PLANT
LIFE

PLANT
LIFE

GENERAL EDITOR
Professor D.M. Moore

New York
OXFORD UNIVERSITY PRESS
1991

CONSULTANT EDITOR
Professor Peter Haggett, University of Bristol, UK

Dr C.D. Adams, The Natural History Museum, London, UK
Central America and the Caribbean

Dr J.R Akeroyd, University of Reading, UK
The British Isles/Italy and Greece

Dr Andrew Byfield, Southampton, UK
The Middle East

Professor Lincoln Constance, University of California
at Berkeley, USA
The United States (Western)

Dr Q.C.B. Cronk, University of Cambridge, UK
Southern Africa

Dr P.J. Garnock-Jones, Canterbury Agriculture and Science Center,
Christchurch, New Zealand
Australasia, Oceania and Antarctica

Professor Dr J. Grau, University of Munich, Germany
Central Europe

Dr Stephen G. Haw, Kingham, UK
China and its neighbors

Charles Jeffrey, Royal Botanic Gardens, Kew, UK
The Soviet Union

Dr S.L. Jury, University of Reading, UK
The British Isles

Dr J.M. Lock, Cambridge, UK
Central Africa

Dr R.E. Longton, University of Reading, UK
Canada and the Arctic

Professor D.M. Moore, University of Reading, UK
The World of Plants/South America/Spain and Portugal

Catherine Olver, University of Reading, UK
Japan and Korea

Dr John Packham, Polytechnic of Wolverhampton, UK
The Nordic Countries

Dr Ian B.K. Richardson, Reading, UK
France and its neighbours

Dr L.P. Ronse Decraene, Catholic University
of Louvain, Belgium
The Low Countries

Martin Walters, Cambridge, UK
Central Europe

Dr S.M. Walters, Cambridge, UK
Eastern Europe

Dr Peter S. White, University of North Carolina, USA
The United States (Eastern)

Dr T.C. Whitmore, University of Cambridge, UK
Southeast Asia

Dr A.J. Whitten, Cambridge, UK
The Indian Subcontinent

Dr G.E. Wickens, Hampton Hill, UK
Northern Africa

AN EQUINOX BOOK

Copyright © Equinox (Oxford) Limited 1990

Planned and produced by
Equinox (Oxford) Limited
Musterlin House, Jordan Hill Road
Oxford, England OX2 8DP

Published in the United States of America by
Oxford University Press, Inc.,
200 Madison Avenue,
New York, N.Y. 10016

Oxford is a registered trademark of
Oxford University Press

Library of Congress
Cataloging-in-Publication Data

Plant life/general editor, D.M. Moore
p.   cm.
"An Equinox book" – Verso CIP t.p.
Includes bibliographical references and index.
ISBN 0-19-520863-3
1. Phytogeography. 2. Plants. 3. Botany.
4. Plant communities.
I. Moore, D.M. (David Moresby)
QK101.P55 1991
581.9 -- dc20   90-47460
                                                 CIP

Volume editor	Barbara Haynes
Assistant editor	Victoria Egan
Designers	Isobel Gillan, Jerry Goldie, Rebecca Herringshaw
Cartographic manager	Olive Pearson
Cartographic editor	Clare Cuthbertson
Picture research manager	Alison Renney
Picture researcher	Linda Proud
Project editor	Candida Hunt
Art editor	Steve McCurdy

ISBN 0-19-520863-3

Printing (last digit): 9 8 7 6 5 4 3 2 1

Printed in Spain by Heraclio Fournier SA, Vitoria

INTRODUCTORY PHOTOGRAPHS
Half title: *Cardinal flower* (Lobelia cardinalis) *in Michigan, USA (NHPA/John Shaw)*
Half title verso: *Hickory* (Carya ovata) *catkins in spring (Bruce Coleman Ltd/Stephen J.
Krasemann)*
Title page: *The grass* Pennisetum villosum, *Asir, Saudi Arabia (Oxford Scientific
Films/Andy Park)*
This page: *Snowbells* (Soldanella alpina) *in the French Alps (Bob Gibbons)*

Contents

PREFACE
7

PREFACE

WE CANNOT ANY LONGER AFFORD TO UNDERESTIMATE THE EXTENT OF our dependence upon plants - they underpin almost all life on Earth. They have the extraordinary ability to convert light energy from the sun into food, and this allows all animals, including humans, to live. Even carnivores benefit at second hand, eating animals that have themselves been nourished by plants.

In the 4,000 million years since life first evolved on Earth, plants have diversified to occupy the planet, become extinct, and further diversified. Some 400 million years ago plants first began to establish themselves on land. Since that time the physical world they have colonized has changed dramatically. The continents we know today have broken away from previous landmasses, and have drifted across the face of the Earth to occupy new positions. The world's climates have changed, the Earth passing through warm phases and cold ice ages. The climate in a particular area may be further changed by the rise of mountain ranges and the opening of seas and oceans. All these factors have contributed to the extraordinary diversity of plants. It is not known for certain how many species of plants now exist, but it is estimated that there are at least 303,000 species, of which more than two-thirds are the flowering plants, with which we are most familiar.

This book offers a journey into the history, diversity, beauty and importance of plants. It begins by examining their origins, the characteristics they have in common and what distinguishes the various groups. Their migration and evolution, and their ability to diversify to colonize almost every part of the world, no matter how hostile the conditions they may find, makes a fascinating story. The requirements that plants have for life is then explained.

As well as providing an enormous variety of more or less delectable foods, plants also give us fuel, shelter, clothing, tools and other everyday items, stimulants and medicines. For thousands of years they have also been cultivated for pleasure – they ornament our houses, gardens and parks. The urgent need to conserve the plants we know about, and to classify and describe those not yet catalogued, is emphasized – for the extinction rate of plant species is frightening, and we do not even know what the potential value might be of some of the plants that are lost.

All these topics are considered at a global level and at a local level, for the various regions of the world. It is hoped that this celebration of the great beauty of the world's plants, and of their many uses, will lead to a greater understanding of how essential it is to ensure their future.

D.M. Moore
UNIVERSITY OF READING

The flamboyant tree *(Delonyx regia)*, a native of Madagascar

Carpet of wildflowers in spring in Texas, USA *(overleaf)*

THE WORLD OF PLANTS

The Diversity of Plants

Plants are found in greater or lesser abundance almost everywhere on Earth, from the warm, moist, equable tropics to the most arid deserts and the coldest polar regions; from the depths of the oceans to the highest mountain peaks. There are some 300,000 species; nearly all other life forms on Earth ultimately depend on the energy they produce. The plants' conquest of so much of our planet is the result of their evolution into an almost bewildering diversity of forms, enabling them to exploit every opportunity for survival.

What is a plant?

Put simply, a plant is an organism that contains a green pigment, chlorophyll. This is used to convert sunlight into energy – a process known as photosynthesis, by which plants obtain food, in the form of glucose, to fuel growth. Plant cells differ from those of other living organisms in that they have a rigid wall composed of a carbohydrate, cellulose. Fungi, which do not contain chlorophyll and whose cell walls are made of a different carbohydrate, chitin, are not generally considered to be plants. Some species that do not photosynthesize but rather obtain their food parasitically from other plants are counted as plants because they have evolved from green, photosynthesizing ancestors.

Plants vary in size and complexity from minute, single-celled algae that measure about 0.003 mm in diameter to giants such as the Californian redwood (*Sequoia sempervirens*) and the Australian gum (*Eucalyptus regnans*). These towering trees, which can reach over 100 m (330 ft) in height, have a complex structure and physiology to sustain their massive size.

But size can also be achieved with a relatively simple structure, as in the case of the giant kelp (*Macrocystis pyrifera*), a seaweed (or alga) whose fronds, which may be as long as the redwoods are high, are supported by the seawater in which they grow.

There is enormous variety in the plants, both above and below the ground. Some stand erect, others are lowgrowing (prostrate), creeping over the ground like strawberries (*Fragaria*), or scrambling over other plants, as blackberries (*Rubus*) do. In dense forests lianas climb up trees to reach the sunlight, and hang like umbilical cords to their roots many meters below on the forest floor. In stark contrast, many plants that grow in harsh, exposed environments form low, dense cushions that at a distance resemble rocks or boulders.

Most plants obtain the water and nutrients necessary for life through their roots, which anchor them in the soil. Some grow on other plants in the angles between the branches and the trunk, and acquire them from the detritus and water that accumulate in these crevices. Such plants are known as epiphytes; some, like Spanish moss (*Tillandsia usneoides*), are virtually free hanging in the humid air of the tropical forests in which they grow, absorbing moisture and nutrients over their entire surface.

Plants have very different lifespans. Some, known as annuals, complete their life cycle in one growing season, which may last only a few weeks. Other (perennials) may live for years. The bristlecone pine (*Pinus longaeva*), found in the southwestern United States, is known to have the longest lifespan, holding the record at about 5,000 years.

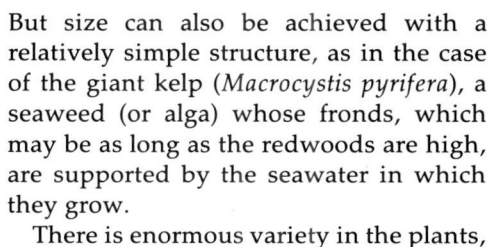

HOW PLANTS ARE NAMED

Plant studies depend on taxonomy – the identification and classification of plants. Each plant is known by its genus and species name: a genus is a group of structurally related species, a species is a group of more or less identical plants, capable of interbreeding. Genera are grouped into larger related categories.

Category	Scientific name of taxonomic group (taxon)	Comments
Division	Anthophyta	Flowering plants (angiosperms)
Class	Dicotyledones	Distinguished by having two seed leaves or cotyledons (dicots)
Order	Asterales	
Family	Compositae	Commonly known as the daisy family
Genus	*Bellis*	
Species	*Bellis perennis*	Daisy
Variety	*Bellis perennis* var. *aucubaefolia*	A naturally occuring variant
Cultivar	*Bellis perennis* 'Double Bells'	A form of double daisy, selected and artificially bred

Massive trees (*left*) Towering above the undergrowth, the Californian redwoods are among the largest and oldest organisms in the world. The General Sherman, in Sequoia National Park, measures 11.3 m (37 ft) round the base. Some of these trees began their lives over 3,000 years ago.

Tropical blossom (*right*) *Hibiscus rosa-sinensis* is just one example of the extraordinary range of glorious flowers found on Earth. The different combinations of color, shape, size and perfume have evolved to ensure the survival of the species, for flowers contain the plant's reproductive organs.

PHOTOSYNTHESIS – THE BASIS OF LIFE

Photosynthesis equation

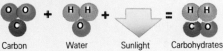

Carbon dioxide + Water + Sunlight = Carbohydrates + Oxygen

Sunlight

Oxygen

Carbon dioxide

Water

Water and minerals

Almost all organisms ultimately derive their energy from photosynthesis. Fundamental to this process is the green pigment chlorophyll, which is carried in microscopic bodies – the chloroplasts – present in the cells of certain tissues (notably leaves) in most plants.

Chlorophyll absorbs light from the sun and transfers this energy to reaction centers in the chloroplasts. The energy is then used in a series of reactions that synthesize adenosine triphospate (ATP) and release oxygen. Plants are unique in their ability to produce oxygen. Almost all living things need oxygen to respire, and give off carbon dioxide as a waste product. Green plants are no exception to this, but they also utilize carbon dioxide: the ATP provides the chemical energy to drive a second series of reactions (the Calvin cycle) in which carbon dioxide from the atmosphere is used to synthesize glucose. The glucose produced is converted into carbohydrates: either starch – the principal storage compound for plants – or other sugars.

This food is stored in various ways, such as swollen roots or underground stems (tubers and rhizomes), and is needed to fuel growth. The food stored in plants is converted back into energy by the animals that feed on them, and thus enters the food chain; ultimately, therefore, all animals depend on the energy created by photosynthesis.

The Origins of Plants

T HE BASIC UNITS OF LIVING MATERIAL – amino acids and the proteins formed from them – were probably present very early in the Earth's history, perhaps as much as 4,000 million years ago. At this stage – and for about another 2,000 million years – the Earth's atmosphere did not contain oxygen, and the creation of these proteins appears to have been the result of the interaction of gases such as ammonia, methane and water vapor.

Photosynthesis is thought to have evolved some time in the next 500 million years: the earliest fossils, which were formed about 3,300 million years ago, are microscopic filaments and spheres, the apparent remains of bacteria. These single-celled organisms (prokaryotes) included the chlorophyll-containing "blue-green algae" (cyanobacteria).

It was not until some 1,000 to 700 million years ago that the nucleic acids (deoxyribose nucleic acid, DNA), which determine inherited characteristics, and associated proteins in the cell became enclosed by a membrane to form a nucleus. The multicellular organisms that developed diversified relatively quickly to populate the waters of the Earth.

The first aquatic plants, the algae, appeared about 500 million years ago. About 100 million years later plants, probably derived from green algae, began to establish themselves on land. They gradually colonized muddy shores and then moved to drier habitats. Some 350 millions years ago primitive precursors of all the present groups of terrestrial plants (bryophytes, pteridophytes and gymnosperms), except the flowering plants (angiosperms), were living on land.

The groups of plants

Biologists divide all the plants on Earth into these five groups. The simplest, the algae, live mostly in marine and fresh water, though some inhabit damp places on land. More than half of the approximately 25,000 species are single celled, while most of the others consist of a number of cells arranged into a filament (as in *Spirogyra*) or a plate (as in *Ulva*). In some algae such as certain seaweeds the cells are differentiated to form a blade, a stalk and a structure for anchoring the plant (known as a holdfast), as well as a rudimentary system for conducting water. All algae contain chlorophyll,

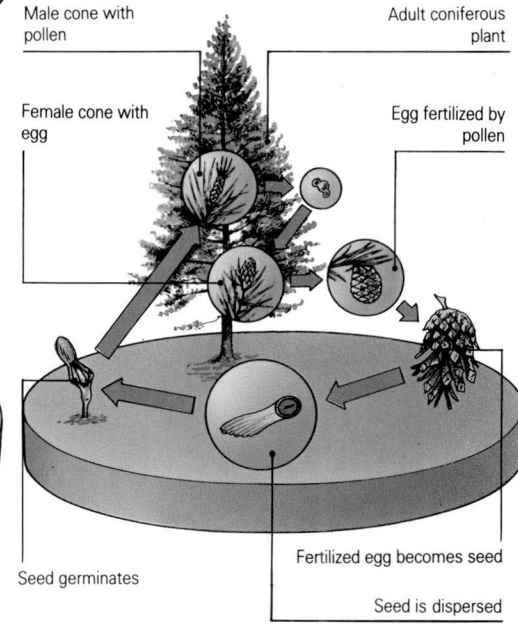

Prehistoric forest (*left*) This is a reconstruction of a forest of 300 million years ago, dominated by seed ferns (long extinct) and seedless trees, giant ancestors of the horsetails, liverworts and ferns.

though the green color tends to be masked by other pigments, carotenoids, in the red algae (*Rhodophyta*) and the brown algae (*Phaeophyta*), which live at depth in the ocean. The reproductive structures of the algae are also single celled.

Mosses, hornworts and liverworts (the bryophytes), of which there are about 16,000 species, rarely exceed 20 cm (8 in) in length. They grow predominantly on land, usually in marshy ground. The life cycle of spore-bearing bryophytes occurs in two stages, one sexual and the other asexual. This method of reproduction is also found in the pteridophytes, plants such as ferns, club mosses and horsetails.

The most highly developed plants – the conifers (gymnosperms) and flowering plants (angiosperms) – are distinguished from all other plants by producing seeds. Seeds consist of an embryo surrounded by a food store for nourishment during germination; a hard outer layer – the seed coat – provides protection. Seeds can remain dormant until conditions for growth are favorable, which greatly increases their chances of survival.

Life cycle of a moss

Moss plant (gametophyte) with mature sporophyte full of spores

New gametophyte develops male and female sex organs

Released spores germinate

Fertilized egg grows into new sporophyte

Fertilization: male sperm swims to female egg

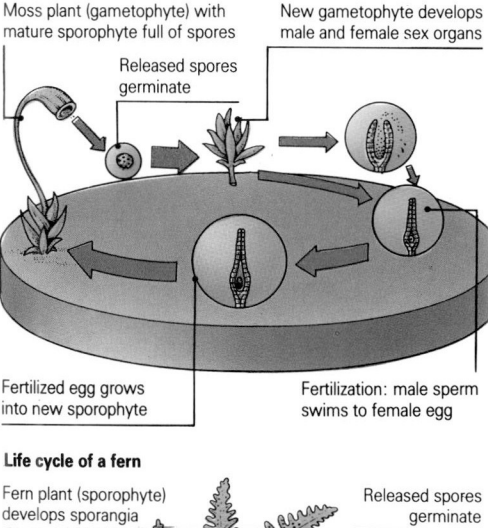

Life cycle of a fern

Fern plant (sporophyte) develops sporangia

Released spores germinate

New sporophyte develops on gametophyte

Gametophyte develops

Fertilization: male sperm swims to female eggs

Mature gametophyte develops male and female sex organs

Life cycle of a gymnosperm

Male cone with pollen

Adult coniferous plant

Female cone with egg

Egg fertilized by pollen

Seed germinates

Fertilized egg becomes seed

Seed is dispersed

Life cycle of an angiosperm

Adult plant bears flowers, male sex organs (anthers) produce pollen

Pollen deposited on female sex organs (stigma), grows down to fertilize egg

Seed germinates

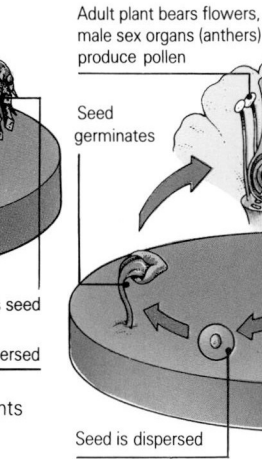

Seed is dispersed

Fertilized egg becomes seed

Reproduction in seedless plants Ferns and mosses have motile sperm that must swim through water to fertilize the spore, restricting them to moist places.

Reproduction in seed-bearing plants In higher plants male pollen grains are carried by wind or animals to female sex organs of other plants.

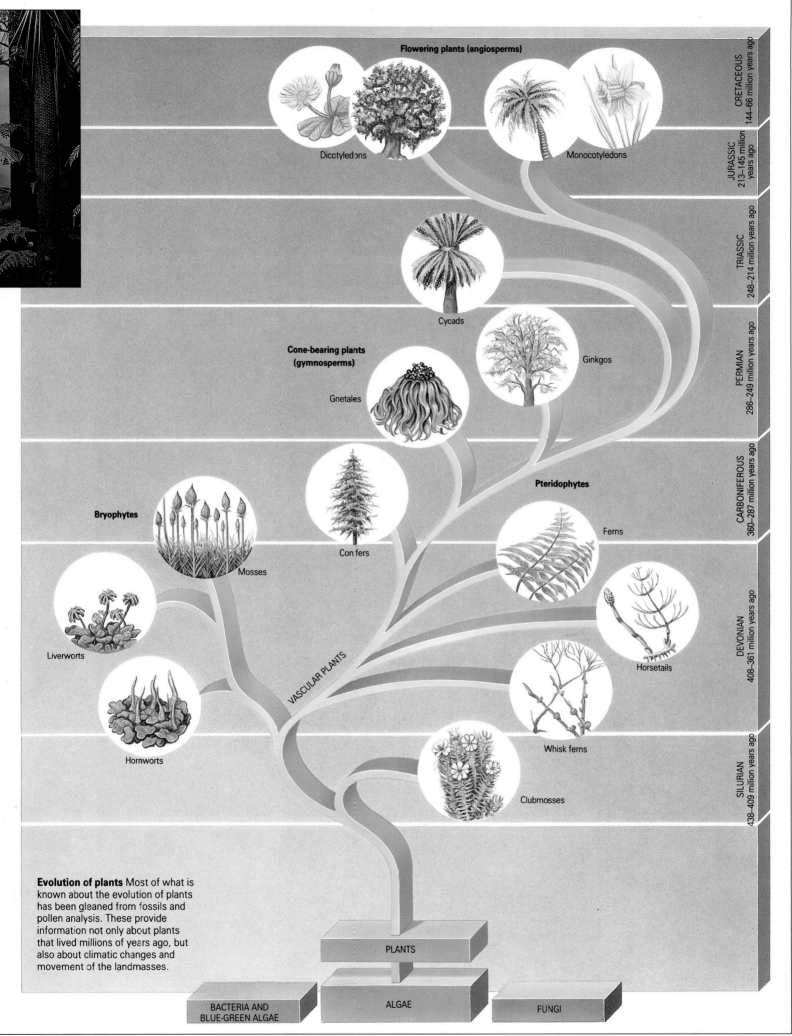

Flowering plants (angiosperms)

Dicotyledons

Monocotyledons

CRETACEOUS 144–66 million years ago

JURASSIC 213–145 million years ago

Cycads

TRIASSIC 248–214 million years ago

Cone-bearing plants (gymnosperms)

Ginkgos

Gnetales

PERMIAN 286–249 million years ago

Pteridophytes

Bryophytes

Mosses

Conifers

Ferns

CARBONIFEROUS 360–287 million years ago

Liverworts

Horsetails

DEVONIAN 408–361 million years ago

VASCULAR PLANTS

Hornworts

Whisk ferns

Clubmosses

SILURIAN 438–409 million years ago

Evolution of plants Most of what is known about the evolution of plants has been gleaned from fossils and pollen analysis. These provide information not only about plants that lived millions of years ago, but also about climatic changes and movement of the landmasses.

PLANTS

BACTERIA AND BLUE-GREEN ALGAE

ALGAE

FUNGI

The Flowering Plants

THE MOVE TO LIFE ON LAND WAS GENERALLY accompanied by the development of vascular systems – strands of tissue that enable water and minerals from the soil to be carried into the plant, and food to be transported around it. The former is called the xylem, the latter the phloem. The systems gave land-dwelling plants a structure to support them in the air as they were no longer supported by water.

Bryophytes do not possess any vascular tissues, though the stems of some mosses contain a central strand of water-conducting cells that resemble cells found in the phloem of primitive vascular plants. The giant horsetails (*Calamites*), tree ferns and primitive gymnosperms that first established themselves successfully on land gave way to forests dominated by gymnosperms, including the earliest conifers, about 280 million years ago. The early pteridophytes evolved into primitive members of the ferns we see today, such as *Osmunda*, *Cyathea* and *Dicksonia*, as well as horsetails (*Equisetum*), about 195 million years ago. Sixty million years later the conifers began to be succeeded by broadleaf flowering plants, which quickly rose to the dominant position they maintain today.

During this long period of time groups of plants evolved, became prominent, and then either died out or became restricted in distribution as other groups replaced them. The British naturalist Charles Darwin (1809–82) first suggested that evolution proceeds by competition between organisms, so those best able to survive and produce offspring gradually become dominant. The study of genetics has revealed a physical basis, through inherited variation, for gradual evolutionary change. Some scientists, however, consider that organisms may remain relatively stable for long passages of time, punctuated with periodic bursts of evolutionary change. The apparently rapid diversification of multicellular aquatic organisms more than 400 million years ago, and the explosive arrival of flowering plants about 135 million years ago, lend strong support to this theory.

Flower evolution

The earliest flowers developed from primitive leaflike structures that gradually folded to form the carpel – the part of the flower that protects the female ovule – and stamens, which protect the male pollen grains. Early angiosperm fossils show incompletely folded carpels, which are still seen in some modern plants such as magnolias (Magnoliaceae), pepper trees (*Drimys*) and the Fijian *Degeneria*. In addition *Magnolia* and *Degeneria* have leaflike stamens.

Subsequent evolution led to the fusion of the carpels and their differentiation into ovary, style and stigma, and the modification of the leaflike stamen so that it has become a filament bearing a lobed anther at its apex. All the parts of a flower are derived from leaves. Leaves below the carpel and stamen lost their chlorophyll to become colored petals, and others – the sepals – which retained their leaflike appearance, protect the upper reproductive parts of the flower.

There is tremendous diversity in the arrangement (inflorescence) of flowers. They may be borne singly or grouped around the stem in different ways. This is largely the result of interaction with the animals, principally insects, that pollinate the flowers – that is, transfer the pollen from the male anther to the female stigma so fertilization can take place. Flowers have developed many different shapes, colors and scents to attract pollinators.

All flowering plants are divided into one of two groups – the monocotyledons and the dicotyledons. This denotes whether the shoot produced by the germinating seed has one or two cotyledons (leaflike organs); the division perhaps shows that they evolved from different ancestors. Dicotyledons include many familiar plants, from dandelions to oak trees; the other group includes grasses, lilies, orchids and palm trees.

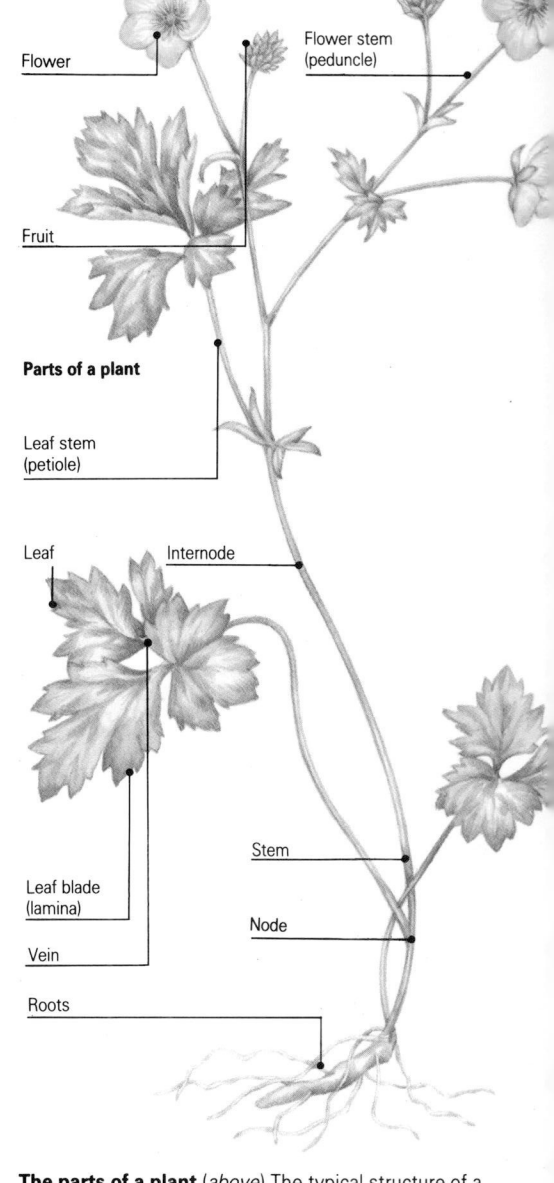

Parts of a plant

Flower

Flower stem (peduncle)

Fruit

Leaf stem (petiole)

Leaf

Internode

Stem

Node

Leaf blade (lamina)

Vein

Roots

The parts of a plant (*above*) The typical structure of a dicot herb. In addition to the parts shown here, the root hairs, which take in water and nutrients, are also important.

Structure of a leaf (*below*) A cross-section through a leaf of a typical dicot. Most of the chloroplasts are concentrated in the upper mesophyll, which is exposed to direct sunlight.

Structure of a leaf

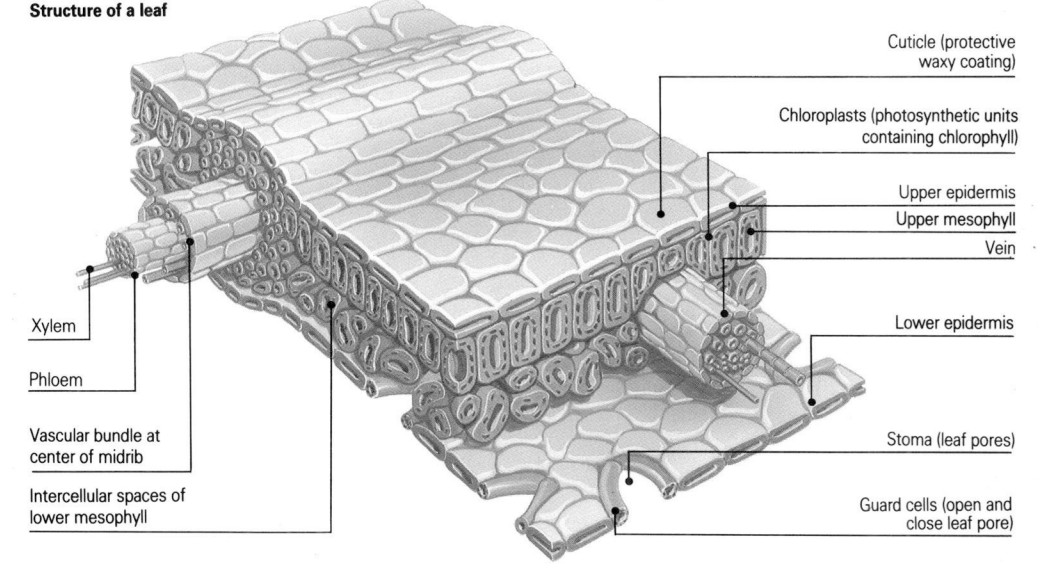

Xylem

Phloem

Vascular bundle at center of midrib

Intercellular spaces of lower mesophyll

Cuticle (protective waxy coating)

Chloroplasts (photosynthetic units containing chlorophyll)

Upper epidermis

Upper mesophyll

Vein

Lower epidermis

Stoma (leaf pores)

Guard cells (open and close leaf pore)

Parts of a flower

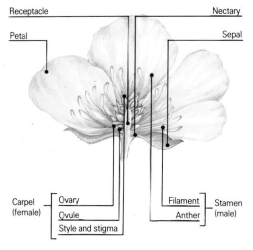

Receptacle

Nectary

Petal

Sepal

Carpel (female)
- Ovary
- Ovule
- Style and stigma

Stamen (male)
- Filament
- Anther

The stem of a herbaceous dicot

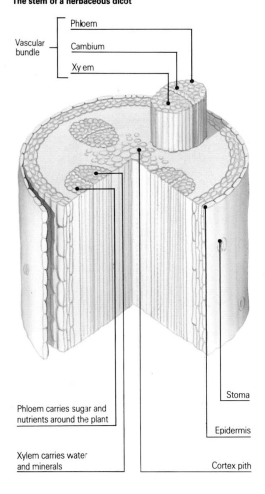

Vascular bundle
- Phloem
- Cambium
- Xylem

Phloem carries sugar and nutrients around the plant

Xylem carries water and minerals

Stoma

Epidermis

Cortex pith

Herbaceous stem and woody trunk Stems and trunks of plants support the leaves and flowers, and also conduct food and water around the plant in specialized vessels. In woody plants, tough protective bark develops from the epidermis and old phloem tubes. The central core and old xylem tubes are compressed and give strength.

The structure of a flower (*left*) A cross-section through a typical angiosperm flower shows how well the embryo seed is protected.

Many different flowerheads (*right*) There are many variations in floral structure and in the way flowers are grouped. They may be large and solitary, like a waterlily, or individually tiny but massed into an eyecatching inflorescence (flowerhead), as in a daisy or a member of the carrot family.

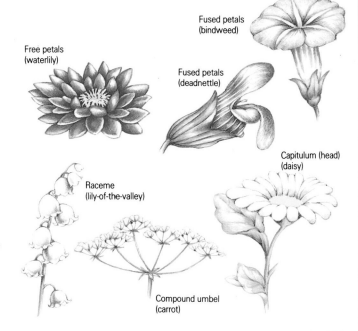

Free petals (waterlily)

Fused petals (bindweed)

Fused petals (deadnettle)

Capitulum (head) (daisy)

Raceme (lily-of-the-valley)

Compound umbel (carrot)

MONOCOTS AND DICOTS

Flowering plants are divided into two groups according to certain characteristics. Some examples give an idea of the differences: lilies and grasses are monocots, beech trees and clover are dicots.

	Monocotyledons	Dicotyledons
Embryo	One seed leaf (cotyledon)	Two seed leaves (cotyledons)
Growth form	Mostly herbaceous; a few treelike	May be either woody or herbaceous
Vascular system	Consists of numerous scattered bundles, without definite arrangement; rarely with cambium	Usually consists of a ring of primary bundles with a cambium, and secondary growth in outer stem layer
Leaves	Usually parallel veined, oblong or linear in shape and often sheathing at base; stalk (petiole) seldom developed	Usually net veined, broad in shape and seldom sheathing at base; usually with stalk (petiole)
Flowers	Parts are usually in three or multiples of three	Parts are usually in fours or fives

---

Development of a woody trunk

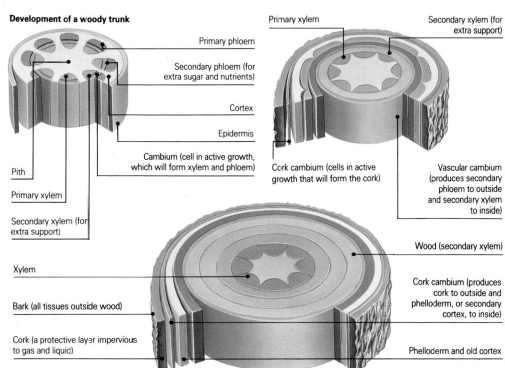

Primary xylem

Secondary xylem (for extra support)

Primary phloem

Secondary phloem (for extra sugar and nutrients)

Cortex

Epidermis

Cambium (cell in active growth, which will form xylem and phloem)

Pith

Primary xylem

Secondary xylem (for extra support)

Xylem

Bark (all tissues outside wood)

Cork (a protective layer impervious to gas and liquid)

Cork cambium (cells in active growth that will form the cork)

Vascular cambium (produces secondary phloem to outside and secondary xylem to inside)

Wood (secondary xylem)

Cork cambium (produces cork to outside and phelloderm, or secondary cortex, to inside)

Phelloderm and old cortex

Plant Migrations

Each of the thousands of species and groups of plants inhabiting the world evolved in a particular area and then spread outward to give the patterns of distribution there are today. Animals are obviously mobile, so it is not difficult to appreciate that species migrate from one place to another. That rooted plants are able to do so may seem more surprising. It was first suggested in the late 18th century that plants migrate when fruits, seeds and spores are dispersed by wind, water or animals. More recent understanding of the nature of the Earth's crust, which consists of tectonic plates that are constantly shifting, has added an additional piece to the jigsaw.

Very similar, or even identical, groups of plants sometimes occur in widely separated areas. There is no evidence to support the possibility that each one evolved independently, and one explanation is that when the single supercontinent of Pangea began to fragment about 190 million to 160 million years ago some species were separated by the movement of the landmasses. An example of this phenomenon, which is known as vicariance, is the southern beech (*Nothofagus*), which had evolved before the continents split and is found in South America and in Australasia.

Ancient events cannot account for the vicariant distribution of other, younger species; for example, sand-spurrey (*Spergularia platensis*), which can be found in widely separated areas of North and South America. In such cases, the seeds were undoubtedly carried over long distances by winds or oceanic currents.

How species spread

It is no easy task to determine where plants first evolved. The presence of several related plant groups in a particular area may be taken to indicate that they originated in that area. But centers of diversity also arise after a species has moved into an area with particular environmental conditions that serve to stimulate evolutionary activity.

Physical features and climate affect the spread of plants: oceans, mountain ranges and deserts provide suitable habitats for some plants but not for others, and they also act as migration routes for plants, filtering out some and encouraging the spread of others. Furthermore, plants

---

MODES OF TRAVEL

Plants have evolved many specialized forms to aid their dispersal from place to place around the world. Some, such as the many kinds of orchids (Orchidaceae), have extremely light seeds, which are carried on the wind in the same way as the spores of ferns and mosses. Others have developed more elaborate structures to make them airborne. The seeds of the willow herb (*Epilobium*) are plumed; those of the dandelion (*Taraxacum*) and thistle (*Cirsium*) have featherlike parachutes. In maples (*Acer*) and ash (*Fraxinus*) the carpels develop prominent wings to carry them on the wind. The seed cases of other plants form natural catapults: when they are split the seed is ejected explosively in all directions. Some, such as vetch (*Vicia*) burst open when the sun dries them; the touch-me-not or balsam (*Impatiens*) is triggered by the movement of the wind or by an animal brushing past.

Seeds and many fruits have waterproof coatings and may be swept for long distances down streams and rivers before germinating. Here again, some plants are able to exploit the possibilities of water dispersal: for example, air trapped within the outer husk of the coconut gives it great buoyancy, and it may be carried across the sea by ocean currents to wash up on distant islands.

The colors and sweet taste of many fleshy fruits such as cherries, grapes and rosehips are particularly attractive to animals. They eat the fruit and the seeds pass unharmed through their digestive systems to be voided some distance from the parent plant. Other fruits and seeds have hooked spines that become attached to the feathers or hair of animals. They can be transported many kilometers before falling to the ground. Burdocks (*Arctium*) and cleavers (*Galium*) are examples of plants that are dispersed in this way.

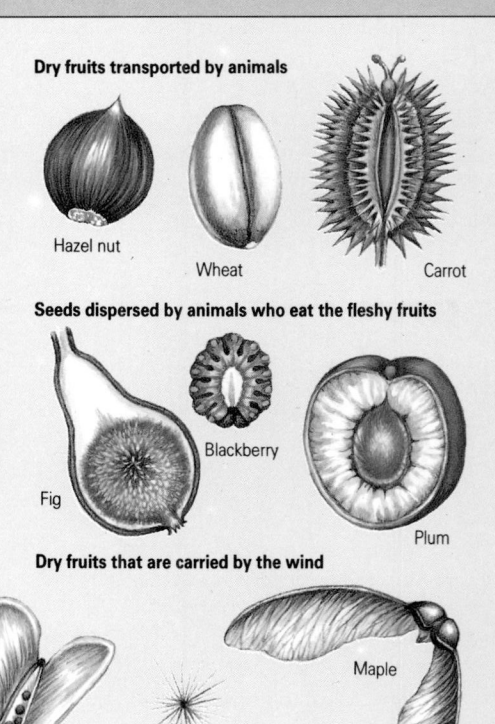

Dry fruits transported by animals

Hazel nut

Wheat

Carrot

Seeds dispersed by animals who eat the fleshy fruits

Fig

Blackberry

Plum

Dry fruits that split open to disperse seeds

Pea

Foxglove

Shepherd's purse

Dry fruits that are carried by the wind

Maple

Dandelion

Dispersal by birds (*above*) Birds are important agents in extending the distribution of fruit-bearing plants. Some birds travel long distances on migration, when they may carry seeds for thousands of kilometers from one landmass to another.

Parachutes in the wind (*left*) The melancholy thistle (*Cirsium heterophyllum*) is one of many plants whose seeds have a fluffy appendage. These catch the wind, so the seeds are transported by air to another site to germinate.

Endemic succulent (*below*) *Didiera trolii* is a strange spiny member of the Didiereaceae. This family of columnar succulents is unique to the dry places of Madagascar. An isolated island with a stable geological history, it is very rich in endemic plants.

moving into a new area face competition from the plants already growing there, which may limit their chances of survival.

It might be expected that the longer the elapse of time since a species evolved, the wider its area of distribution. But there are many exceptions to this apparently simple proposition. Some of the widest distributions have been shown to be among the most recent. For example, purple saxifrage (*Saxifraga oppositifolia*) and mountain avens (*Dryas octopetala*) extend throughout northern Eurasia and North America in areas that were left as open habitats when the ice sheets retreated about 12,000 years ago. Since there were few competing species they were able to establish themselves widely and rapidly.

Sometimes a species of plant is restricted to an unexpectedly small area. The plant may require a specialized habitat that occurs only sporadically or – in the case of plants on isolated islands – it is unable to disperse from its area of origin. Such plants are known as endemics.

Endemics may be very old (when they are known as paleoendemics). These are the surviving remnants of groups that were formerly more widely distributed but have become extinct elsewhere. The maidenhair tree (*Gingko biloba*), now restricted to a small area in central China, belongs to a group that was widespread in the northern hemisphere 100 million years ago. Neoendemics, on the other hand, are likely to be geographically restricted because they have not yet had time to spread more widely.

Factors for Survival

THE ENVIRONMENT, OR SURROUNDINGS, IN which plants grow is the key to their survival. The complex interaction of factors such as climate, altitude, soils and the presence of other living organisms affects the conditions available for growth and determines which plants can or cannot establish themselves in a particular locality. On a global scale, the limiting factors for plant distribution are influenced by latitude, which affects seasonal and daily temperatures, the duration of the growing season and length of daylight.

Sunlight is the ultimate source of most plant energy. Its efficiency depends on the amount of the Earth's atmosphere it has to pass through; it is consequently lower in temperate regions, where it strikes the Earth more obliquely, and at sunrise and sunset, than it is close to the Equator. At high intensities sunlight has a heating effect, which plants must be able to cope with. If its intensity is low, they must adapt themselves to utilizing all the lower amounts of energy available. They do this by increasing their leaf size.

The duration of light and dark has a profound effect on plant growth. In the tropics, day and night are of roughly equal length throughout the year; the inequality between them increases the nearer to the poles, with longer days and shorter nights in summer and the reverse in winter. The length of the period of daylight is important in initiating flowering: for example, the willow herb *Epilobium hirsutum* will flower only after it has been exposed to about 16 hours of light a day, whereas chrysanthemums require the shortening days of late summer and early fall to flower.

The importance of temperature

Most of a plant's life processes depend on the activity of certain complex protein molecules, or enzymes. These are destroyed at temperatures above 40–50°C (104–122°C). If temperatures suddenly fall below 0°C (32°F) the water and dilute salt solutions within plants freeze and expand, damaging cell structures. Within these limits plant groups and species

Groundsels on Mount Kenya (*above*) High on the equatorial mountains plants have evolved to survive in a climate that has high temperatures every day and freezing conditions every night. Similar giant plants can be found in the Andes.

Climate zones and ocean currents (*below*) Climate may be modified by ocean currents. Western Norway and southwestern Greenland, for example, are warmed by the North Atlantic Drift, allowing plants normally found farther south to grow there.

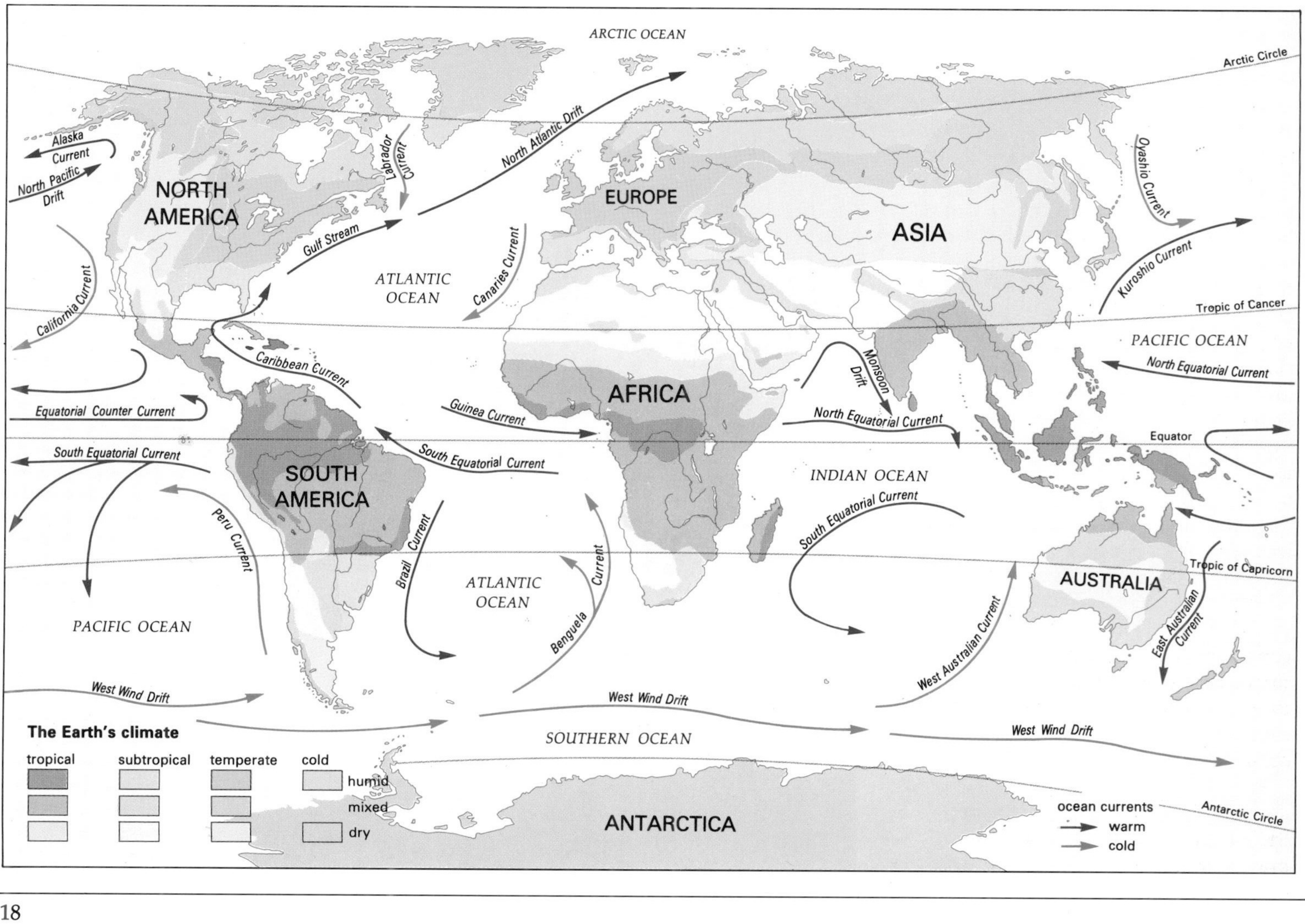

ARCTIC OCEAN

Arctic Circle

Alaska Current

North Pacific Drift

NORTH AMERICA

Labrador Current

North Atlantic Drift

EUROPE

ASIA

Oyashio Current

Gulf Stream

ATLANTIC OCEAN

Canaries Current

Kuroshio Current

California Current

Tropic of Cancer

PACIFIC OCEAN

North Equatorial Current

Caribbean Current

AFRICA

Monsoon Drift

Equatorial Counter Current

Guinea Current

North Equatorial Current

Equator

South Equatorial Current

South Equatorial Current

SOUTH AMERICA

INDIAN OCEAN

South Equatorial Current

Peru Current

Brazil Current

ATLANTIC OCEAN

Benguela Current

AUSTRALIA

Tropic of Capricorn

PACIFIC OCEAN

West Australian Current

East Australian Current

West Wind Drift

West Wind Drift

West Wind Drift

SOUTHERN OCEAN

The Earth's climate

tropical	subtropical	temperate	cold	
				humid
				mixed
				dry

ANTARCTICA

ocean currents

→ warm

→ cold

Antarctic Circle

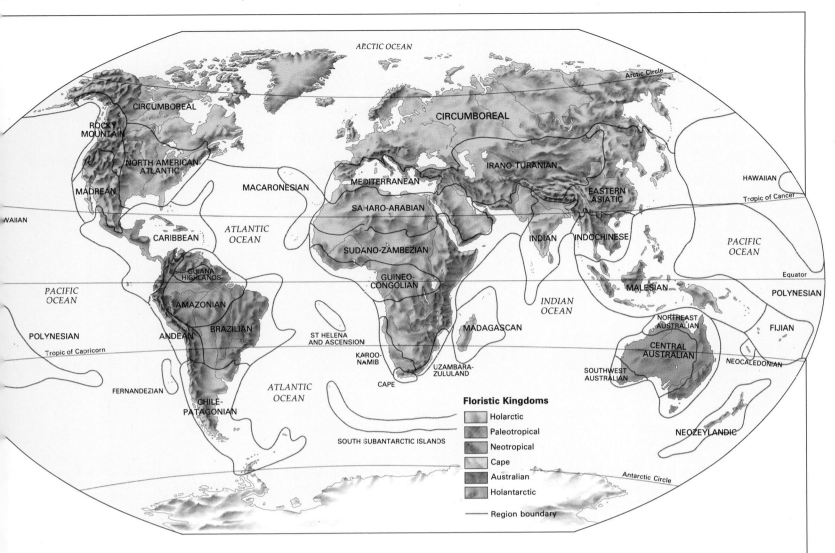

Map showing floristic kingdoms and regions with labels including:

ARCTIC OCEAN

CIRCUMBOREAL

ROCKY MOUNTAIN

NORTH AMERICAN ATLANTIC

MADREAN

MACARONESIAN

MEDITERRANEAN

SAHARO-ARABIAN

IRANO-TURANIAN

EASTERN ASIATIC

HAWAIIAN

Tropic of Cancer

CARIBBEAN

ATLANTIC OCEAN

INDIAN

INDOCHINESE

PACIFIC OCEAN

GUIANA HIGHLANDS

SUDANO-ZAMBEZIAN

GUINEO-CONGOLIAN

Equator

PACIFIC OCEAN

AMAZONIAN

BRAZILIAN

ST HELENA AND ASCENSION

MADAGASCAN

INDIAN OCEAN

MALESIAN

POLYNESIAN

POLYNESIAN

ANDEAN

KAROO-NAMIB

UZAMBARA-ZULULAND

NORTHEAST AUSTRALIAN

FIJIAN

Tropic of Capricorn

CAPE

CENTRAL AUSTRALIAN

NEOCALEDONIAN

FERNANDEZIAN

CHILE-PATAGONIAN

ATLANTIC OCEAN

SOUTHWEST AUSTRALIAN

SOUTH SUBANTARCTIC ISLANDS

NEOZEYLANDIC

Floristic Kingdoms

- Holarctic
- Paleotropical
- Neotropical
- Cape
- Australian
- Holantarctic
- ——— Region boundary

Antarctic Circle

have developed their own ranges of temperature tolerance.

As you ascend a mountain, temperatures fall by about 1°C for every 200 m (660 ft). This profoundly affects plant distribution, giving rise to zonational patterns of vegetation: from the bottom to the top of a mountain, broadleaf trees and woodland flowers give way to conifers, then to tussocky grasses and lowlying cushion plants – a pattern that broadly corresponds to the vegetation zones that run from the Equator to the poles. Plants growing on slopes facing the Equator are at temperatures between 5°C and 6°C higher than those facing the poles, and so aspect, too, exerts considerable influence on what plants grow where.

Water is essential for plant survival, forming up to 90 percent of their tissue weight. Its presence, or lack of it, is a key limiting factor to growth, with some plants being much more tolerant of arid conditions than others. Some plants even thrive best when their roots are standing in water; a few are able to withstand the salinity of coastal waters, which the tissues of most plants cannot absorb.

Plants are also affected by the chemical makeup of the soil in which they grow. Soils formed from igneous rocks such as granite are acid, those from limestone or chalk are alkaline; each type supports its own community of specialized plants. Nutrients, vital for growth, are present in soil when there is a large amount of humus – that is, organic matter broken down by the activities of fungi, bacteria and animals such as worms and beetles. They are easily washed out of well-drained soils, such as sands and gravels, by a process known as leaching, so these soils are able to support only a limited number of plants.

Map of floristic kingdoms and regions This map reflects the divisions into communities of the world's plants. The major divisions, kingdoms, are subdivided into regions. They reflect the cohesion of plants in their communities as a result of both historical and climatic factors.

The difference between woodland and grassland, or heathland and bog, is easy to see. The distinctions are to do with the structure of plant communities, in the same way that human communities can be recognized as cities, towns, villages and hamlets. We are also accustomed to recognizing that human populations differ in their racial, cultural and historical composition. This is paralleled in the plant world as we move over the surface of the planet. Particular combinations of plant groups – species, genera or families – characterize particular areas of the world. For example, gum trees (*Eucalyptus*) are native only to Australia and adjacent regions; a combination of Californian lilac (*Ceanothus*), *Clarkia* and wellingtonia (*Sequoiadendron*) would indicate the western United States, while the mountain avens (*Dryas octopetala*) in parts of upland Britain points to links with the Arctic and alpine parts of continental Europe, resulting from a common heritage wrought by the glaciations some 16,000 years ago. Such floristic differences between various parts of the world reflect their differing amounts of isolation, the topographic upheavals they have sustained, and the extent of migration between them.

Plants in the water (*left*) Plants have evolved in different ways to exploit a watery habitat. The bald or swamp cypress (*Taxodium distichum*) has developed raised "knees" (pneumatophores) from its roots to take in oxygen, not available in the waterlogged soil. The tiny duckweeds (*Lemna*) that float on still water are little more than a single leaf with a few roots.

Adapting to Surroundings

ADAPTATION – THE WAY IN WHICH ALL living organisms alter their pattern of behavior, or even their structure and physical makeup, so that they are better fitted to cope with their environment – is usually a gradual, evolutionary process. It comes about through the transmission, by breeding, of particular genetic characteristics that allow individual plants or animals to utilize the available resources of their environment more efficiently. The offspring of individuals possessing these characteristics are better able to withstand the challenges of the environment, so more of them survive, passing the characteristics on to an ever increasing number of descendants.

When plants move into a new area, the first species to take over are those able most rapidly to evolve characteristics that enable them to survive, so they are those that can reproduce most quickly. Later arrivals tend to put more energy into their vegetative growth; they take up more of the light and squeeze out the early pioneers. This process creates a succession of colonizers, each adapting in turn to the series of biological changes that colonization brings about.

Studies of widely distributed plant species show that they are composed of a series of populations that have adapted to the diverse environments they have encountered and colonized. This sometimes results in a series of genetically distict populations (ecotypes). Genetically unrelated species often produce ecotypes that have similar characteristics when exposed to similar environments. For example, most species that grow on sandy coasts develop narrow, tough or fleshy leaves to withstand the stress of reduced water availability, and often produce shoots from the stem (adventitious roots) to anchor them in the drifting sand.

Many species show a gradual variation, or cline, in characteristics as they move from one set of environmental conditions to another. The sea plantain (*Plantago maritima*) grows on coasts all the way from the tideline to grasslands lying behind the shore. Those plants that live in or near the tidal zone, where salinity is high, have low, prostrate growth, short leaves and small seeds. Away from the sea, where conditions are less saline, they grow relatively erect, with longer leaves and larger seeds.

These changes may come about both because of the environmental influences and also as a result of breeding behavior. A plant growing in one place may pollinate individuals lying some distance away. This reduces the inherited differences between individuals within species, so they grade into each other.

The gradual (clinal) variation of characteristics enables many species to

ENLISTING THE HELP OF ANIMALS

Flowering plants have evolved a set of features – the flower – that allows them to "move" to seek a mate, thereby enhancing species diversity. In the most simple flowers, pollen is shed from one flower and carried on the wind to the stigmas of other, often far distant plants. This process is wasteful, as much of the pollen is lost or is carried to plants of other species that cannot accept it. The same is true of indiscriminate animal pollinators that visit plants at random in search of nectar.

Mutual adaptation (or coevolution) between plants and animals has resulted in a much more advantageous system – plants offer their nectar to specific insects, which then carry it directly to another plant of the same species. Scent and distinctive patterns of colored petals act as "honey guides" to lead particular insects to the nectar; the structure of the flowers allows entry to certain insects and denies it to others. Even the time at which the flowers open is significant – the flowers of the various species of evening primrose (*Oenothera*), for example, open either in the evening or in the early morning, and each group has its own clientele of insects.

Attracting bats Flowers that rely on bats for pollination have a plentiful supply of nectar and a musky scent.

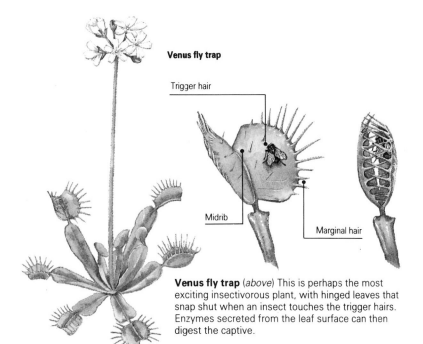

Venus fly trap

Trigger hair

Midrib

Marginal hair

Venus fly trap (*above*) This is perhaps the most exciting insectivorous plant, with hinged leaves that snap shut when an insect touches the trigger hairs. Enzymes secreted from the leaf surface can then digest the captive.

Pitcher plant

Insects attracted to rim, slip down slippery walls into a pool of digestive fluid

Insect eaters (*above*) These specialized plants usually live in boggy soils where vital minerals are scarce. To make up their nutrient requirements they have evolved mechanisms to trap and digest insects. Some plants have leaves like pitchers, slippery pitfalls full of digestive fluid; while the sticky tentacles of sundews surround and hold an insect visitor.

Living on other plants (*left*) This large lobelia (*Lobelia bequaerti*) is an epiphyte. Epiphytes live on other plants. They are not parasites, for their roots take no nourishment from the host; they merely perch on a branch to gain more light and avoid the competition of the forest floor.

adapt to more than one environment. This has enabled many plants to extend their range into suitable environments worldwide. The mountain sorrel (*Oxyria digyna*), for example, lives in cool northern areas in Eurasia and North America with generally low temperatures. It also grows farther south at high altitudes in the Caucasus and the Rocky Mountains, where similar temperatures prevail. To survive at these latitudes it has had to adapt to considerable differences in the duration of daylight.

Ecotypic variation

When one environment changes abruptly to another – for example, from woodland to grassland, or from acid to alkaline soil – the change in the ecotype of plants each supports may be equally abrupt. The hawkweed *Hieracium umbellatum*, for example, has the usual characteristics of coastal populations when it grows on sandy coasts; it is also found in woodland, where it has developed broad, thin leaves suited to the higher humidity and lower levels of sunlight.

Another example of abrupt change in the pattern of adaptation is provided by plants that grow in soil containing toxic metals. Outcrops of serpentine rock produce soils that are full of toxic metals such as nickel, chromium and magnesium. Nevertheless, certain plants have been able to colonize even such an apparently hostile environment. These outcrops are usually sharply differentiated from the surrounding soils, which are free of heavy metal deposits, and plants are unable to cross the boundary.

Living in the Oceans

As on land, plant life in the world's oceans depends on the availability of light and nutrients. In the oceanic depths, where there is no light, no plants can live. But the accumulated skeletons of past plants and animals at the bottom of the ocean are a rich source of nutrients that remain unused until upwelling ocean currents bring them to the surface. Where this happens, as along the west coast of South America, for example, the abundance of plant life, and consequently of animal life too, is enormous.

The microscopic floating algae (phytoplankton) that are the descendants of the primeval one-celled algae from which all plant life originated form the basis of the marine food chain. Phytoplankton can be very abundant: 3 cubic cm (1 cubic ft) of seawater may contain as many as 12.5 million organisms. There are two main groups: the Chrysophyta (consisting of diatoms, golden-brown and yellow-green algae) and Pyrrophyta (dinoflagellates).

The diatoms have a crystalline, siliceous skeleton formed in two parts that fit neatly together and are frequently sculpted into fantastic and beautiful shapes. Some diatoms are not floating (or planktonic) but live on the sea bed in shallow water. Others are loosely held together in flexible chains linked by fine threads or strands of slime. Many of the golden-brown algae are mobile, powered by whiplike organs (flagella).

The dinoflagellates have a cell wall composed of small cellulose plates that fit together to form a mosaic. They are moved by two flagella, one working transversely in a groove almost girdling the body, the other projecting backward in a small longitudinal groove. The beating of the flagella causes the organisms to spin like tops, while their red "eye-spot" (stigma) guides them toward the light.

From time to time "red tides" build up on the surface of the sea, caused by accumulations of reddish dinoflagellates. The dinoflagellates contain toxic substances that do not harm the water-filtering animals, such as mussels and clams, that feed on them, but can nevertheless prove lethal to the animals, including humans, that eat the shellfish.

Chains of diatoms (*right*) *Diatoma elongatum* lives in fresh water; other diatoms – microscopic single-celled algae – inhabit oceans and even soil. Each species has a distinctive pattern of ridges and markings by which it is classified.

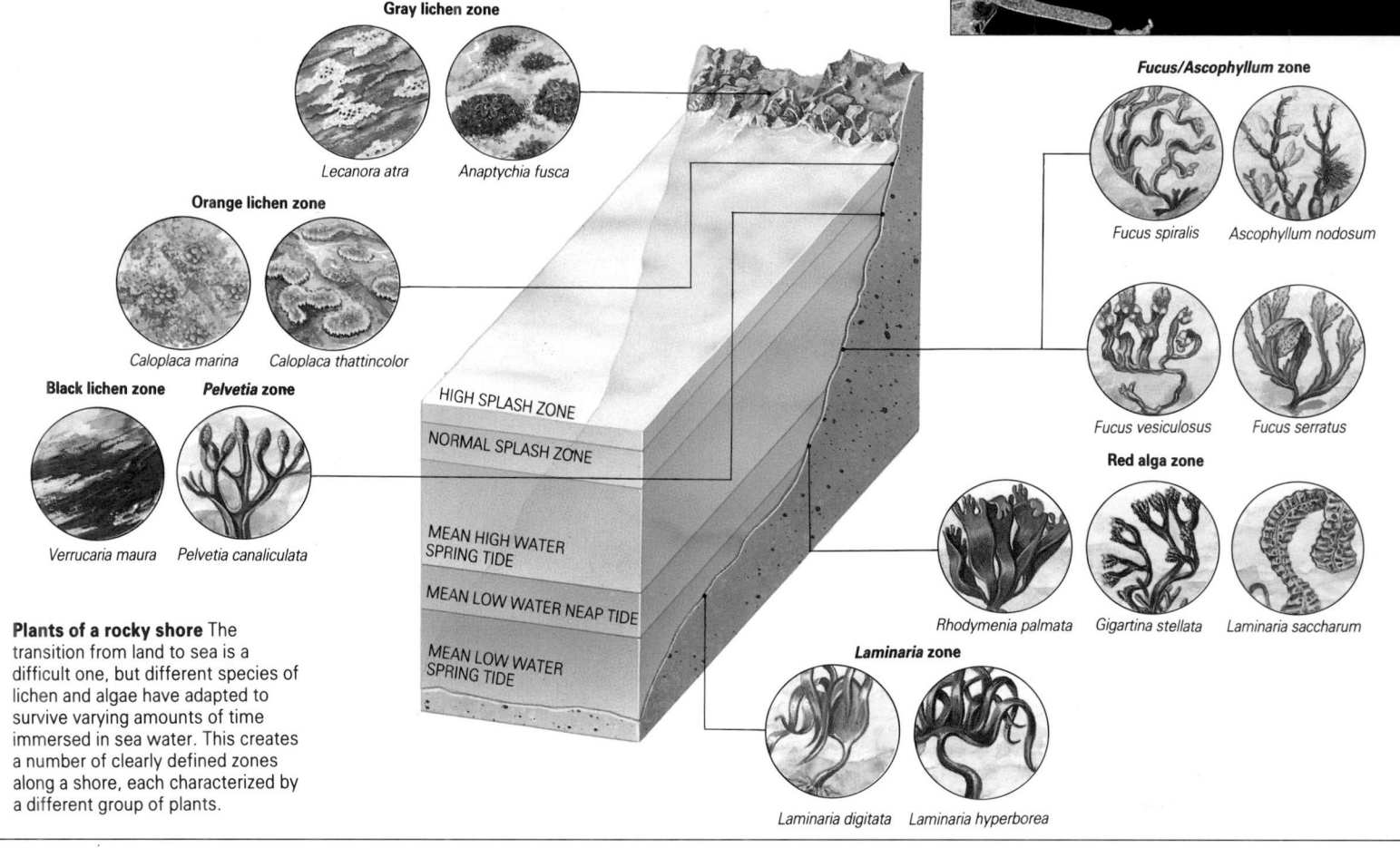

Gray lichen zone

Lecanora atra   *Anaptychia fusca*

Orange lichen zone

Caloplaca marina   *Caloplaca thattincolor*

Black lichen zone   ***Pelvetia* zone**

Verrucaria maura   *Pelvetia canaliculata*

***Fucus/Ascophyllum* zone**

Fucus spiralis   *Ascophyllum nodosum*

Fucus vesiculosus   *Fucus serratus*

Red alga zone

Rhodymenia palmata   *Gigartina stellata*   *Laminaria saccharum*

***Laminaria* zone**

Laminaria digitata   *Laminaria hyperborea*

HIGH SPLASH ZONE

NORMAL SPLASH ZONE

MEAN HIGH WATER SPRING TIDE

MEAN LOW WATER NEAP TIDE

MEAN LOW WATER SPRING TIDE

Plants of a rocky shore The transition from land to sea is a difficult one, but different species of lichen and algae have adapted to survive varying amounts of time immersed in sea water. This creates a number of clearly defined zones along a shore, each characterized by a different group of plants.

Kelps at low tide The kelps (*Laminaria*) are large brown algae with a broad, tough lamina and strong stem. A rootlike holdfast anchors them to the rocky shore. At low tide they often form a band along the edge of the sea.

Anchored to the sea bed

The most prominent plants of the shoreline are the multicellular algae, or seaweeds and kelps. These have organs (holdfasts) to attach them to the sea bed. The green seaweeds (Chlorophyta), which require unfiltered sunlight, live on the upper parts of the shore, between the tidelines or just below it. Some, such as *Cladophora*, form erect tufts of slender, branched fronds up to 15 cm (6 in) long, whereas the sea lettuce (*Ulva*) has flat fronds two cells thick and up to 1 m (3 ft)

long. Both seem to tolerate daily exposure to air but not savage battering by waves.

The pigmentation of brown (Phaeophyta) and red (Rhodophyta) seaweeds allows them to absorb the green, violet and blue wavelengths of light that can penetrate deep water. Some red seaweeds have been found at depths of as much as 175 m (575 ft) – deeper than any other living plant. Despite this ability, brown seaweeds, such as the fronded *Fucus*, grow most thickly in the intertidal zones, and are able to withstand the constant battering of the surf. In less turbulent coastal waters underwater jungles of *Laminaria* and the gigantic *Macrocystis* and *Mereocystis*, with fronds measuring up to 100 m (330 ft), are found.

Some groups of flowering plants, whose ancestors evolved on land, have now partially returned to the oceans. Conspicuous among them are the eel grasses (Zosteraceae). There are about 36 species in the 9 genera that make up this family, and they are found in coastal shallows from cool to tropical waters. They have no vascular systems; as far as is known they all have small flowers, which are usually unisexual. They produce pollen filaments up to 2 mm long, which lack an outer coat (exine) and are carried by water to the feathery stigmas. The adaptation of these flowering plants to an aquatic environment has been so successful that they have been found living at depths of 50 m (165 ft).

Islands of Plants

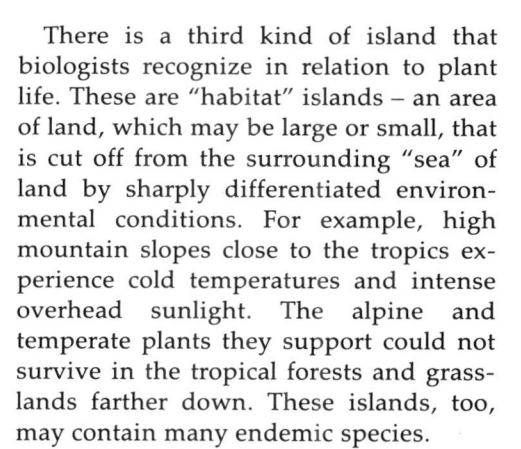

THE PLANT LIFE OF MANY OCEANIC ISLANDS IS of particular interest to biologists because they are inhabited by species that have evolved without competition from invading species, except when these have been introduced by people. This means that the plants they contain have been able to adapt and diversify to fill all the environmental opportunities (niches) the island offers. They therefore contain a high proportion of what are known as endemic species – those that are found nowhere else in the world.

Islands lying on a continental shelf are close enough to their adjacent landmasses not to have developed a discrete plant life. They experience the same weather patterns, and may once have been joined to the neighboring landmass, so they will have shared a common plant life before they were separated by rising sea levels. If they have been formed by volcanic activity, or as part of a coral reef, they will have been colonized by seeds dispersed from the nearby mainland by wind, water or animals. The plant life of the British Isles, for example, differs little from that of northwestern Europe, or that of Sri Lanka from the Indian Subcontinent.

There is a third kind of island that biologists recognize in relation to plant life. These are "habitat" islands – an area of land, which may be large or small, that is cut off from the surrounding "sea" of land by sharply differentiated environmental conditions. For example, high mountain slopes close to the tropics experience cold temperatures and intense overhead sunlight. The alpine and temperate plants they support could not survive in the tropical forests and grasslands farther down. These islands, too, may contain many endemic species.

Some oceanic islands are very old: many of those in the Pacific Ocean are fragments of the ancient supercontinent of Gondwanaland, which began to break up about 80 million years ago. New Zealand and New Caledonia are examples: the plants that grow on them include some of the most ancient groups known – plants such as the monkey puzzles (*Araucaria*) and podocarps.

The remote islands on the Kerguelen Ridge in the Indian Ocean are ancient remnants of Gondwanaland. It is unlikely that they were covered by ice, as they are home to the peculiar endemic Kerguelen cabbage (*Pringlea antiscorbutica*).

The plants of young oceanic islands have arrived more recently, by dispersal. For example, the islands of the Tristan da Cunha group arose in the middle of the South Atlantic about 20 million years ago. They were colonized by plants from South America, following the direction of the prevailing winds and ocean currents. Only 40 percent of the plants on these islands are endemic, compared with 81 percent on New Zealand.

Geological age is not an infallible indicator of an island's plant origins. The earliest plants of Macquarie Island, as of many other subantarctic islands, were wiped out during the last ice age, at its height some 20,000 years ago. The islands were recolonized by plants that reached them between 12,000 and 10,000 years ago. They have not yet had time to diversify, so there are relatively few endemic species.

Diversity and environmental change

An island like New Caledonia, which seems to have experienced relatively little environmental change since its formation, retains a high proportion of its ancient plants, with only a few modern families. New Zealand, on the other hand, has had a far more turbulent history. Between 38 and 26 million years ago a rise in sea level submerged much of its land area, and many ancient groups, such as most proteas, died out. In more recent geological times, between 7 and 2 million years ago, its mountain ranges were formed; this dramatic alteration of the environment triggered evolutionary activity among its plants.

The volcanic Hawaiian islands provide a further illustration of the correlation between environmental change and species diversity. Volcanic activity has never ceased here, and has created a great variety of environments from high peaks to deep valleys. There are both recent and ancient lava flows; the latter have gradually eroded to gravels and sands. These offer a wide range of environmental niches, and the number of species filling them now totals about 1,800. These have evolved from about 275 species that originally colonized the islands by dispersal from elsewhere in the Pacific.

High-altitude islands East Africa has several mountain islands rising above a lowland sea of tropical vegetation. The upper slopes are covered with a distinct vegetation, which includes shrubby *Hebenstretia dentata* and members of the heath family.

Plants as Food, Shelter and Clothing

IT IS AS FOODSTUFFS THAT PLANTS ARE CHIEFLY, and most obviously, used by people; human history is considered to begin when people learned to domesticate and cultivate plants. The essential ingredients of the human diet – carbohydrates, fats, proteins and vitamins – can all be obtained from plants.

The staple food of nearly all the world's people comes from those plants that provide carbohydrate starch. In grain plants such as wheat, maize and rice, starch is stored in the seeds as nutrient for the developing embryo; it is stored underground in the tuberous roots of the tropical sweet potato (*Ipomoea batatas*) and the cassava or manioc (*Manihot esculenta*), as well as the stem tubers of the temperate potato (*Solanum tuberosum*) and tropical yam (*Dioscorea*).

The carbohydrate sugar is gained in quantity from the stems of sugar cane (*Saccharum officinale*) and the thickened roots of sugar beet (*Beta vulgaris*). Sugars are also provided by the fleshy fruits of innumerable temperate and tropical plants; these are also rich in vitamins, as are the green leaves of many plants.

Oils (fats) are obtained from the fruits or seeds of the olive (*Olea europaea*), sunflower (*Helianthus annuus*), maize (*Zea mays*), soybean (*Glycine maxima*), peanut (*Arachis hypogaea*), rape (*Brassica campestris*) and sesame (*Sesame indicum*). The richest sources of plant protein are the seeds of the many members of the pea family (Leguminosae), such as mung, kidney, lima and runner beans (all species of *Phaseolus*), broad beans (*Vicia faba*), soybeans, jack beans (*Canavalia ensiformis*) and peas (*Pisum sativum*).

Of lesser importance for nutrition, but greatly valued for their enhancement of flavor, are the many spices derived from the seeds, fruits and roots of various plants, as well as culinary herbs from their leaves. The leaves of the tea plant (*Camellia sinensis*) and the beans of coffee (*Coffea arabica*) and cocoa (*Theobroma cacao*) provide stimulating beverages.

Rolling wheatfields (*above*) Wheat is one of the many grasses that has had its cropping ability greatly improved over that of wild species. Grasses not only provide a staple food for humans, but are also grown to provide nutritious grazing for their beasts. Among the most important forage crops are ryegrass (*Lolium*), bluegrasses (*Poa*), buffalo grasses (*Buchloe*) and cocksfoot (*Dactylis*).

Preparing cassava (*left*) Native to South America, cassava (*Manihot esculenta*) has become an important food crop throughout the tropics, particularly in Africa. The large, swollen tuberous roots are peeled, boiled and mashed into a sustaining meal. Tapioca is a refined product of the plant.

Woven shelters (*right*) The stems of reeds and grasses have a multitude of uses. They can be woven into baskets, used for bedding, and make an insulating, weatherproof roof. These plaited grainstores are both functional and decorative.

The many uses of wood

For almost as long as people have been cultivating food plants they have been using the wood of trees and shrubs to provide shelter, as handles for their implements, to make barrels and cases, boats and waggons, and in countless other ways. Until recently wood has been the principal source of fuel, and still is in many parts of the world today.

The qualities of toughness, durability and elasticity for which wood is valued derive from the type and size of the cells of which it is composed, the thickness of the cell wall and the nature of the chemicals deposited in them. The principal cells that form wood have their cells strengthened by lignin. Fiber cells give the plant elasticity, so that it does not snap under stress.

Wood-producing flowering plants are known as hardwood trees: the cells of the xylem are joined end to end to form vessels. They are tougher than the tracheids, cells that permit the flow of sap in conifers, consequently known as softwoods. The terms are confusing: parana pine (*Araucaria angustifolia*), a conifer, has tough wood suitable for shelving; the tropical balsa (*Ochroma*), a hardwood, produces a lightweight wood used for rafts and model airplanes. The colors of hardwoods obtained from tropical rainforest trees such as teak (*Tectona grandis*) and mahogany (*Swietenia*), ranging from deep yellow to deep red-brown, and also their decorative grain, make them especially prized for furniture making.

Many of the fiber cells in plants are extracted, combed and then spun into strands that can be woven to make cloth and textiles. Cotton is obtained from the bundles of hairs that cover the seeds of the cotton plant (*Gossypium*); the kapok tree (*Ceiba pentandra*) produces similar fluffy fibers, especially useful for insulation. The stem fibers of flax (*Linum usitatissimum*) are extracted by soaking the plant so that the other tissues decompose; they produce linen, perhaps the oldest textile in use. The stem fibers of jute (*Corchorus*) are similarly treated to make rope and sacking, and the leaves of *Agave sisalana* yield sisal, used to make binding twine and string. Paper is made by putting woodpulp in a water suspension and laying down the fibers on a screen – though much paper is recycled, the industry is a major consumer of the world's coniferous forests.

Plants in Ritual and Healing

FROM EARLIEST HUMAN TIMES CERTAIN TREES seem to have had particular mystical meaning for certain peoples – perhaps enshrining a common folk memory of when our ancestors lived in the forests as hunter–gatherers. Examples of the reverence given to particular trees include the oak, worshipped by the Celtic peoples of pre-Christian northwestern Europe; the ash tree, Yggsdrasil in Scandinavian mythology, whose branches and roots extend through the universe and support it; the fig tree revered by Romulus, one of the twin founders of the city of Rome; the sacred peepul (*Ficus religiosa*) of Buddhists; and the sausage tree (*Kigelia pinnata*), regarded with awe in much of tropical Africa.

Plants in ritual

Plants have long been associated with religious ritual. At least 5,200 years ago in Egypt the bodies of important people were embalmed to preserve them from decay. Aromatic spices such as myrrh (*Commiphora abyssinica*), cassia and cinnamon (*Cinnamomum*), cumin (*Cuminum cyminum*), anise (*Pimpinella anisum*) and others were used to plug the body cavities. The Egyptian pharaohs burned incense (plant resins mixed with spice) and poured an offering of wine into a bowl of sacred lotus flowers (*Nelumbo nucifera*) in front of an image of the god Sekhmet in the hope of receiving strength and bravery. The gifts present by the magi, or three wise men, to the infant Jesus included two plant resins, frankincense and myrrh, both of which were vested with religious significance. Incense is still burned in many religious ceremonies today.

Plants that induce visions and hallucinations have also long been used both in religious ceremony and as a pleasure stimulant to take the edge off human pain and misery. The peyote cactus (*Lophophora*) produces mescal, a stimulant drug used ritualistically by the indigenous Amerindians of Central and North America; the leaves of the coca plant (*Erythroxylum coca*) were similarly used by the peoples of the Andes. They contain an alkaloid, cocaine, which acts as a stimulant by raising the heart rate and blood pressure. The leaves are chewed by Andean people to ease hunger and the burden of everyday living; in its processed form as cocaine or crack, the substance is potentially much more harmful.

Cannabis (*Cannabis sativa*) – hashish or marijuana – is another widely used stimulant. It was featured in the rituals of the Assassins, a fanatic Muslim sect of the 11th and 12th centuries.

Alcohol, produced from the fermented sugars of many plants, is the most widely used stimulant of all: much human ingenuity has gone into ways of obtaining and distilling it. Bananas, maize and barley give beers; wheat and rye, whiskey and vodka; grapes, wine; the agave, pulque and tequila – the list is seemingly endless. Alcohol has symbolic importance in many religions, one well-known example being the offering of wine that plays a central role in Christian ritual.

Cures for human ills

In very many cultures healing has been associated with religion, and the curative properties of many plants have long been known to – and exploited by – priests, shamans and medicine men. Western orthodox medicine in recent centuries has increasingly tended to dismiss the use of healing plants as "folk medicine" of little practical value, but traditional remedies have always had widespread popular use. To give a few examples, feverfew (*Chrysanthemum parthenium*), as

THE PLANT PHARMACOPOEIA

Modern medicine relies on many plant products in the control of disease and the relief of pain. Among them is the bark of the *Chinchona* tree, found in the Amazonian forest, which is the source of quinine, widely used to prevent malaria and to reduce fever. The dried juice or latex exuded by the opium poppy (*Papaver somniferum*) provides the substances for the painkillers morphine and codeine. The poppy is a native of the Mediterranean, but is now intensively cultivated in South and Southeast Asia: much of the harvest supplies the illegal drug trade. Another painkiller, aspirin, originally derived its active ingredient salicin from the bark of the willow (*Salix*); the inner bark of the tree *Liquidambar orientalis*, from the eastern Mediterranean, exudes storax resin, used in inhalants for bronchial infections.

Substances manufactured by certain plants to defend themselves against predators can be highly toxic to humans. In controlled doses, their physiological effects are beneficial in medicine. For example, curare, a substance obtained from the tree *Strychnos toxifera*, was used by Amazonian Indians to poison the tips of their arrows: it paralyzed their victims. Today it is used as a muscle relaxant in surgical operations. The leaves of the common foxglove (*Digitalis purpurea*) are the source of digitoxin, which stimulates the heart to beat very rapidly, causing death. It is used to treat heart failure. Atropine, obtained from the highly poisonous deadly nightshade (*Atropa belladonna*), is used to dilate the pupil of the eye for surgical examination.

Many people both within the medical profession and outside it are coming to realize how little we have tapped the full potential of plants as a tool in overcoming disease. It is only in very recent years that the Madagascar periwinkle (*Catharanthus rosea*) was discovered to be a source of leurocristine, found to be effective in the treatment of certain childhood leukemias. The periwinkle is endangered by the destruction of its natural habitat in Madagascar. The fear is that there may be many other plants with equally important life-saving characteristics that have already become extinct.

A chapel in a yew tree (*above*) The early Christians frequently chose to build their churches on sites that had been used for pagan worship, often by a yew tree. These dark spreading trees, an ancient symbol of death, are still grown in graveyards.

Medicinal plants for sale (*right*) Plant material continues to be widely used for medicinal and superstitious purposes. In many parts of the world plants are collected and sold in markets to people living in the towns.

its name suggests, is used to bring down fever; arnica (*Arnica montana*) reduces bruises and swellings; slippery elm (*Ulmus fulva*) has been resorted to by women to promote miscarriages.

Alternative or complementary systems of medicine – those that are not officially recognized by the practitioners of orthodox medicine – are nevertheless gaining wide acceptance among people in the Western world. They lay great stress on plant remedies. Herbal medicines play an essential part in the remedies of homeopathy, for example, which treats disease by administering minute doses of substances that would, in a healthy person, produce symptoms of the disease that is being treated.

Plants for Pleasure

THE CULTIVATION OF PLANTS FOR THEIR beauty alone has long appealed to the aesthetic side of human nature. The earliest surviving evidence of gardening comes from a plan showing the estate of an Egyptian official. It dates from about 1400 BC, but it is likely that designed pleasure gardens existed some time before that. The garden is laid out on symmetrical lines with ponds, treelined avenues and pavilions. Similar pleasure gardens, where shade and cool water could be enjoyed, were important in the Assyrian, Babylonian and Persian cultures of the Middle East. They were later taken up by the Greeks and Romans, and spread into Europe with the expanding Roman empire.

With the disintegration of the Roman empire in the 4th and 5th centuries AD, the tradition of gardening disappeared from Europe for several centuries, upheld only in the monasteries where herbs were grown for medicinal purposes. Gardening was kept alive in the Mediterranean and the Middle East by the Arabs. From the 7th century onward the idea of the pleasure garden, based on the Persian model, spread throughout the Islamic world – westward into Egypt, North Africa, Sicily and Spain, and eastward as far as Mughul India. Water was a major features of these Arab gardens – splashing fountains cooled the air and soothed the ear, providing for desert dwellers an oasis of peace in an often turbulent and arid environment.

The garden reappeared in Europe with the 15th-century Italian Renaissance. The return to ideas of classical proportion meant that villas and gardens were planned to complement each other: statuary moved outdoors, and planting patterns were formal and symmetrical. As these ideas spread through Europe gardens became more ornate, particularly in France. The Dutch added topiary (elaborately trimmed trees and shrubs) and adopted the Turks' expertise in breeding bulb plants, particularly tulips.

During the 18th century there was a move away from formal gardens toward more "natural" vistas. Groups of trees and

An exotic collection of plants The subtropical garden of Tresco Abbey, in the Scilly Isles off southwest Britain, is rich with plants collected for more than 150 years from countries bordering the Mediterranean, from Australia, New Zealand and South Africa.

shrubs were planted to form clumps in open, landscaped parkland that apparently owed nothing to human artifice. The Englishman Lancelot "Capability" Brown (1716–83) was renowned for producing some of the most notable examples of this art, which involved large-scale engineering of the landscape to create artificial lakes and hills.

At this time European plant collectors began to scour the world for plants of horticultural value, introducing new families and species to Europe, and transmitting them to other places in the world. Very many of the ornamental plants introduced to European gardens in the 18th

Paradise garden (*above*) Persian gardens were usually formal in layout, with pools and canals representing the waters of life. They were planted at random with many flowers, including bulbous plants and roses, and shade trees to create a perfumed oasis.

A living collection (*left*) The Royal Botanic Gardens, at Kew in England, have nearly 30,000 species of plants that have been collected from all over the world either for their beauty, for their economic importance or their potential usefulness.

(*Alcea rosea*), petunia and lobelia, which are not. A garden in southern California might combine native lilacs (*Ceanothus*), Australian kangaroo-paw (*Anizoganthus*) and Spanish red valerian (*Centranthus*).

The international character of horticulture is reflected today in the mingling of Western styles with those from other parts of the world. The gardens of China and Japan have always reflected a keen pleasure in natural scenery: symmetry is avoided, and garden features such as bridges and garden buildings are designed to harmonize with the trees and other plants around them, according to strict rules. In Japanese gardens the precise placing of rocks, carefully selected for their color and shape, in decorative patterns that have symbolic meaning, is also of considerable importance.

The search for plants for the world's gardens still continues. Plant collectors today have to satisfy stringent international legal requirements passed to protect the world's natural environments, but valuable plants continue to be cultivated by plant breeders. By selecting characteristics from naturally occurring plants, and mixing them with those of artificially produced hybrids, the gardens of the world continue to be enriched with new, improved varieties.

and 19th centuries came from the Himalayas and from China. Trees and shrubs from all over the world were brought together in botanical gardens and arboretums. Not only decorative plants were sought: naturalists also visited remote regions in South America such as the Amazon to hunt for plants of agricultural and commercial value.

As a result of these endeavors gardens now contain a rich mixture of native and introduced garden plants. The traditional English "cottage" garden will be planted with foxglove (*Digitalis*), columbine (*Aquilegia*) and larkspur (*Consolida*), which are native, and with hollyhock

THE ROLE OF THE BOTANICAL GARDEN

When the plant collectors brought their new plants home, they were usually deposited in botanical gardens – living museums, often furnished with glasshouses, where rare and special plants could be displayed and the relations between the plant groups and species studied. These gardens had their origins in the "physic gardens" of university medical schools, where plants were grown to make medicine. The earliest of these, still extant, was established at Padua, Italy in 1542.

By the end of the 18th century there were, according to one reckoning, 1,600 botanical gardens in Europe, reflecting the current scientific interest in botany. Many were founded as places to assess and "acclimatize" the new plant introductions from abroad, and were closely linked with colonial exploration and expansion. The botanical garden at Orotava in the Canary Islands, established in 1792, was a receptacle for plants from the Spanish empire. The world famous Royal Botanic Gardens at Kew, England were established as a

national collection in 1841; the gardens there date back to the 17th century and include plants from all over the British empire. Kew was particularly important as a center both for studying plants of economic value, and also for establishing them in suitable environments throughout the world; it was responsible for the worldwide distribution of rubber (*Hevea brasiliensis*), cocoa (*Theobrama cacao*), bananas, pineapples, tea and coffee.

As the botanical gardens grew and became more varied, they were increasingly popular as places to visit. Their living collections of plants continue to provide a valuable resource for plant classifiers and scientists. Botanical gardens have reacted to the recent emphasis on the importance of conservation by studying ways of propagating species that have become extinct in the world, or are in danger of doing so. Much current work is concerned with establishing taxonomic connections between plants using the most up-to-date methods of genetic research.

Disturbing the Habitats

WHEN HUMAN COMMUNITIES WERE SMALL and scattered, the selective removal of particular plants for food, shelter or clothing did not unduly disturb the natural balance of the habitats in which they grew. But as human populations increased, people's exploitation of plants soared too, with devastating effect on their natural communities.

Disturbance has been particularly massive in the forests. Some 400 years ago vast numbers of the oaks of Britain (*Quercus robur*, *Q. petraea*) and of Spain (*Q. rotundifolia*) were felled to satisfy those countries' seemingly insatiable demand for ships to fight their wars, at the same time opening up the land for grazing and crops. Much of the forest that once covered eastern Northern America was similarly exploited for timber for construction and export, while creating open, cultivated land. The gum tree (*Eucalyptus*) and red cedar (*Toona australis*) forests of southeast Australia were cleared for the same reasons. Today, tropical rainforests around the world are being felled to meet the international demand for valuable hardwood timer.

In general, trees such as ebony (*Diospyros*), teak (*Tectona*), and mahogany (*Swietenia*), which provide the most desirable hardwoods, grow as scattered individuals or in small groups. The felling of a single tree normally causes little disruption to the habitat, but once logging is carried out as a commerical operation it creates widespread damage to the forest around. In Cameroon, for example, logging has destroyed many bush mangoes (*Cardyla pinnata*), whose fleshy fruit-pods are an important item in the diet of local people.

However carefully undertaken, commercial logging means that roads or tracks have to be built to give access for machines. This exposes soil, which then becomes eroded – a particular problem in the tropics, where underlying laterites (rocks containing a high amount of iron and other oxides) make the soil increasingly unproductive. In addition, the removal of certain species from a plant population, no matter how sensitively carried out, inevitably alters its genetic structure: the "best" are harvested, so succeeding generations are reproduced from poorer quality members.

It is more profitable to clear-fell an area of trees and then extract those species that are especially desirable; the rest are normally burned. In theory, the cleared areas should be replanted. If they are, however, it is usually with non-native single species, and the diversity of the original forest is lost forever.

From grassland to grazing land

The natural grasslands of the world were the first areas on which people grazed their newly domesticated animals. The prairies of North America, the pampas of South America and the steppes of central Asia had clearly been grazed by wild herbivores long before domestic herds appeared on the scene. But since then the impact of domestic animals has been very great indeed.

Not unnaturally, animals selectively graze the plants they find most palatable, choosing softer-leaved grasses and herbs (forbs) rather than hard-leaved "bunch" and "tussock" grasses. This selective grazing has considerable effects, which go beyond the alteration of the balance between the plant species. Most of these softer-leaved plants grow in areas where water is abundant. The animals that gather in large concentrations to eat them compact the soil by the constant trampling of their hooves, so it can no longer retain the water to support the grasses.

Some animals are not particular about what they eat. Goats, for example, have become the scourge of dry Mediterranean or African hillsides, where their overgrazing has stripped the land of all plant cover, exposing the soil to widespread gullying and erosion.

Threatened bulb *Cyclamen mirabile* is a plant at risk through overcollection. The Convention on International Trade in Endangered Species (CITES) is now enforced by many governments to protect such plants, though it is difficult to monitor the trade.

Sweeping away (*left*) All plants, and trees in particular, stabilize the soil with their network of roots and protect it from being washed away by heavy rain. When forests are cleared there is nothing to prevent water and wind from eroding the soil.

The forest is destroyed (*above*) Every year tens of thousands of square kilometers of the Brazilian rainforest are cleared. The trees are not felled to extract saleable timber but burned simply to clear the land for ranching.

PLANTS AS PLUNDER

The gardens of the world have always been enriched by bringing in plants from the wild, which are then crossbred to produce hybrid garden species. Bulb plants such as tulips, hyacinths and crocuses, which grow naturally in southern Europe and the eastern Mediterranean, were introduced to western Europe in the middle of the 16th century from the court of the Ottoman ruler Suleiman the Magnificent (1520–66), where they were highly valued. Their rise in popularity made tremendous inroads into the natural plant populations; their sale to collectors provided a welcome source of income for local people. Cyclamen, another horticultural favorite native to the same region, suffered a similar fate.

Other major plant groups have been seriously depleted by the attention of enthusiasts. In Southeast Asia and South America, orchids belonging to such genera as *Cattleya*, *Dendrobium*, *Paphiopedilum* and *Vanda* have been collected almost to extinction. The determination of some collectors to acquire rare specimens at any cost is so great that this plunder continues in spite of internationally enforceable protective legislation, with the result that there is now a considerable smuggling trade in endangered species. Devotees of succulent plants have contributed to a serious reduction in the population of New World cacti. Depredations have been especially marked in Mexico, whose natural plant life is exploited to supply an enormous market in the United States.

Pressure from People

IT ALL STARTED WITH THE STONE HAND-AX AND with fire, which gave our early ancestors power to clear the forests. Later they used hoes and digging sticks to turn the soil for cultivation, and then adopted wooden plows that they either pulled themselves (as some still do today) or harnessed to oxen or horses; later still the plows were made of iron or steel. Today the tools people use are far more efficient and devastating in their effects – vast mechanical plows, chainsaws and bulldozers. In open land, plowing breaks up the grassland turf that stabilizes the soil. Where the climate is dry and windy, the plowing up of soils can have disastrous effects. Natural prairie in the central and southern United States, converted into wheatlands in the 1930s, soon turned into "dustbowls" instead – arid, devastated landscapes capable of supporting very little plant life at all.

As populations grow, more and more land is need for housing, transport, industry and leisure, as well as for farming. The construction of huge new highways and of sprawling townships with their commercial and industrial fringes removes great swathes of natural plant cover. In clearing the land, the bulldozers and other earthmoving machines create areas of open or disturbed ground. When it is not immediately used for building, it is rapidly colonized by so-called "weedy" plant species. Weeds are, in effect, plants that are out of place. Gardeners and farmers open up land to cultivate their favored flowers or crops, and then wage war against those "unwanted" plants that exploit the open soil.

New highways, too, provide open ground for colonization along their margins. In cool, temperate regions these are rapidly taken over by plantains (*Plantago*), dandelions (*Taraxacum*) and other flowering plants. The yardgrass (*Eleusine indica*) has similarly spread along roadside verges in western Africa and other tropical regions.

Highways and railroads have had other modifying effects on natural plant communities. In cold parts of the world, where salt is thrown on the roads during winter to prevent the formation of ice, the roadside soils become salty. This makes them suitable for salt-tolerant plants, usually found in coastal environments, and these have consequently been able to spread far inland. Oxford ragwort (*Senecio squalidus*), a native of the volcanic soils of Etna in Sicily, was introduced to the Oxford Botanic Garden in the 19th century. It soon established itself as an alien weed species, spreading rapidly along railroad embankments, where it grows in the clinker that lines the tracks. The California poppy (*Eschscholzia californica*) similarly follows the railroad and highway verges of central Chile. Wherever suitable niches are created, there are plants to exploit them.

Pressure on coastal environments

In recent decades human activity has pressed particularly heavily on the plants of coastal environments. All around the coasts of the Mediterranean, in California and Florida – and in other places favored by vacation-makers in search of the sun – hotels, apartment blocks, villas and trailer parks have sprung up, causing widespread havoc to plant communities.

In addition, the sewage waste from these developments, which is at best only partially treated, is pumped into the adjacent oceans, with serious consequences for seaweeds and other forms of marine life. Noxious "algal blooms" are an increasing problem along many stretches of coastline. The construction of barrages across estuaries, which use river currents and tidal surges to generate electricity, has also seriously affected coastal habitats. So has the widespread drainage of coastal wetlands for reclamation and cultivation.

Pressure from tourists (*above*) The increasing interest in the wild countryside has resulted in many more people visiting conservation areas. These reserves need to be carefully managed in order to avoid the destruction of the often fragile vegetation because of the sheer number of visitors.

Destroyed by nomads (*left*) This acacia tree was badly damaged in order to collect honeycomb. It is difficult to control destruction of plants in poor, arid lands by an increasing number of people who subsist from day to day.

Poppies by the roadside (*right*) In developed countries native plants are being pushed to the margins, to areas of land that cannot be exploited. In this way highway verges and railroad embankments have become havens for wildflowers, especially in countries where local authorities no longer use herbicides as a matter of course.

Damage from Pollution

ENVIRONMENTAL DISASTERS NOW CAPTURE the headlines. In the 1980s there was a series of incidents, such as the massive oil spillage from the tanker *Exxon Valdez* off the coast of Alaska, the death of vast stretches of coniferous forest in Germany and Scandinavia as the result of acid rain pollution, and the widespread and long-lasting damage caused by the release of radiation from the damaged Chernobyl nuclear reactor in the Soviet Union. These drew public attention to the harm that increased industrial activity inflicts on plant and animal communities, as well as on humans. The scale of the damage caused to plant life by the constant release of noxious chemicals into the atmosphere is much more widespread and unrelenting than these single incidents suggest.

Pollutants reach plants as solid particles or gases in the air, or suspended in water solution. It is in this latter form that they are absorbed through the soil. In industrial and urban areas particles of carbon in smoke, deposited indiscriminately as unsightly black soot, are the most obvious airborne pollutant. In many countries they have been the target of "clean" air legislation to eliminate coal burning in towns and cities. Deposits of soot undoubtedly reduce the photosynthetic efficiency of leaves. However, the unseen damage from sulfur dioxide, associated with such particles, is of greater consequence for plant life, and this pollutant,

along with ozone and various fluorides, continues to be poured into the atmosphere by coal and petroleum-burning industries and by automobile exhaust.

It is these substances that are the cause of acid rain. Released into the atmosphere, they are oxidized and then dissolve to form acid droplets in the rain clouds. They fall as snow, hail or rain onto vegetation and into lakes and rivers, causing widespread damage to plant life. Leaves lose their green pigment, chlorophyll, so the plant cannot photosynthesize, and then fall.

Heavy metals such as copper, zinc and lead, which are released into the soil from the spoil heaps thrown up by mining, are also harmful to plant life. They inhibit root growth, eventually causing the plant to die. Lead is also given off in automobile exhaust, and is responsible for killing roadside trees and shrubs. Other poisonous substances, such as cyanide and mercury, are pumped into waterways as industrial waste.

Coping with poisons
Strong concentrations of all these chemicals result in the almost total loss of plant cover. However, some plant species are more resistant to poisons than others, which means that areas that are denuded of plants will eventually be reclaimed. By studying these resistant species, we are able to gain some understanding of how

Plants in the city (*left*) European blackthorn (*Prunus spinosa*) flourishes by a motorway. Some plants are better able than others to cope with urban pollution by restricting the movement of metals from their roots to the rest of the plant.

Metal indicators (*above*) It has long been known that certain wild plants, such as Jacob's ladder, indicate metal deposits. Such indicator plants have evolved to take advantage of the relative freedom from competition that such toxic soils provide.

plants are affected by pollution, and how they are able to cope with it.

There seems to be no scientific evidence to support the suggestion that the leaves of resistant plants contain fewer openings, or stomata, than others, so reducing the exchange of gases. What seems clear is that these substances reduce the plant's ability to photosynthesize and that they are more harmful in fluctuating concentrations than if they are constantly present – in other words, plant cells are able to adapt to steady toxic pressure. How they do so is not understood.

Pollutants appear to interrupt the metabolic processes that take place within the plant cells, altering rates of photosynthesis and growth. In some plants, such as spruce (*Picea*), vetches (*Vicia*) and tobacco (*Nicotiana*), these processes are increased above their normal rate; in others, such as maize (*Zea*), beet (*Beta*) and oranges (*Citrus*), they are lowered or inhibited altogether. Both effects are harmful to the functioning of the plant.

Serpentine endemic *Garrya elliptica* is one of a limited number of plants that grow on serpentine soils. Although they thrive in the presence of heavy metals, even these tolerant plants grow more successfully on nonserpentine soils.

There is considerable evidence to show that the key to a plant's ability to withstand pollution lies within its cellular structure. In the old days of metal prospecting, the presence of certain wild plants in an area was taken as a sure sign that metal deposits lay in the rocks beneath. In California, for example, Jacob's ladder (*Polemonium*) and California phlox indicate the presence of serpentine rocks, which are rich in nickel, chromium and magnesium. Plants like these appear to adapt to the presence of heavy metals in the soil by the action of certain proteins within the cell walls. These bond with the pollutants to stabilize them, and in this way they are prevented from entering the cells themselves to disrupt the plant's life processes.

Completing the Inventory

IN THE PAST, THE IMPORTANCE OF CONSERVING and protecting plants in their natural environments took second place in the popular imagination to the need to save endangered animals. The plight of animals such as pandas, koalas, seals and elephants had more immediate appeal, and encouraged people to reach into their purses to support programs to protect individual species, rather than whole ecosystems. In the long run, however, these schemes have played an important part in fostering a deeper understanding of conservation issues generally. Other factors, such as television wildlife programs and the media attention given to such problems as the greenhouse effect and rainforest depletion, have made many people much more aware of the vital role played by plants in maintaining the Earth's ecological balance.

Today it is being increasingly stressed by scientists, conservation organizations and funding agencies that plants, which provide energy for nearly all the other living organisms on Earth, are the underpinning for the world's fragile ecosystems. But reasoned decisions about what to conserve, and how to do it, have been hampered by incomplete knowledge of what plants exist, and to what degree they are threatened by extinction.

Recording the data in time

In the last 20 years the business of gathering information about plants has been tackled with tremendous vigor. The first list of threatened plant species was published in 1970. Since then *Plant Red Data Books*, published by the International Union for Conservation of Nature and Natural Resources (IUCN), have been produced for most regions of the world. These give information about plant species considered to be rare, vulnerable, threatened or probably extinct, based on detailed plant surveys; nearly three-quarters of the land area of the United States, most of Western Europe and Australasia, a quarter of the Soviet Union, a sizable amount of Central America, and parts of northern and central Africa, the Indian subcontinent and Southeast Asia have been covered. The data for many other areas are not yet available, and are particularly lacking in many of the

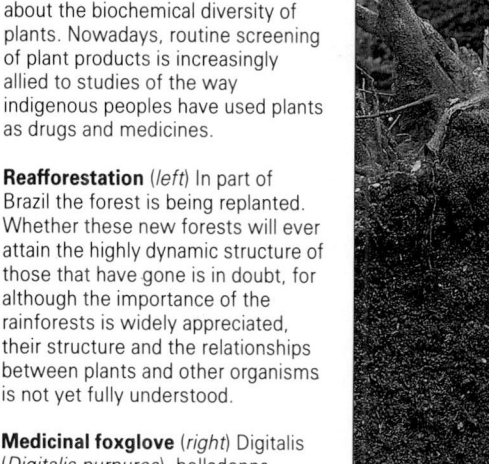

Secrets of the forests (*above*) As forests are being destroyed many plants of potential value go with them. In some lands species are being extinguished before they have been described. Too little is known about the biochemical diversity of plants. Nowadays, routine screening of plant products is increasingly allied to studies of the way indigenous peoples have used plants as drugs and medicines.

Reafforestation (*left*) In part of Brazil the forest is being replanted. Whether these new forests will ever attain the highly dynamic structure of those that have gone is in doubt, for although the importance of the rainforests is widely appreciated, their structure and the relationships between plants and other organisms is not yet fully understood.

Medicinal foxglove (*right*) Digitalis (*Digitalis purpurea*), belladonna (*Atropa belladonna*) and quinine (*Cinchona*) are examples of well-established plant-derived drugs. At least 20,000 chemical compounds are known from plants, many of them of considerable importance in the pharmaceutical industry.

world's tropical regions, such as South America, where the greatest diversity of plant species is found.

The size of the threat to plant life on the planet cannot realistically be assessed until we know what plants exist. It is estimated that 10 percent of the flowering plants have still not been described. This is the best-known group; the situation is worse for the other plant groups. Taxonomy – the recognition and description of species – a branch of science that has a long and distinguished history, has fallen somewhat out of favor in relation to other areas of plant biology in recent years. Yet such studies are crucial if the world's invaluable plant resources are to be charted and analyzed. The current impressive and encouraging interest in plant conservation is now spearheading endeavors to explore and describe all the Earth's plant life, but most especially in the tropical regions where information is needed so badly.

Can the inventory be completed? Almost certainly not. It has been estimated that about a thousand plant species a year are becoming extinct, or could become extinct by the middle of the next century. One such species is a relative of the ebony tree, *Diospyros hemiteles*, which grows only on Mauritius. It was first described in 1980, from preserved specimens. In the wild only one, female, tree is known to exist, so the survival of the species in its natural environment is doomed.

On the other hand, increasingly sophisticated detailing of plant resources can reveal that species known to be extinct in the wild still exist in botanical gardens. A single specimen of Easter Island's only tree, *Sophora tormiro*, was found growing in Göteborg in Sweden; the extinct British grass *Bromus interruptus* survives in the Edinburgh Botanical Garden in Scotland, and the Romanian endemic primrose, *Primula wulfenia* subsp. *baumgarteniana* is found only in the Botanic Gardens at Meise in Belgium. Even such apparently small additions to our knowledge are vital to build up our understanding of the world's plant life.

Safeguarding the Future for Plants

THE RECOGNITION THAT PLANTS SUSTAIN ALL life on Earth is at the heart of the conservation movement. However, in many parts of the world, where population growth is soaring, the immediate use of plants to supply food, heat and shelter today understandably seems more important than concern about what might happen to the planet tomorrow. Against the overwhelming need to feed all the people of the world into and beyond the next century, what can be done to ensure that plants – and, by extension, humans as well – have a future?

The surest way to preserve plants is to protect their natural environments – a strategy known as *in situ* conservation. Yet the designation of national parks and other protected reserves in which plants (and the animals that depend on them) are sheltered from all forms of exploitation, though of enormous value, does not wholly solve the problem. Protective legislation is rarely totally inflexible in the face of economic considerations such as logging and mineral extraction. People living within the reserved areas need to pursue their livelihood through farming or some other means of support.

One way to solve these problems is to introduce buffer zones around the reserves, in which varying degrees of exploitation are allowed under certain restraints. Restrictions are total at the center; yet even here human pressure can impinge. Many reserves are visited by people for purpose of scientific research or for recreation – and even the best-intentioned visitors will upset the biological balance of the ecosystems. Some reserves may restrict visitors altogether, others limit daily quotas.

There is still no general agreement among conservationists about the minimum area of land that is needed to protect all the species living within it. Requirements will vary according to the region and the type of plants. Much research is currently being carried out into this aspect of conservation.

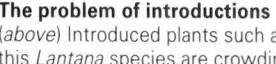

The problem of introductions (*above*) Introduced plants such as this *Lantana* species are crowding out the native species in parts of Hawaii. Alien weeds aggravate the problems caused by grazing animals on the Marquesas Islands in the southern Pacific by colonizing disturbed habitats. These are two of many instances where introduced weed species have adapted very successfully, thriving in habitats where the native plants are under increasing pressure as a result of the activities of humans.

Replanting trees in Africa (*left*) Projects are being funded to encourage local people in many parts of the world to replant trees. This is a longterm strategy, as a forest takes a thousand years to evolve and cannot be reestablished in a decade.

Soil erosion (*right*) When Philip Island was discovered in 1774 it was densely forested. A few years later pigs were introduced, then goats and rabbits. Between them these animals have steadily destroyed much of the vegetation. Recent experiments have, however, shown that native tree seedlings planted in the bare earth will grow well, so regeneration of the island's natural vegetation seems possible.

Virgin areas are not the only targets for protection; vegetation in areas formerly opened up by human development can be regenerated, with careful management. This is not simply a matter of sowing the seeds of native plants along the verges of highways. Alien species introduced by humans must be eliminated – first of all domestic livestock, and then other animals such as rabbits and goats.

Control of these animals, which have frequently become pests, can be very difficult, but there have been some successes. On Philip Island in the South Pacific, vegetation was restored on denuded land within six months of the island's population of rabbits being destroyed. The same has happened on Campbell Island, south of New Zealand, though here a small number of alien plants have become integrated into the plant cover.

Alien plants, introduced deliberately or inadvertently, can be just as destructive to the native plant life as animals. Much research is now being carried out into the best methods of clearing them out. In the Galapagos Islands off the coast of Ecuador invasive plants such as the quinine-related *Cinchona succirubra* and passion fruit (*Passiflora edulis*) are cut back regularly. Selective herbicides may eventually provide a more effective remedy.

Conservation in botanical gardens

When the native habitat is totally destroyed, plant conservation in botanical gardens (*ex situ* conservation) may be the only course available. However, the worldwide distribution of gardens bears little relation to where the areas of richest plant life are or which are at greatest threat. Of the 1,500 botanical gardens around the world, half are in temperate Europe. Of the remainder, about 120 are in tropical Asia, 60 are in South America, and 30 in tropical Africa.

The value of these gardens does not lie just in their role of preserving specimens of endangered plants. Some plants have been successfully reintroduced into the wild from plants propagated in botanical gardens from seed, cuttings or tissue cultures. One example is provided by the she cabbage (*Senecio redivivus*), which has been reestablished on the Atlantic island of St Helena.

REGIONS OF THE WORLD

CANADA AND THE ARCTIC
Canada, Greenland

THE UNITED STATES
United States of America

CENTRAL AMERICA AND THE CARIBBEAN
Antigua and Barbuda, Bahamas, Barbados, Belize, Bermuda, Costa Rica, Cuba, Dominica, Dominican Republic, El Salvador, Grenada, Guatemala, Haiti, Honduras, Jamaica, Mexico, Nicaragua, Panama, St Kitts-Nevis, St Lucia, St Vincent and the Grenadines, Trinidad and Tobago

SOUTH AMERICA
Argentina, Bolivia, Brazil, Chile, Colombia, Ecuador, Guyana, Paraguay, Peru, Uruguay, Surinam, Venezuela

THE NORDIC COUNTRIES
Denmark, Finland, Iceland, Norway, Sweden

THE BRITISH ISLES
Ireland, United Kingdom

FRANCE AND ITS NEIGHBORS
Andorra, France, Monaco

THE LOW COUNTRIES
Belgium, Luxembourg, Netherlands

SPAIN AND PORTUGAL
Portugal, Spain

ITALY AND GREECE
Cyprus, Greece, Italy, Malta, San Marino, Vatican City

CENTRAL EUROPE
Austria, Germany, Liechtenstein, Switzerland

EASTERN EUROPE
Albania, Bulgaria, Czechoslovakia, Hungary, Poland, Romania, Yugoslavia

THE SOVIET UNION
Mongolia, Union of Soviet Socialist Republics

THE MIDDLE EAST
Afghanistan, Bahrain, Iran, Iraq, Israel, Jordan, Kuwait, Lebanon, Oman, Qatar, Saudi Arabia, Syria, Turkey, United Arab Emirates, Yemen

NORTHERN AFRICA
Algeria, Chad, Djibouti, Egypt, Ethiopia, Libya, Mali, Mauritania, Morocco, Niger, Somalia, Sudan, Tunisia

CENTRAL AFRICA
Benin, Burkina, Burundi, Cameroon, Cape Verde, Central African Republic, Congo, Equatorial Guinea, Gabon, Gambia, Ghana, Guinea, Guinea-Bissau, Ivory Coast, Kenya, Liberia, Nigeria, Rwanda, São Tomé and Príncipe, Senegal, Seychelles, Sierra Leone, Tanzania, Togo, Uganda, Zaire

SOUTHERN AFRICA
Angola, Botswana, Comoros, Lesotho, Madagascar, Malawi, Mauritius, Mozambique, Namibia, South Africa, Swaziland, Zambia, Zimbabwe

THE INDIAN SUBCONTINENT
Bangladesh, Bhutan, India, Maldives, Nepal, Pakistan, Sri Lanka

CHINA AND ITS NEIGHBORS
China, Taiwan

SOUTHEAST ASIA
Brunei, Burma, Cambodia, Indonesia, Laos, Malaysia, Philippines, Singapore, Thailand, Vietnam

JAPAN AND KOREA
Japan, North Korea, South Korea

AUSTRALASIA, OCEANIA AND ANTARCTICA
Antarctica, Australia, Fiji, Kiribati, Nauru, New Zealand, Papua New Guinea, Solomon Islands, Tonga, Tuvalu, Vanuatu, Western Samoa

North America

Central and South America

CANADA AND THE ARCTIC

THE UNITED STATES

CENTRAL AMERICA AND THE CARIBBEAN

SOUTH AMERICA

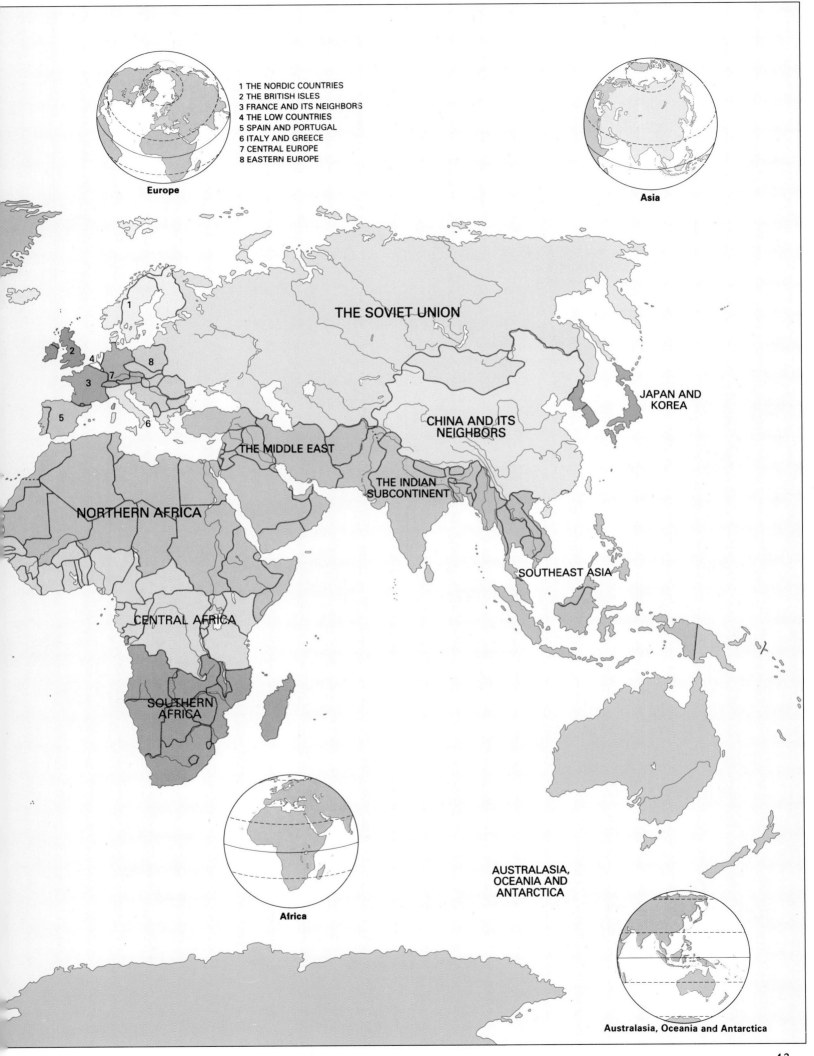

1 THE NORDIC COUNTRIES
2 THE BRITISH ISLES
3 FRANCE AND ITS NEIGHBORS
4 THE LOW COUNTRIES
5 SPAIN AND PORTUGAL
6 ITALY AND GREECE
7 CENTRAL EUROPE
8 EASTERN EUROPE

Europe

Asia

THE SOVIET UNION

JAPAN AND
KOREA

CHINA AND ITS
NEIGHBORS

THE MIDDLE EAST

THE INDIAN
SUBCONTINENT

NORTHERN AFRICA

CENTRAL AFRICA

SOUTHEAST ASIA

SOUTHERN
AFRICA

Africa

AUSTRALASIA,
OCEANIA AND
ANTARCTICA

Australasia, Oceania and Antarctica

PLANTS OF THE COLD NORTH

TUNDRA, FOREST AND PRAIRIE · THE NORTHERN FORESTS · PLANTS AND PEOPLE

Canada's plants reflect the size of the country in their distribution and their variety. They have found ways to survive in the very different conditions of temperate and polar latitudes, humid coastal areas and the dry continental interior. Much about Canada is vast – the expanse of the central grasslands, the frozen stretches of the tundra, and the towering heights of stands of coniferous and deciduous forest. In the Arctic, tundra plants withstand icy temperatures, frozen soil and short summers. Farther south are forests of evergreen conifers and mixed forests, which cover about half the country. Deciduous forests in the southeast enjoy long, mild summers, and herald the cold winters with spectacular displays of color in the fall. Only grasslands survive the hot, dry summers and freezing winters of the arid interior.

COUNTRIES IN THE REGION

Canada

DIVERSITY

	Number of species	Endemism
Canada	4,000	low
Greenland	497	3%

PLANTS IN DANGER

	Threatened	Endangered	Extinct
Canada	8	7	–
Greenland	–	–	–

Examples *Armeria maritima* subsp. *interior*; *Cypripedium candidum* (small white lady's slipper); *Isotria medeoloides* (small whorled fogonia); *Limnanthes macounii*; *Pedicularis furbishiae* (Furbish's lousewort); *Phyllitis japonica* subsp. *americana*; *Plantago cordata*; *Salix planifolia* subsp. *tyrrellii*; *Salix silicicola*; *Senecio newcombei*

USEFUL AND DANGEROUS NATIVE PLANTS

Crop plants *Acer saccharum* (sugar maple), *Oxycoccus macrocarpus* (cranberry), *Vaccinium myrtilloides* (blueberry)
Garden plants *Hamamelis virginiana*, *Lilium canadense* (Canada or meadowlily), *Syringa reflexa*, *Syringa villosa*, *Tsuga canadensis*
Poisonous plants *Aconitum columbianum*; *Cicuta maculata* (water hemlock); *Robinia pseudacacia* (black locust); *Sanguinaria canadensis*; *Toxicodendron radicans* (poison ivy); *Veratrum viride* (false hellebore)

BOTANIC GARDENS

Montreal (20,000 taxa); Royal Botanic Gardens, Hamilton; University of Western Ontario (15,000 taxa)

TUNDRA, FOREST AND PRAIRIE

About 40 million years ago the climate in Canada was warmer than it is now, and much of the land was covered with temperate mixed forest of both deciduous broadleaf trees and evergreen conifers. Gradually the climate in the north cooled, and the vegetation zones shifted southward. The deciduous species of the mixed forests were displaced farther than the conifers as they were less able to withstand cold conditions, leaving the northern zone inhabited only by pine (*Pinus*) and spruce (*Picea*).

As the Rocky Mountains formed they increasingly deprived the interior of rain. Grasslands developed in the dry areas lying in the rain shadow to the east of the mountains, and tundra plants evolved in the far north as forest and grassland species adapted to the extreme cold.

Lowgrowing Arctic shrub Mountain avens is typical of many boreal plants, growing very slowly in the low temperatures and northern light. It may take up to a hundred years to spread 1 m (3 ft). The mosses and lichens grow even more slowly.

Plants of the north
The tundra plants of the Arctic and the coastal strip of Greenland consist mostly of grasses, sedges and rushes, and herbaceous species such as lousewort (*Pedicularis*) and yellow marsh saxifrage (*Saxifraga hirculus*). There are also lowgrowing shrubs such as Arctic white heather (*Cassiope tetragona*), cushion plants such as mountain avens (*Dryas integrifolia*) and purple saxifrage (*Saxifraga oppositifolia*), and plants with creeping woody stems such as the willow *Salix arctica*. Mosses and lichens are also widespread. Many flowering plants, such as *Ranunculus sulphureus*, are found only in the Arctic; some, like alpine sorrel (*Oxyria digyna*), extend southward to alpine habitats.

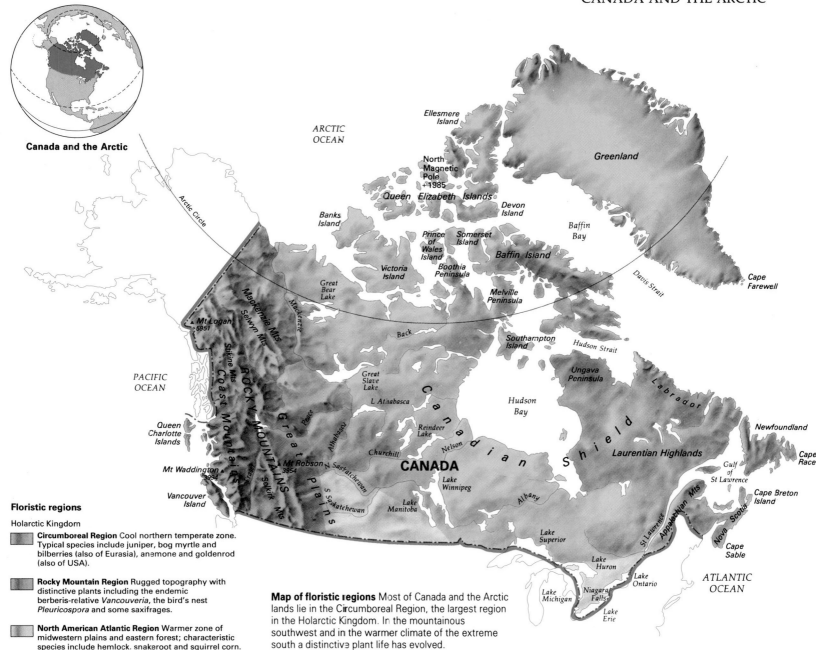

Floristic regions

Holarctic Kingdom

Circumboreal Region Cool northern temperate zone. Typical species include juniper, bog myrtle and bilberries (also of Eurasia), anemone and goldenrod (also of USA).

Rocky Mountain Region Rugged topography with distinctive plants including the endemic berberis-relative *Vancouveria*, the bird's nest *Pleuricospora* and some saxifrages.

North American Atlantic Region Warmer zone of midwestern plains and eastern forest; characteristic species include hemlock, snakeroot and squirrel corn.

Map of floristic regions Most of Canada and the Arctic lands lie in the Circumboreal Region, the largest region in the Holarctic Kingdom. In the mountainous southwest and in the warmer climate of the extreme south a distinctive plant life has evolved.

In the coniferous forests tree species such as black and white spruce (*Picea mariana* and *P. glauca*) and the one-flowered wintergreen (*Moneses uniflora*), range across the continent and into Eurasia. Other important trees include tamarack (*Larix laricina*) and jackpine (*Pinus banksiana*) in the east, and lodgepole pine (*Pinus contorta*) and alpine fir (*Abies lasiocarpa*) in the north and west. White birch (*Betula papyrifera*), aspen (*Populus tremuloides*) and other poplars are distributed throughout the region.

Temperate forests and grasslands

Temperate mixed forest covers much of the southeast, with conifers such as white pine (*Pinus strobus*) and eastern hemlock (*Tsuga canadensis*) growing together with deciduous trees such as sugar maple and red maple (*Acer saccharum* and *A. rubrum*), basswood (*Tilia americana*), white elm (*Ulmus americana*) and oaks (*Quercus*).

Other conifers, such as red spruce (*Picea rubens*), thrive in the wetter forest areas. Deciduous forests flourish in southern Ontario, in the southernmost part of the region; their range extends into the eastern United States.

The temperate forests of the west coast contain only a few deciduous species. They are dominated instead by conifers such as western hemlock (*Tsuga heterophylla*) and douglas fir (*Pseudotsuga menziesii*). These trees form towering evergreen rainforests along the coast and on the western slopes of the Coast Mountains. Similar species occupy the western slopes of the Rocky Mountains, giving way farther east to Engelmann spruce (*Picea engelmannii*) and eventually to alpine fir, lodgepole pine and white spruce, as the trees of the upland (montane) forests mingle with those of the far north.

Grassland and ponderosa pine (*Pinus ponderosa*) savanna have developed in the

drier lands between the mountains. Alpine meadows rich in grasses and herbaceous perennials occupy the high summits throughout the western mountains.

In the arid south-central plains of Canada the temperate forests give way to grasslands that form part of the northern American prairies. In the driest areas, close to the Rocky Mountains, the most abundant species include blue grama grass (*Bouteloua gracilis*), which is short and densely tufted. It grows together with shrubs that have adapted to dry conditions, such as sage (*Artemisia*) and even cacti (*Opuntia*).

Farther east the mixed-grass prairies contain other short, tufted grasses, the taller wheat grass (*Agropyron*), sedges (*Carex*), shrubs and broadleaf herbs. Larger grasses, such as big bluestem (*Andropogon gerardi*), which grows to 2 m (6.5 ft), predominate in tall-grass prairie, which is found on heavy clay soils.

THE NORTHERN FORESTS

About half of Canada is covered with forests composed of species that have adapted to the long, cold winters and short, but relatively warm summers.

Black spruce is one of the most abundant trees, partly because it is evergreen and can withstand dry, cold conditions. It is one of the few trees able to survive in the waterlogged, oxygen-deficient soils of *Sphagnum* bogs. In the north it grows on drier soils, and it even persists in climates that are too severe to allow seed production. Instead, it reproduces vegetatively: the lowest branches grow close to the ground and become covered by moss and litter, in which they form roots and eventually give rise to new trees. Along the northern boundary of the forest black spruce survives in a shrublike form often less than 1 m (3 ft) high. Its low, compact shape and a blanket of snow protect it from the desiccating winter winds.

Lightning fires

The accumulation of dead lower branches, and the resins of many of the other plants in a spruce forest, make it particularly vulnerable to fire. Perhaps surprisingly, fires caused by lightning are of major importance in these forests, opening the way for succeeding plant communities and maintaining the diversity of species in the ecosystem.

Black spruce is well adapted to fire. Its cones can remain on the trees for five years or more without opening; when fire strikes, the heat triggers the release of large quantities of seed, which enables the species to reestablish itself. Pines produce similar cones, and they too may release abundant seeds after a fire. Aspens seldom reproduce by seed; instead they sprout freely from their roots. After fires, dense stands of aspen saplings spring up, but they decline as the growing conifers cut out the light and the soil becomes depleted of nutrients.

Labrador tea (*Ledum groenlandicum*), a lowgrowing shrub 30–80 cm (12–30 in) tall, which is characteristic of northern forests, is also adapted to survive fire. It sprouts from charred stem bases, and the young shoots grow rapidly with the flush of nutrients that are released from the burnt vegetation.

Mosses, coral-roots and lichens

Much of the ground in northern forests is densely covered with the feather mosses *Hylocomium splendens* and *Pleurozium schreberi*. These mosses slow down the evaporation of moisture from the soil. Unlike higher plants, they are able to absorb mineral nutrients directly from rain. The nutrients are later released at the base of the slowly decomposing moss layer, where the fine roots of black spruce absorb them.

Branched lichens (*above*) Cladonia, like all lichens, is a partnership between a fungus and an alga. The fungus surrounds and protects a layer of green alga, which manufactures sugars by photosynthesis. Water is absorbed through the surface of the organisms.

Vegetation pattern of a bog (*below*) Peatlands and *Sphagnum* bogs cover large areas of the north. Trees and lowgrowing shrubs colonize the raised areas of peat, while in the surrounding wet fens sedge (*Carex*) and cattail (*Typha*) dominate.

Tamarack (*Larix lariciana*), labrador tea and dwarf juniper (*Juniperus communis* var. *depressa*) form forests on raised peat

On gently sloping ground, birch (*Betula pumila*) and a variety of peat-forming plants occupy low ridges of peat known as "strings". Sedges (*Carex*) fill shallow pools between the ridges

Sedges and cattail (*Typha*) grow on the wet fen and marsh marigolds (*Calla palustris*) may also be found

Sedges, pitcher plants (*Sarracenia*) and bogmosses occupy the pools

Coral-root orchids (*Corallorhiza*) are also typical of northern forests. These plants are saprophytic – they live off dead matter. They consist only of an underground rhizome or stem – a yellowish, many branched coral-like structure. The rhizomes have no roots, but are heavily infected by a fungus that absorbs organic nutrients from the soil and supplies them to the orchids. When the plant has accumulated sufficient food, aerial stems develop measuring up to 40 cm (16 in) in height, each ending in an inflorescence of small cream to purplish flowers. There are no leaves on these stems, but to compensate, both the green stalk of the inflorescence and the plant's ovaries contain chlorophyll; their photosynthesis presumably supplements the food derived from the fungus.

Like coral-roots, lichens consist of an association between plants and fungi. The partners in this symbiotic relationship are algae and fungi. The algae, which contain chlorophyll and photosynthesize, supply the organic nutrients for both partners. Lichens have no roots to draw water from the soil, but the fungal hyphae are able to absorb both water and mineral nutrients from precipitation. They also form a protective sheath, contracting around the algal cells to reduce metabolic activity when sunlight, which dries the

White-flowered Labrador tea The tough leaves and rust-colored hairs on the undersides restrict water loss. Such precautions are necessary because the roots cannot function efficiently in the anerobic conditions of the boggy soils where the shrub grows.

lichen, penetrates the open woodland canopy, and expanding again when conditions become more moist and photosynthesis can be resumed.

Reindeer lichen (*Cladonia rangiferina*) and related species cover the ground in open woodland in the northern forests. These pale brown or whitish, branched structures, up to 15 cm (6 in) tall, resemble tiny shrubs. Above them, fine strands of old man's beard lichen (*Usnea*) festoon the tree branches.

Pale lichens grow in areas that receive heavy snow in winter, as a result of which they are inactive. In summer the lichens reflect the sun's rays, reducing water loss and extending their periods of metabolic activity. Black lichens reverse this pattern of seasonal activity. *Alectoria*, for example, grows in exposed places with only a thin snow cover. The black coloration increases their absorption of radiation from the sun, enabling them to melt the snow immediately surrounding their stems, and stimulating photosynthesis in spring and fall when meltwater is available. The black lichens are dry and inactive for long periods in summer.

TREES IN A COLD CLIMATE

Forests develop in areas where conditions are neither too cold, as in the tundra, nor too dry, as in the prairies. Even so, the conditions they have to withstand may be extreme. Some areas, such as southern Ontario, may have long, mild summers, but these alternate with winters cold enough to freeze the soil, which restricts the amount of water the trees can take up. These areas are characterized by broadleaf deciduous trees, which survive by shedding their leaves in winter. This minimizes the amount of water lost through transpiration.

In northern and upland areas, where winter temperatures are as low as −30°C (−22°F), evergreen conifers predominate. Here the growing season is very short, so deciduous trees, which can photosynthesize only when new leaves develop in spring, and have to invest considerable resources in leaves that function for only a few months, are at a severe disadvantage. In contrast, the needlelike leaves of evergreen trees, such as pine and spruce, remain active for several years whenever the weather is favorable. The trees are able to conserve water during the long, very cold winters because their leaves have a thick, protective outer layer and a reduced surface area.

PLANTS AND PEOPLE

Plants feature prominently in Canadian culture. The maple leaf on the national flag proclaims the distinctive nature of the Canadian landscape, particularly in the fall. Each of the twelve provinces and territories has adopted a characteristic plant species as an emblem. Newfoundland, for example, has chose the carnivorous pitcher plant (*Sarracenia purpurea*); the Northwest Territories selected the mountain avens.

The Canadian forests are famous for the berries they yield, such as blueberry (*Vaccinium myrtilloides*) and cranberry (*Oxycoccus macrocarpus*). These fruits are picked every summer for local domestic use, and are also harvested commercially on a small scale. The woods of Ontario are also well known for their maples. Following the ancient custom of the Iroquois Indians, the trunks of sugar maples are tapped in March, as the sap rises, to collect the sugary fluid that is refined and sold as maple syrup.

Native Canadian plants that are widely cultivated in gardens include the shrubby goldenrod (*Solidago canadensis*) and the dogwood tree (*Cornus alternifolia*) from the southeast, creeping wintergreen (*Gaultheria procumbens*) from the northern forests, the western temperate trout lilies (*Erythronium*) and Arctic alpines such as purple saxifrage. However, it is not always clear whether the cultivated stocks came from Canadian populations or from elsewhere within the range of the species.

Commerce and conservation

The exploitation of forest resources, both in natural stands and increasingly in plantations, is a major industry. Important timber species include hemlock, douglas fir, ponderosa pine and red cedar in the west, and white pine, sugar maple and white oak (*Quercus alba*) in the east. Across the country various species of spruce are logged, and are particularly important in the pulp and paper industry. The western coastal rainforests, the central deciduous forest, and the mixed forests in much of the east have mostly been felled, either for timber or to make way for agriculture and urban development. Regrowth forests develop rapidly, whether by succession in abandoned fields or following replanting. However, it will take a long time for them to match the grandeur of the few remaining primeval stands; those of western hemlock reach 40-50 m (130-165 ft) in height and have taken up to 500 years to develop.

The best way to conserve the forests is to increase the extent of protected reserves, and practice selective rather than clear-felling (in which all the trees in an area are felled). The last strategy has the added advantage of retaining some forest cover, which benefits some orchids and other plants of the forest floor.

Maples in the fall (*right*) The brilliant reds and yellows of the maples now attract tourists. Sugar maples were once the only source of sweetness in North America; a large tree can produce up to 115 litres (25 US gallons) of sap in a single spring.

Blueberries (*below*) Growing on boggy soil, these fruits were an important part of the diet of the Indian people, who dried any surplus for the winter. Blueberries are also eaten in vast quantities by birds and other animals, especially bears.

Pinus strobus
white pine

Picea pungens

Thuja occidentalis
American or white cedar

Tsuga heterophylla
western hemlock

Conifers of Canada These are primarily timber trees, but they and their decorative varieties are planted for their beauty. The foliage and cones vary considerably from genus to genus.

PLANTS IN LOCAL TRADITION

In a land of limited natural resources, the indigenous people demonstrated great ingenuity in their use of local plants. Pacific Coast Indians fashioned western red cedar into massive square houses, totem poles and dugout canoes, as well as boxes and bowls. They also used strips of cedar bark and spruce roots in weaving and as twine. Red cedar and other softwoods are easily carved, and were fashioned into elaborate ceremonial masks.

Indian women capitalized on the absorbent, antibiotic properties of sphagnum moss to diaper their babies, and to stem menstrual blood. In the east and north the men built canoes from strips of white birchbark stretched over a cedar frame, sewing them in place with the fine, pliable roots of white spruce. They could later be repaired with pine resin. Birchbark was also used to make barkcloth, which served as a fabric for clothing.

Many of the indigenous people have always relied on plants to nourish the animals that are their principal source of food. In the past the prairie grasses supported bison, the mainstay of the Plains Indians, while in the Arctic many Inuit relied on the annual migrations of caribou, one of the few animals able to digest the lichens of the northern woodlands.

It is not only the forests that have suffered from economic pressure. Most of the natural grasslands have been replaced by fields of Eurasian cereal grasses. Many of the native prairie species are now restricted to a few scattered reserves and to roadside verges, where they are increasingly threatened by herbicides and other agricultural chemicals. No native plants are used significantly in agriculture. This is not a new trend; even the traditional crops of the Iroquois Indians (maize, pumpkins, beans and tobacco) were introduced from farther south.

The state of the far north

In contrast to those of the south, the plants of the northern and Arctic areas, including Greenland, have so far not been much disrupted by agriculture. The plants here are, however, particularly susceptible to pollution. The emission of sulfur dioxide and heavy metals by nickel smelting activities at Sudbury, Ontario, has had a devastating effect on nearby forest. Local disruption of tundra vegetation has also been caused by the oil industry, both through oil spillage and because the heavy vehicles used in oil exploration erode the soil so the permafrost begins to melt.

The northern territories are so vast, inhospitable and remote that there is little prospect of largescale human settlement, and little imminent danger of the native vegetation being seriously disturbed.

More significant is the potential threat of acid rain and other pollutants spreading to the north from existing centers of population and industry. Global warming and the depletion of ozone in the upper atmosphere could also have a major impact. Any significant change in climate is likely to disrupt existing vegetation patterns, particularly in the Arctic, and plants are damaged by increases in ultraviolet radiation. These are international problems. The causes are largely not of Canada's making, while the effects could seriously damage ecosystems throughout the world.

Survival in the Arctic

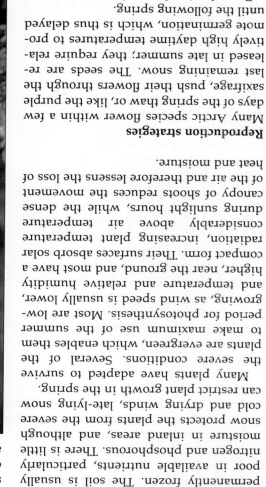

The Arctic is one of the most rigorous environments on Earth. The summers are short and cool, with average temperatures above freezing (0°C–6°C/32°F–43°F) for less than three months during the year. Summer days are long, with continuous daylight in midsummer, but the intensity of the sunlight is low. In winter there is a corresponding period of continuous darkness, and prolonged severe frost. Only a thin surface layer of the soil thaws in summer – beneath this the soil is permanently frozen. The soil is usually poor in available nutrients, particularly nitrogen and phosphorous. There is little moisture in inland areas, and although snow protects the plants from the severe cold and drying winds, late-lying snow can restrict plant growth in the spring.

Many plants have adapted to survive the severe conditions. Several of the plants are evergreen, which enables them to make maximum use of the summer period for photosynthesis. Most are low-growing, as wind speed is usually lower, and temperature and relative humidity higher, near the ground, and most have a compact form. Their surfaces absorb solar radiation, increasing plant temperature considerably above air temperature during sunlight hours, while the dense canopy of shoots reduces the movement of the air and therefore lessens the loss of heat and moisture.

Reproduction strategies

Many Arctic species flower within a few days of the spring thaw or, like the purple saxifrage, push their flowers through the last remaining snow. The seeds are released in late summer; they require relatively high daytime temperatures to promote germination, which is thus delayed until the following spring.

Reproduction by seed is unreliable in the Arctic, where both flower and fruit development is very slow, there is a shortage of insect pollinators, and the cold climate means that seeds do not always succeed in germinating. There are consequently very few annuals in this region, and many of the perennials re-produce vegetatively as well as by seed. Arctic communities of alpine sorrel are particularly well adapted, developing both flowers and rhizomes (the under-ground plant stems used for food storage). In contrast, communities that live in the less severe alpine habitats farther south do not produce rhizomes.

Apart from rhizomes, plants such as creeping willows and cushion plants re-produce by putting out roots where the stems come into contact with the ground. New shoots arise as the roots become established; eventually the stem rots and breaks, and the new plant becomes inde-pendent. This process is known as layer-ing. Some plants even replace a few or all of their flowers with vegetative bulbils (small bulbs). These include nodding saxifrage (*Saxifraga cernua*) and grasses such as *Poa alpina*, where the bulbils develop into young plants with leaves and partly developed roots before they

are shed – an effective means of ensuring successful reproduction.

Many members of the daisy family (Compositae), rose family (Rosaceae) and other groups overcome the unreliability of seed production by producing flowers and seed asexually. By one method, a cell in the ovule repeatedly divides to form an embryo without fertilization having taken place. There is an unusually high proportion of such apomictic plants in the Arctic; they include many species thought to have evolved recently in order to occupy new habitats formed since the last ice age ended some 10,000 years ago.

Arctic bells (*above*) *Cassiope tetragona* is a dwarf shrub with tiny leaves pressed close to the stem. These are produced very slowly, but are capable of photosynthesizing for many seasons. It has been known for them to function for as long as 15 years.

Kidney-shaped leaves (*left*) The growth of the mountain sorrel (*Oxyria digyna*) is governed by daylength. As the days get longer growth begins; as the daylength decreases to 15 hours, buds begin to develop. These will not open until the days lengthen and the next growing season begins.

Spider plant (*right*) The extraordinary appearance of *Saxifraga flagellaris* is due to the numerous runners, creeping stems capable of producing tiny bulbils that will grow into daughter plants. This additional method of reproduction is particularly useful in the Arctic tundra, where pollinating flies are infrequent visitors.

PLANTS OF THE NEW WORLD

The huge diversity of plant life in the United States reflects the continent's size and the extraordinary variety of its environments. The plant life can be roughly divided into two regions – the plants that lie to the west of the Rocky Mountains and those that lie to the east. The mixed forests of the humid east gradually give way to tall-grass prairie and short-grass plains in the dry interior. In Florida and the southeast subtropical plants flourish. The west is typified by dark forests of majestic conifers along the Pacific coasts and on the parallel mountain ranges (or cordillera) that run from north to south down the west coast, while drought-resistant gray scrub and green succulents are typical of the deserts. The exceptions are California and the Hawaiian islands, botanically two of the most interesting areas in the world.

COUNTRIES IN THE REGION

United States of America

DIVERSITY

	Number of species	Endemism
USA (mainland)	20,000	high
Hawaii	950	very high

PLANTS IN DANGER

	Threatened	Endangered	Extinct
USA (mainland)	1,000	300	90
Hawaii	197	646	270

Examples *Agave arizonica*; *Asimina rugelii*; *Cladrastis lutea* (American yellowwood); *Conradina verticillata* (Cumberland rosemary); *Fritillaria liliacea* (white fritillary); *Hudsonia montana* (mountain golden heather); *Malacothamnus clementinus* (San Clemente Island bush-mallow); *Pediocactus knowltonii* (Knowlton cactus); *Prunus gravesii* (Graves's beach plum); *Rhapidophyllum hystrix* (needle palm)

USEFUL AND DANGEROUS NATIVE PLANTS

Crop plants *Carya illinoensis* (Pecan); *Helianthus annuus* (sunflower); *Helianthus tuberosus* (Jerusalem artichoke); *Zizania aquatica* (Indian wild rice)
Garden plants *Catalpa bignonioides*; *Ceonothus dentatus*; *Clarkia elegans*; *Cornus florida* (flowering dogwood); *Cupressus macrocarpa*; *Fremontodendron californicum*; *Magnolia grandiflora*; *Yucca gloriosa*
Poisonous plants *Argemone mexicana* (prickly poppy); *Euphorbia maculata* (spotted spurge); *Hippomane mancinella* (manchineel); *Karwinskia humboldtiana* (coyotillo); *Toxicodendron radicans* (poison ivy); *Toxicodendron quercifolium* (poison oak)

BOTANIC GARDENS

Brooklyn Botanic Gardens and Arboretum (14,000 taxa); Huntington Botanic Gardens, San Marino (12,000 taxa); Kennet Square Longwood Gardens (13,000 taxa), Missouri; New York (15,000 taxa)

FORESTS AND GRASSES OF THE EAST

The eastern United States has a huge variety of environments for plants. The highest altitudes in the northeast support a tundra vegetation, which gives way on the lower slopes to coniferous and deciduous forests. In contrast, the southern tip of Florida in the extreme south supports broadleaf tropical evergreen forest. In the drier interior, flat or rolling grasslands stretch westward to the foothills of the Rocky Mountains. There are also bogs, swamps, salt marshes, dunes and cliff faces in the region.

The eastern forests

There is one narrow belt of rugged mountains in the eastern United States – the Appalachians. They are not very high, reaching only 2,000 m (6,500 ft) at their highest point. This means that, unlike the broad expanses of alpine vegetation in the west, the alpine tundra in the east is restricted to only a small area. It grows above the treeline, which lies at about 1,500 m (4,900 ft) in the north. The lower mountain slopes are covered with dark forests of spruce (*Picea*) and fir (*Abies*).

The conifers include hemlock (*Tsuga*), pine (*Pinus*) and cedars, which grow both as dominant trees and within upland deciduous forests. Among other conifers found in the region are the baldcypress (*Taxodium*), which grows in southwestern swamps, Atlantic white cedar (*Chamaecyparis*), found in coastal swamps, northern white cedar (*Thuja*), which occurs in northern swamps and on limestone cliffs, and spruce (*Picea*), fir (*Abies*) and larch (*Larix*), found on high mountain slopes and in northern bogs. In general, coniferous forests prefer sites with acid soil poor in nutrients; they can be wet or dry, steep, cold, very protected or exposed.

The heart of the eastern forest is dominated by broadleaf deciduous trees, the most important of which are oak (*Quercus*), hickory (*Carya*), maple (*Acer*),

Wild rice in a Louisiana swamp *Zizania aquatica*, a staple food of the Amerindians, grows in shallow water among the buttress roots of baldcypress (*Taxodium disticum*). This deciduous conifer belongs to a small genus found only in the United States and Mexico.

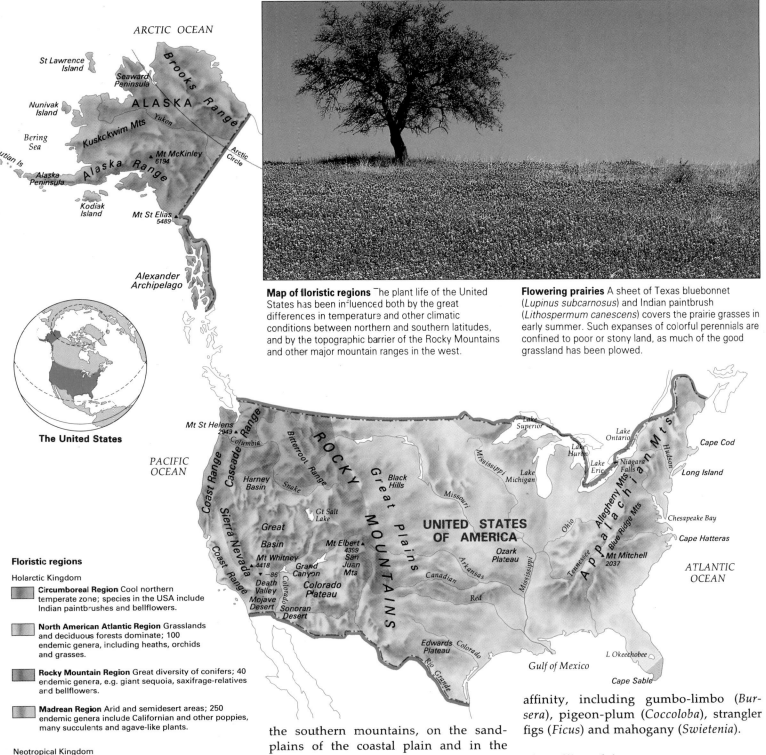

Map of floristic regions The plant life of the United States has been influenced both by the great differences in temperature and other climatic conditions between northern and southern latitudes, and by the topographic barrier of the Rocky Mountains and other major mountain ranges in the west.

Flowering prairies A sheet of Texas bluebonnet (*Lupinus subcarnosus*) and Indian paintbrush (*Lithospermum canescens*) covers the prairie grasses in early summer. Such expanses of colorful perennials are confined to poor or stony land, as much of the good grassland has been plowed.

Floristic regions

Holarctic Kingdom

Circumboreal Region Cool northern temperate zone; species in the USA include Indian paintbrushes and bellflowers.

North American Atlantic Region Grasslands and deciduous forests dominate; 100 endemic genera, including heaths, orchids and grasses.

Rocky Mountain Region Great diversity of conifers; 40 endemic genera, e.g. giant sequoia, saxifrage-relatives and bellflowers.

Madrean Region Arid and semidesert areas; 250 endemic genera include Californian and other poppies, many succulents and agave-like plants.

Neotropical Kingdom

Caribbean Region Tropical climate; rich plant life. Covers Caribbean and much of Central America.

---

birch (*Betula*), elm (*Ulmus*), magnolia (*Magnolia*), ash (*Fraxinus*) and buckeyes (*Aesculus*). They are found on a wide variety of upland sites, as well as at lower altitudes and on river floodplains. With them grows a host of other broadleaf trees, including beech (*Fagus*), American hornbeam (*Carpinus*), tuliptree (*Liriodendron*), sweetgum (*Liquidambar*), sycamore (*Platanus*), hackberry (*Celtis*), and dogwood (*Cornus*).

Pines (*Pinus*) thrive within the deciduous forest and on its margins; they are particularly common on the dry ridges of the southern mountains, on the sandplains of the coastal plain and in the glaciated region to the north. Their survival in the past depended on fires (natural fires and those caused by Indians and settlers) to keep down competing vegetation. Now that fires are rarer, the pine forests are being taken over by oaks.

On the southeastern coastal plain subtropical broadleaf evergreens such as *Persea* and native palms grow among the pines and temperate deciduous trees. Mangroves (*Rhizophora, Avicennia, Conocarpus,* and *Laguncularia*) take the place of salt marshes along the southernmost coasts of Florida and the Gulf of Mexico. The extreme southern tip of Florida has small areas of broadleaf evergreen and deciduous trees of a tropical Caribbean affinity, including gumbo-limbo (*Bursera*), pigeon-plum (*Coccoloba*), strangler figs (*Ficus*) and mahogany (*Swietenia*).

The rolling plains

On its western flank the deciduous forest originally merged into prairies that were dominated by big bluestem (*Andropogon*) and Indian grass (*Sorghastrum*), but these have largely been replaced by fields of corn and soybeans. With the decrease in rainfall toward the west, the tall-grass prairies yield first to mixed-grass prairie, and finally to short-grass plains dominated by buffalo grass (*Buchloe*) and little grama grass (*Bouteloua*), which stretch to the foothills of the Rockies. The short-grass plains were the home of the American buffalo before the herds were devastated and the land was used as grazing for commercial livestock.

STRATEGIES FOR SURVIVAL

The many different kinds of plants in the eastern United States are closely associated with the wide range of environments to which they have adapted. Even within a single habitat there can be enormous variety: just one hectare of deciduous forest may contain as many as 30 species of trees and 150 species of vascular plants.

Taking advantage of spring
In early spring in a deciduous forest, before the leaves come out, full sunlight can penetrate what is usually a dense and shaded habitat. The temperature fluctuates greatly, giving warm days and cold nights (though near the ground it varies less widely). In addition, a fresh supply of plant nutrients is suddenly made available in the soil – the result of bacterial action on plants that have died during the previous fall and winter. These conditions result in a carpet of herbaceous wildflowers that covers the forest floor: dutchman's britches (*Dicentra*), spring beauty (*Claytonia*), toothwort (*Dentaria*), trout lily (*Erythronium*), blood root (*San-*

guinaria), dwarf ginseng (*Panax*) and showy orchids (*Orchis*).

Some of the plants that emerge in early spring are strict ephemerals – they are active and above ground only in spring, lying dormant in the ground for the rest of the year. Others are longer lived, and have to adjust as the unfurling leaves of the tree canopy above cut out the sun's light and warmth.

The availability of sunlight, warmth and nutrients explains why understory plants produce leaves at this time, but not why they flower. A few woodland herbs come into leaf and bloom at separate times, such as wild ramps (*Allium tricoccum*). The leaves appear in early spring but then drop off, and the plant blooms in summer without any leaves.

There may be a cost to a plant that puts out leaves in spring but stores the carbohydrates it manufactures for later use in flowering and fruiting; however, no one knows what that cost is. It is possible that many woodland ephemerals flower and fruit in spring because they rely on ants to disperse their seeds; it may be that ants travel mainly in spring and are thus an effective way of ensuring the successful dispersal of seeds.

Rival seeds and saplings
Some trees in deciduous forests cannot tolerate shade and therefore cannot live for long in the understory – but they nevertheless survive. Other species are shade-tolerant, and often have numerous

Dodecatheon meadia f. *alba*
shooting star

Trillium sessile

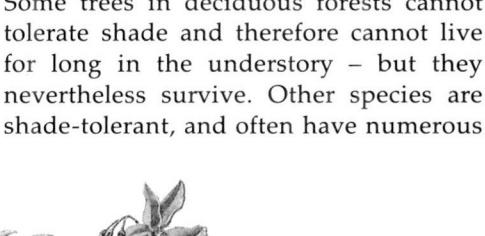

Mertensia virginica
Virginia bluebell

Erythronium grandiflorum

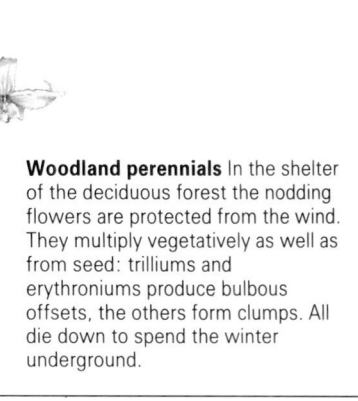

Woodland perennials In the shelter of the deciduous forest the nodding flowers are protected from the wind. They multiply vegetatively as well as from seed: trilliums and erythroniums produce bulbous offsets, the others form clumps. All die down to spend the winter underground.

The longleaf pine savannas of the southeast coastal plain owe their existence to sandy soil and fire. They contain one of the highest densities of carnivorous plants in the world. This wealth of carnivorous plants is partly due to the fires that once swept the region, removing the taller woody plants that would have prevented light from reaching the smaller carnivorous plants in the understory.

Carnivorous plants are able to thrive on nutrient-poor sites, including these sandy soils. They may use insects as an extra source of nutrients when they are flowering and fruiting. The southeast coastal plain has a high density of such plants, with five genera and 35 species. The Venus fly trap (*Dionaea*), the butterworts (*Pinguicula*) and the sundews (*Drosera*) capture insects by moving their leaves. The fly trap closes its leaves on its prey, while the sticky hairs on the sundews' leaves hold and digest the prey. The bladderworts (*Utricularia*) close a trapdoor over a pouch to prevent the prey escaping.

In contrast, the leaves of the pitcher plants (*Sarracenia*) are modified into a rosette of pitchers with downward-pointing hairs. These prevent insects from escaping, so they drown in the digestive fluid at the bottom of the pitcher, and their nutrients are absorbed by the plant.

seedlings and saplings growing in the shady lower levels of the forest, waiting for an opportunity to enter the upper canopy and gain access to the light.

Pin cherry (*Prunus pensylvanica*) dislikes shade, but its seeds live for 40 or more years in the soil. If a large disturbance (such as a hurricane or fire) clears part of the forest, these seeds will sprout to establish a new generation of trees. Another opportunist is yellow birch (*Betula lutea*). Its seeds do not survive in the soil for long, but it grows so quickly in spaces where a canopy tree has fallen that

Hooded pitcher plant Insects are lured by honey glands into the domed head. Lit by a mosaic of clear windows, it appears to be an exit, but instead it is the pitcher's mouth.

it often manages to secure a position in the upper story.

If the disturbance to the canopy is small and the added light a minor, short-lived change, shade-tolerant species usually win the race. However, where the disturbance is larger and the amount of light that penetrates to the understory is greater, less shade-tolerant species often increase. Trees in old forests are larger than those in young forests, and leave larger gaps when they fall. The coexistence of shade-tolerant and intolerant species may therefore work best in the oldest forests.

Out of the forest, trees also employ elaborate strategies to establish their seeds. The floodplains are inhabited by trees such as silver maple (*Acer saccharinum*), willow (*Salix*), cottonwood (*Populus*) and elm (*Ulmus*). They are among the first to bloom in early spring, and the seeds from their flowers mature and fall within weeks. The seeds themselves are unusual in that they do not lie dormant; they die if they do not germinate within a matter of weeks. The seeds, which float, are carried by the high spring waters, and as the waters recede they are deposited on the muddy banks. They then germinate, establishing a strong root system to anchor themselves before the next winter and spring raise the water level again.

A CHANGED LANDSCAPE

The destruction of land for farming began in the eastern United States in the 1600s. By the mid-1800s the more fertile soils of the Midwest and the prairies lured settlers westward; only those farms on the richest soils in the east continued to be farmed. The farms on the rich prairie soils are still there today, and areas of native prairie vegetation are highly restricted. One of the few places that prairie plants can be seen is along railroads that cut through tracts of the original vegetation. The short-grass plains have mostly been taken over for grazing.

The fall of the old forests
By 1920 most old growth forests of the eastern United States had been cut down, either to make room for agriculture or for logging. The logging industry expanded at the same time as farming was moving westward. It was aided by increasing mechanization. Trees were often harvested without regard for the future. Second growth forests are now widespread, particularly on hilly or otherwise less productive land, and only a few, mostly small, old growth forests remain. On the best sites, their trees may reach 2–3 m (6.5–10 ft) in diameter, and include species such as white pine (*Pinus strobus*) and red spruce (*Picea rubens*).

The forests have been altered in many other ways too. For example, they have been invaded by foreign plants and affected by the diseases that came with them. Among the exotic invaders, the most destructive was the chestnut blight (*Endothia*), an Asian fungus that was accidentally imported in the early 1900s on Chinese chestnut trees. In just a few decades this disease spread throughout the forests of the very susceptible American chestnut (*Castanea dentata*). A dominant forest tree, it provided the food source (beech mast) for many woodland animals so its destruction had a disastrous effect. Efforts are now being made to breed resistance into the American chestnut, and to search for less virulent strains of the fungus to build up immunity in the trees.

The cardinal flower (*Lobelia cardinalis*) This shortlived perennial of moist places and watersides has become a popular plant for bog and water gardens. It is easily increased by removing the offshoots that grow from the base of the plant.

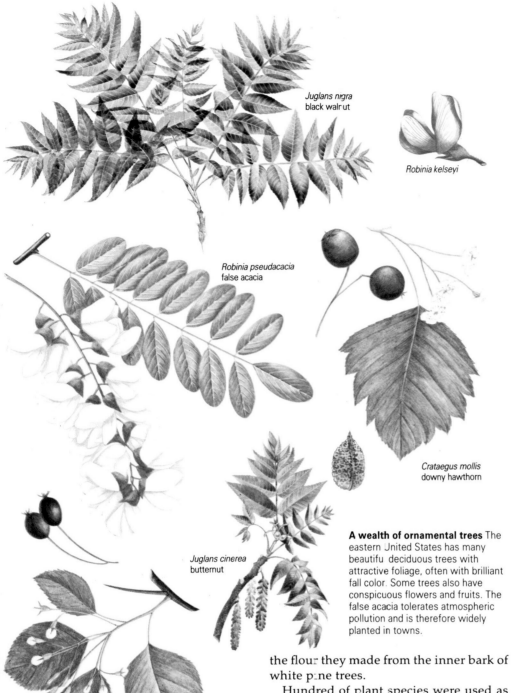

Juglans nigra
black walnut

Robinia kelseyi

Robinia pseudacacia
false acacia

Crataegus mollis
downy hawthorn

Juglans cinerea
butternut

Crataegus crus-galli
cockspur thorn

A wealth of ornamental trees The eastern United States has many beautiful deciduous trees with attractive foliage, often with brilliant fall color. Some trees also have conspicuous flowers and fruits. The false acacia tolerates atmospheric pollution and is therefore widely planted in towns.

Air pollution in the form of acid rain is a major problem in the eastern United States, stunting and even killing the trees of the forests. Current research is under way to study the effects of pollutants on the growth and death of eastern trees.

Nurturing plants to make money

For some 10,000 years before European settlement, many Indian cultures in the eastern United States made use of a wide range of plants. The dependence of one tribe on natural resources was commemorated in their name: the Adirondacks, or "tree-eaters", were named for the flour they made from the inner bark of white pine trees.

Hundred of plant species were used as medicines, a tradition that was continued by white settlers, particularly in the southern Appalachian Mountains. Trade in herbs continues to this day, and the two largest herb companies in the United States are both located in the North Carolina mountains. One of the best-known medicinal plants grown in this region is ginseng (*Panax quinquefolium*), whose roots are gathered in huge amounts for export to Asia. Today the ginseng trade is a multimillion dollar business, and is regulated by both federal and state governments to prevent over-exploitation of the plant in the wild. Other important regional crops of native plants include blueberries and cranberries (*Vaccinium*) and pecans (*Carya*).

While many of the traditional medicinal uses of mountain herbs have no currently documented scientific basis, they cannot be disregarded. A common deciduous forest understory herb, mayapple (*Podophyllum*), is today the source of an important drug used to treat cancer.

In a number of northeastern and mountain states the glorious fall colors of the decidous trees attract many visitors. Among the most colorful are maples (*Acer*), ashes (*Fraxinus*), birches (*Betula*), sweetgum (*Liquidambar*), black gum (*Nyssa*) and sourwood (*Oxydendrum*). Spring wildflower displays are also the basis of local tourism in some areas.

Conservation in the eastern states now focuses on protecting the last wilderness or semiwilderness areas, unique habitats and rare species. Increasing emphasis is being placed on research and management to ensure that natural diversity survives. As in other parts of the country, a growing interest in native plants has led to the creation of many new plant societies and botanical gardens. In addition, many of the plants in the eastern United States are important in horticulture. Among the most frequently planted species in the eastern states is flowering dogwood (*Cornus florida*). The wide array of native azaleas (*Rhododendron*) are also very popular.

Plants of the Great Smoky Mountains

One of the world's best preserves of plant diversity is the Great Smoky Mountains National Park on the border between North Carolina and Tennessee in the southern Appalachian Mountains. An area of 200,000 ha (494,200 acres), it contains some 1,500 species of vascular plants, including more than 100 species of trees. In 1934, when the park was created, one-third of the land had never been disturbed by people. Now protected, it contains the largest block of old growth spruce–fir forest in the south, along with hardwood and hemlock forests with trees that have a girth of up to 2.5m (8ft).

Mixed forest canopy in the fall (*above*) Among the many trees can be found eastern hemlock, yellow buckeye, birch, beech and tulip trees, which can soar to more than 55 m (180 ft) in the rich soil.

Conserving a rich resource

The conservation of this rugged landscape saved more than just the old growth trees. It preserves plants whose resources can be studied and put to use in many spheres outside the park. A recent project showed that 60 percent of the national park's plants have some use, ranging from foods to medicines, dyes, inks, building materials, chewing gum, oils, soaps, perfumes and industrial chemicals. Almost 1,000 species growing in the park have been used for medicinal purposes, including some 800 that were used by American Indians throughout North

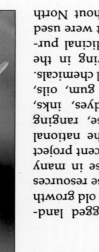

Purple wake-robin (*right*) *Trillium erectum* has flowers with a fetid smell that attract carrion flies to act as pollinators. Like many woodland plants, it flowers in spring, before the leaves have fully unfurled.

America. However, it is not economically viable to use all these plants for medicinal purposes today. Only some 60 species are grown commercially, among them the well-known ginseng and golden seal (*Hydrastis*), the physiological benefits of which are still under investigation.

More than 400 species of the plants in the park have toxic properties, many of which the Amerindians put to use against insects and to stun fish. Of them, the red cedar (*Juniperus virginiana*), is used commercially for its insecticide and repellent properties. Sumac (*Rhus*) has an irritant sap sometimes used in tanning; several species of this genus and the related poison ivy (*Toxicodendron radicans*) are poisonous. Bloodroot (*Sanguinaria canadensis*) contains an alkaloid that affects the heart, muscles and nervous system.

Many of the plants are the source of natural dyes used by local craftspeople: yellow and gold come from yellowwood (*Cladrastis*) and sumac; orange from coreopsis (*Coreopsis*) and Osage orange (*Maclura pomifera*) – an ornamental tree with hard orange wood; red from bloodroot and cardinal flower (*Lobelia cardinalis*), widely grown in gardens; blue from false indigo (*Baptisia*) and *Inula*; and purple from greenbrier (*Smilax*) and various oaks.

The multiple uses of some of these plants is exemplified by the sugar maple (*Acer saccharum*). Its spring run of sap is used to make maple sugar and syrup. Its hard wood provides top quality flooring, including that used in bowling alleys, as well as excellent firewood. The inner bark was ground and used for flour by Amerindians, who also boiled and ate the seeds. The tree also produces spectacular fall colors that help to draw large numbers of sightseers to the area, contributing to tourism. It is widely planted as a lawn and street tree.

Plants cannot be economically exploited within a national park. But as a genetic resource they can be investigated and even used with minimum impact to their habitat. Seeds and cuttings can be the subject of research and used to propagate and increase the numbers of plants outside the park. The many different properties of the plants within the park remain a potential resource for the future.

***Iris cristata* is a plant** of streamsides and hills. The ridged markings in the center of each petal are honey guides to lure insects to the nectaries and reproductive organs hidden between the petal and the crest.

WOODS AND DESERTS OF THE WEST

Running from Alaska and Canada in the north down the west coast as far south as Mexico, coniferous forest clothes the parallel ranges of the western mountains of the United States. It consists of familiar trees such as fir (*Abies*), spruce (*Picea*), hemlock (*Tsuga*), pine (*Pinus*) and larch (*Larix*), as well as Douglas fir (*Pseudotsuga*). In California the forests also contain the giant redwoods (*Sequoia* and *Sequoiadendron*) and the aromatic incense cedar (*Calocedrus*). The moister Pacific forests that extend northward to Alaska are dominated by cypresses such as *Thuja* and *Chamaecyparis*, whereas in the drier woodlands they are replaced by *Cupressus* and juniper (*Juniperus*).

The dominant trees determine what plants are able to grow beneath them. In the moist, dark, dense coastal forests swordfern (*Polystichum*), huckleberry (*Vaccinium*), salal (*Gaultheria shallon*), *Oxalis oregona* and mosses cover the fallen logs and the forest floor. Where daylight can filter through more easily, alder (*Alnus*), oaks (*Quercus*), willow (*Salix*), poplar (*Populus*), birch (*Betula*) and dogwood (*Cornus*) mingle with the conifers to form an understory.

In drier areas, where snow provides most of the moisture, the forests gradually open into parklike woodland. This has a hardier underbrush of pygmy forests or chaparral formed by hard-leaved shrubs such as *Ceanothus*, manzanita (*Arctostaphylos*), chamise (*Adenostoma*) and various shrubby oaks. These forests also contain maple (*Acer*), aspen (*Populus tremuloides*) and larch, with their spectacular fall colors of red, pink and yellow.

Colorful southern scrub

In much of the Great Basin (which lies between the mountains of the Sierra Nevada and Cascade Range in the far west) and the Rockies farther east, a thin vegetation barely covers the bony skeleton of the underlying rock. Piñon juniper woodland borders the lower edge of the forest. This gives way to arid shrub communities dominated by *Artemisia* or, on alkaline and saline soils, by *Atriplex*.

Californian poppies The state flower of California extends to the horizon along the lowland plain. Glorious flowerfields such as this are a brilliant herald of the summer to come.

The monotony is relieved in spring by the pink flush of *Phlox*, the red to orange *Castilleja*, blue lupins (*Lupinus*) and yellow monkey flowers (*Mimulus*).

The arid deserts of the southwest support even more drought-resistant plants. Among them are bright-flowered cacti, the ocotillo (*Fouquieria*) and, most commonly, spiny legumes such as *Prosopis*. After the rare occasions on which rain falls the plants burst into magnificent displays of color, with sand verbena (*Abronia*), evening primroses (*Oenothera*), blazing star (*Mentzelia*) and an endless variety of yellow daisylike flowers.

Intermingling and isolation

The range of plants found in the western United States is a result of the intermingling of northern plants with species from the drier Mexican plateau to the south. As the landscape and climate changed over the millennia many species died out. Their places were taken by plants more suited to the new conditions, either immigrants or newly evolved species.

This process is most evident in California. Cut off from the continental interior

The giant tree *Sequoiadendron giganteum*, the tallest tree in the world, has found a refuge on the relatively humid western slopes of the main forest belt of the Sierra Nevada, where it forms scattered groves. The effects of glaciation in the past may have contributed to its present isolation.

by the Sierra Nevada for the last million years, California now has nearly 5,000 native plant species, almost half of them unique (endemic). From sea level to the treeline, and from the rainforest in the northwest to the deserts of the southwest, the plant life is extremely diverse. The lowland fields in the foothills of the mountains are spectacular, with intermixed lupins and golden California poppy (*Eschcholzia*) stretching to the horizon.

The Hawaiian Islands also have an extraordinary number of endemic species – the result of specialization by the plant species from both sides of the Pacific that colonized the islands. These plants were carried on the wind, by ocean currents and most of all by birds. Many of the plants have small populations that are confined to a single mountain slope, valley, or lava-encircled sanctuary, and are therefore extremely vulnerable.

COPING WITH A HOSTILE ENVIRONMENT

From Pacific coasts to mountain summits, volcanic peaks and continental deserts, the western United States has a remarkable range of ecological niches to which plants have adapted.

Alpine plants

Alpine environments are notable for their very short and unpredictable growing season, their low summer temperatures and hostile winters. Adaptability to these conditions means being able to respond easily to rapid changes in temperature. Most of the alpine plants – dwarf perennial herbs such as saxifrages (*Saxifraga*), *Sibbaldia*, *Draba*, campions (*Silene*), buttercups (*Ranunculus*), gentians (*Gentiana*), *Primula* and *Phlox* – have well-developed underground systems for storing carbohydrates, which enables them to respond quickly when the snow melts in summer. Many alpine plants also form shoots and flower buds a year or more before flowering, as in *Oxyria*, *Oreonana* and *Orogenia*, so that they can quickly flower and set seed during the short summer.

Prostrate or dwarf shrubs such as willows and heaths (*Cassiope* and *Phyllodoce*) are abundant above the treeline; most of these have small, tough evergreen leaves that prevent water loss. The few true cushion plants that are found include *Arenaria*, *Eritrichium*, *Eriogonum* and a number of grasses. Their rounded shape, composed of crowded, often hairy, stems and leaves, protects the plant from the wind and cold, and raises the temperature inside the cushion slightly higher than it is in the surrounding air. The drying effect of wind is a constant threat, countered by the low stature and compact shape of the plant; the buds are protected by a covering of old leaves.

Most alpine plants are pollinated by insects whose activity is often limited by low temperatures, so the amount of seed set is often irregular and meager. To attract pollinators the flowers are frequently large and brightly colored in comparison with the size of the plants.

Plants of the desert

The desert plants of the southwest are all subject to extreme drought and often to greatly fluctuating temperatures. Despite this, what is striking about them is their complete lack of uniformity in size and habit. The desert plants have found radically different ways to thrive in these hostile conditions.

One example is the community of trees and shrubs growing along watercourses that experience flash floods. Prominent among them are woody leguminous species such as paloverde (*Cercidium*),

Hugging the ground (*above*)
Bearberry (*Arctostaphylos alpina*) grows on the central plateau of Alaska. It is a tough, woody shrub, lowgrowing and strong-rooting.

Spines in the Sonoran Desert
(*right*) Among the desert plants are the ocotillos (*Fouquieria splendens*) and various cacti. The tall saguaro cactus stores water in its fleshy stem; it is toxic to most animals.

VERNAL POOLS

One of the most remarkable features of the grasslands in the valleys and foothills of California are vernal pools, or "hog wallows". These are depressions that have formed over hardpan – an impermeable layer of soil – in which water accumulates during winter storms and then slowly evaporates during the rising temperatures of spring. Any plant that survives in a vernal pool must either be a fast-maturing annual, or be able to start life as an aquatic and complete its life cycle under increasingly dry conditions. A number of unrelated plants have adapted to these limitations, among them flowering plants and many grasses, the latter including *Orcuttia* and *Neostapfia*.

The plant groups that flourish in vernal pools have different temperature and moisture requirements for flowering. As a result they often form a series of colored bands around the deepest parts of the pools – yellow (*Lasthenia* and *Mimulus*); blue (*Downingia*), lavender (*Sidalcea* and *Boisduvalia*), purple (*Orthocarpus*) and white (*Limnanthes* and *Plagiobothrys*). The pools are in effect islands in which new species may evolve in isolation, so divergent populations may develop very close to each other. In particularly wet years, however, the individual pools may flood and join up, in which case the diverging populations are thrown back together again.

Acacia, smoketree (*Dalea*) and desert willow (*Chilopsis*). These plants have seeds or nutlets with very hard coats that are highly resistant to germination. When the rain comes, in a torrential downpour, it becomes a rushing stream laden with sand and gravel. The abrading action of the rapidly moving debris erodes the seedcoats, allowing the seeds to germinate in the newly moistened soil. The seedlings immediately send down a deep taproot to provide secure anchorage, ensuring that the plants are not swept away in future floods.

Succulents, such as cacti and agaves, reduce water loss but continue to photosynthesize by storing water in special tissues. They conserve the scarce moisture by recycling it. Succulents vary very widely in form, and different species have developed different ways of retaining water. Agaves produce dense rosettes of succulent leaves containing water-retaining tissue. Yuccas store water in the trunk or in the leaves, or both. Ocotillos shed their thin green leaves very quickly in dry periods, and grow new ones equally rapidly when water is available.

Many shrubs depend on modifications to their leaves to control the loss of water – small or dissected leaves, hidden stomata (pores on the underside), a thick cuticle or outer leaf layer, various kinds of hairiness, and ultimately seasonal leaf shedding (abscission). These features are also found to the north in the shrubby sagebrush and other communities of the Great Basin and the mountains.

Frequently interspersed with these strenuous water-savers, and often putting them to shame with their bright colors and sheer abundance, are the various ephemeral annuals that last for only a few days; they appear in early spring or late fall, or at other times when favorable moisture and temperature conditions coincide. Ephemerals are able to respond quickly to these conditions because they have no period of seed dormancy. This means that the seeds that were dispersed during the previous rains can immediately germinate and grow into plants. These plants complete their life cycle quickly so that they can hide their vulnerable embryos within protective seed coats before the next drought. Their brilliant colors probably serve to encourage their pollinators to work fast as well.

EXPLOITING A RICH RESOURCE

Before the arrival of Europeans and their east coast descendants in western North America, the plant resources there were being harvested and put to use by the indigenous Indian peoples. Today many of these plants are of use to modern agriculture and industry; some are commercially cultivated, while others are being exploited to extinction.

The devastation begins

The Amerindians of the west consisted primarily of three groups: the settled cultivators of the southwest, the fishing cultures of the Pacific northwest, and the nomadic hunter–gatherers of the cordillera and the Great Basin. The Klamaths and some other lake and wetland tribes based their cultures on the giant tule (*Scirpus*) – a large bullrush – which provided them with protective shelter as well as a raw material for weaving mats and fashioning baskets. Basket-making was developed to a high level by Californian Indians, as pottery played little part in their culture.

The nomadic Indians knew about and utilized a broad range of native plants for food, medicine and religious rituals. Early travelers reported that they used a number of plants for food, particularly lily bulbs such as quamash (*Camassia*) and *Calochortus,* corms, and tubers such as cous (*Lomatium*) and yamp (*Perideridia*). Acorns were also a staple food for the Californian Indians.

The settlers adopted the Indian diet and no major crop plants were introduced. The cultivation and use of tobacco (*Nicotiana*) was widespread in the west, but it is not known to what extent imported cultivars were used.

The later landowners did much to devastate the local plant life, turning it over to pasture for their livestock, draining and irrigating the land for agriculture, and exploiting the forests of the cordillera and the Pacific northwest, which have since supported a rapacious timber industry. The rate of cutting still exceeds that of replacement, and local pressure is mainly in favor of speeding up the felling process to avert unemployment and the economic problems associated with it. These include an illicit marijuana culture that has allegedly become one of the most profitable growth industries in some parts of the Pacific coast.

The western United States contains a large number of national parks that offer legal protection to substantial areas of land, including national forests. Elsewhere, the supply of lumber continues to diminish at an alarming rate; the pressure for multiple use of public lands and the encouragement of clear-felling large areas of the dwindling forests that remain may prove politically irrestible.

There is a campaign to save the last remnants of virgin forest, including any stands of redwood located outside parks or private reserves. There is also the problem of off-the-road vehicles, which can wreak havoc in the mountains and deserts. Oak woodland, too, is threatened, having proved to be a good indicator of a climate suitable for vineyards.

Gardeners' favorites

As the native plant life loses species at an accelerating rate, public concern mounts. A natural consequence is the patronage of gardens and arboretums as safe havens for endangered species. Botanical gardens have never enjoyed more active support.

The increased interest in the native plant life has led to the introduction of a greater number and variety of native species into local horticulture. *Ceanothus* and manzanita (*Arctostaphylos*), two of the most ubiquitous and attractive genera of western flowering shrubs, are regularly used for landscaping. Flannel bush (*Fremontodendron*) and Matilija poppy (*Romneya*) are increasingly cultivated. Rock gardens teem with members of the family Cruciferae, such as *Arabis* and *Draba*, together with wild buckwheats (*Eriogonum*) and native succulents (*Sedum, Dudleya, Lewisia*). Red, blue or purple lupins, *Delphinium* and *Pentstemon* often provide a definitive touch.

Clarkia elegans

Lupinus polyphyllus
Washington lupin

Calochortus uniflorus
mariposa lily

Polemonium carneum

Flowers of California (*left*) Four of the myriad native plants that have been adopted by horticulture. Some annuals, including *Clarkia*, are in an active state of evolution; of 43 species only one is not native to the golden state.

Pseudotsuga macrocarpa
large-coned Douglas fir

Pinus contorta
shore pine

Chamaecyparis lawsoniana
Lawson cypress

Abies magnifica
red fir

Abies procera
noble fir

Timber trees of the west Many of these noble trees have made a contribution to horticulture, none more so than Lawson cypress. This tree, which can reach 60 m (200 ft) in the wild, has dwarf forms so small they are suitable for a rock garden.

Candles in the desert (*above*) The curious inflated stems of squaw cabbage (*Streptanthus inflatus*) are topped by a halo of purple flowers. When young the stems are sweet tasting, and were eaten both raw and cooked by the Amerindians.

In addition, native plants are being cultivated for their byproducts. Jojoba (*Simmondsia chinensis*), for example, a plant endemic to the Sonoran Desert of the southwest, has large edible seeds that contain nearly 50 percent liquid wax. This can easily be hydrogenated into a hard white wax used as a substitute for sperm whale oil. Highly productive strains are now successfully being cultivated.

Another oil plant is *Limnanthes*, some species of which are dubbed "meadow foam" because of their masses of white, yellow or rose-colored flowers. These delicate, often succulent annual herbs occur naturally to the west of the Sierra Nevada–Cascade Range. Their large flowers each produce one to five large seeds, which are unique in the high molecular weight of the fatty acids they contain. *Limnanthes* oil can be readily converted into a product very like the liquid wax of jojoba. The plants show considerable variability, and would probably be susceptible to selection and cultivation without many of the problems encountered with jojoba.

GUAYULE: AN ALTERNATIVE TO RUBBER

It has long been known that many desert shrubs contain rubber. The Paiute Indians of the southwest were well aware of the fact – they used it as chewing gum. The rubber occurs as deposits in the cells, particularly in the tissues of the vascular system (phloem) and the inner bark. To extract it the plant must be cut, ground and softened by soaking.

The only North American plant that has been used commercially for rubber is guayule (*Parthenium argentatum*), which grows in north-central Mexico and Texas. The plants are long lived, slowgrowing perennials, usually less than 1m (3ft) tall. Their great advantage over other potential sources is that they have a rubber content of some 10 percent, four or five times that of other native rubber-bearing plants. Guayule was first exploited at the beginning of the century, when German interests built several extraction plants in Mexico. The American government investigated guayule during World War II, and learned that various species of *Parthenium* can hybridize readily. This raised hopes that a program of crossbreeding and selection might increase the rubber content and make the extraction process easier. However, the rubber is of inferior quality because of its high resin content, and cannot compete with the high quality modern synthetic rubber that is now being produced in vast quantities.

Opportunists under threat

The Hawaiian Islands rose as a chain of volcanoes from the deep ocean floor as recently as 5 million years ago. The steep mountain slopes combined with the effects of strong winds have resulted in many microclimates and an abundance of ecological niches. The islands were colonized from both sides of the Pacific, but overwhelmingly from the east. The result is a rich flora, about 95 percent of which is endemic, one of the highest levels of endemism in the world.

Most of the colonizing species had some feature that favored long-distance movement – airborne spores (such as the ferns), seeds carried by birds (*Plantago, Vaccinium, Bidens, Coprosma, Pittosporum*), or seeds that floated in the sea – morning glory (*Ipomoea*), screw pine (*Pandanus*) and *Erythrina*. Once they had arrived, they faced a variety of vacant niches that could be exploited only if their offspring were capable of adaptive radiation – the process whereby plants evolve and spread into new environments by changing their structure.

The Hawaiian tarweeds (such as the silverswords, *Argyroxiphium*) provide a striking example of plants that have been able to adapt in this way. There are 41 species in 3 genera, 23 of them endemic to one or more of the 6 principal islands. All are believed to have descended from a single Pacific Coast tarweed, yet they now include cushion plants, mat-forming subshrubs, rosette shrubs, trees and lianas. They fill a range of habitats, from near sea level to high altitudes and from the very driest to the very wettest of regions. Their evolution is so recent that they have not yet developed effective barriers to hybridization with each other.

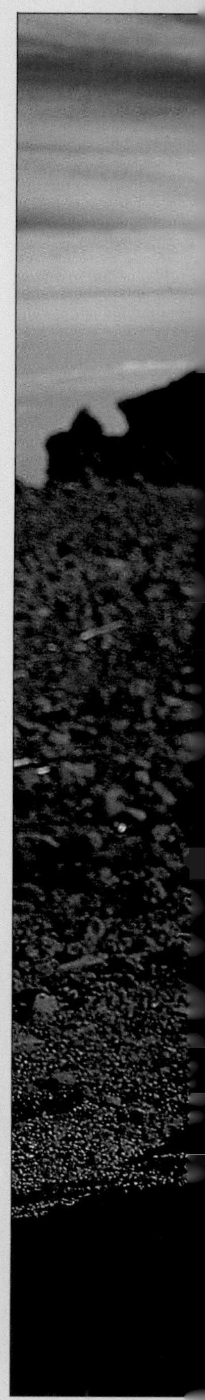

Moisture and altitude

Hawaiian vegetation is divided into wet forest, on the leeward side of the islands, dry forest, on the windward side, and a poorly defined alpine zone.

The characteristic plants of the wet forests are of Indo-Malaysian origin, many of them ferns or tree ferns. Those that came from the Andes or western North American sources such as Mexico are mostly plants from dry areas that have adapted to the wetter conditions.

Above about 2,000m (6,600ft) is an alpine zone of sorts with scattered bogs. Most typical alpine species are not found here; instead the habitat supports high-altitude genera of temperate groups, such

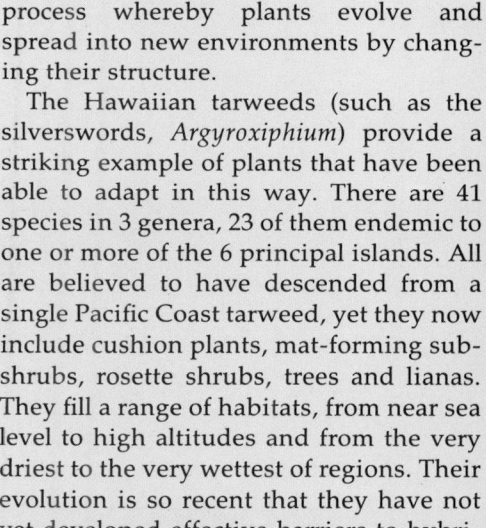

Silversword on lava (above) *Argyroxiphium sandwichense* grows on the edge of a volcanic crater and at high altitude, where it withstands heat and snow. It has deep maroon flowers on a stout flower spike that is covered with a sticky secretion similar to that found on its continental relatives, the Californian tarweeds. It is known by the Hawaiians as *ahinahina*, which means extremely gray.

Red brushlike flowers (above left) These showy flowers of *Metrosideros collina* var. *polymorpha* were once used to make *leis* (Hawaiian ceremonial wreaths), and the tree itself was dedicated to Pele, the goddess of volcanoes. It dominates the lush vegetation found on the lower slopes of the Alakai swamp, one of a decreasing number of unspoilt areas that are still rich in native plants.

Hibiscus flower (left) Many island species of *Hibiscus* have floating seeds that are dispersed by the sea. This capability has been lost by the endemic species on Hawaii, probably as they evolved in moist upland areas. However, these do have seeds covered in stiff hairs that cling to the feathers of birds.

as *Geranium, Fragaria, Silene* and *Vaccinium* from the north, and *Coprosma, Santalum* and *Wikstroemia* from the south.

The destruction of lowland flora

The introduction of grazing and browsing animals was catastrophic to the plant life of the Hawaiian Islands, particularly at lower altitudes. It was accompanied by the introduction of many aggressive weeds. The limited wetlands have been depleted by draining, filling or dredging, and areas that were cleared encouraged further invasion by introduced species (exotics). The lumber industry not only reduced the stands of native timber but also encouraged the establishment of Australian trees such as *Eucalyptus* to restore the damaged watersheds. When it

shifted its emphasis to wood production, trees such as the red cedar *Toona*, eucalypts and various pines were introduced. The extinction of many birds must have adversely affected numerous native plants, which evolved in parallel with the birds and were pollinated by them.

The result is that most of Hawaii's native plant species have been classified as extinct, endangered or threatened. The lowland plant life has virtually disappeared, and the native plants can only be glimpsed in the uplands. Some remedial steps have been taken – nearly half the remaining public land has been zoned for conservation, for example – but policies for multiple use of the land frequently conflict with the conservation of individual species.

The oldest plants in the world

The oldest living plants in the western United States – and in the world – are the bristlecone pines, *Pinus longaeva.* These extraordinary trees grow in scattered groves on dolomite at the dry, cold treeline of the White Mountains on the western edge of the Great Basin. In these high altitudes and isolated places the trees grow extremely slowly. Over the centuries they become increasingly gnarled and twisted, with most of their trunk stripped of bark and bleached almost white in the sun and wind. Many look more dead than alive.

In common with the giant redwoods, *Sequoia* and *Sequoiadendron*, bristlecone pines probably have their origins in the rich mixed forests that occupied western North America well before the ice ages, and have subsequently become ecologically diversified and floristically depleted. The rise of the coastal mountain ranges disrupted the equable climate, concentrating rainfall into the winter months and leading to high summer tempeatures. As seasonal extremes increased in the Great Basin, redwood forests were pushed westward towards the Pacific coast, while the bristlecone pines moved northward to higher elevations.

Although *Sequoia sempervirens* and its close associates show remarkable powers of vegetative regeneration, which has enabled them to survive in the face of extreme pressure from the lumber industry, neither *Sequoiadendron giganteum* nor the bristlecone pine is reproducing successfully. It seems likely that the species may not survive the lifespan of their oldest living individuals.

Trees to measure the past Bristlecone pines not only set records for longevity of more than 5,000 years, but also provide an accurate guide used in dating archeological finds (dendrochronology) and for measuring climatic change.

LUSH FORESTS AND DESERT CACTI

DESERT THORNS AND PALM-FRINGED ISLANDS · PLANTS AND POLLINATORS · A STOREHOUSE OF FOOD AND FLOWERS

From subtropical forest to desert scrub and cloudcapped islands, the plant life of Central America and the Caribbean islands is extremely diverse. In the luxuriant forest grow many species of bromeliads, ferns and orchids; some are ground-dwelling, while others are epiphytic, growing on the bark of trees. Numerous species of cacti thrive in the desert areas of Mexico and Cuba. Most major groups of plants such as orchids, palms and grasses grow throughout the region, but mainland Central America and the islands of the Caribbean have many different species. The least tropical plants of the Greater Antilles are related to those of North America; those of the Lesser Antilles have relatives in South America. The islands support about 200 endemic genera, most with only a few species.

COUNTRIES IN THE REGION

Antigua and Barbuda, Bahamas, Barbados, Belize, Costa Rica, Cuba, Dominica, Dominican Republic, El Salvador, Grenada, Guatemala, Haiti, Honduras, Jamaica, Mexico, Nicaragua, Panama, St Kitts-Nevis, St Lucia, St Vincent and the Grenadines, Trinidad and Tobago

DIVERSITY

	Number of species	Endemism
Central America (mainland)	30,000	very high
Caribbean islands	12,000	high

PLANTS IN DANGER

	Threatened	Endangered	Extinct
Mexico	248	72	8
Costa Rica	258	53	4

(Little information available for other countries)

Examples Ariocarpus agavoides; Auerodendron glaucescens; Carpodiptera mirabilis; Eupatorium chalceorithales; Freziera forerorum; Guzmania condensata; Ipomoea villifera; Lincania retifolia; Lycaste suaveolens; Streblacanthus parviflorus

USEFUL AND DANGEROUS NATIVE PLANTS

Crop plants Gossypium species (cotton); Ipomoea batatas (sweet potato); Persea americana (avocado); Psidium guajava (guava); Theobroma cacao (cocoa); Vanilla planifolia (vanilla); Zea mays (maize)
Garden plants Echeveria setosa; Fuchsia fulgens; Lobelia fulgens; Mammillaria elegans; Plumeria rubra (frangipani), Sedum allantoides
Poisonous plants Abrus precatorius; Comocladia dodonaea (Christmas bush); Croton betulinus (broom bush); Daubentonia purica (purple sesbane); Euphorbia pulcherrima (poinsettia); Jatropha curcas (physic nut); Lophophora williamsii

BOTANIC GARDENS

Royal Botanic Gardens, Kingston (1,000 taxa); University National Autonoma of Mexico (3,000 taxa); University of San Carlos of Guatemala

DESERT THORNS AND PALM-FRINGED ISLANDS

The Central American mainland, from Mexico to Panama, is home to more than 300 endemic genera. Some, such as the dahlias (Dahlia), and the tuberous agave-relatives (Polianthes) have a limited natural range, while others, such as the true agaves (Agave) extend into the Caribbean and the southwestern United States.

Mexico contains a great diversity of cacti; these grow from the dry central plateau at 2,000 m (6,500 ft) down to the lowland deserts. Armed legumes such as mesquite (Prosopis juliflora), and species of Acacia and Pithecellobium make up a scrubby thorn forest. Higher up there are mixed forests of pine and oak; they are dominated by Aztec pine (Pinus teocote) and Pinus patula.

Toward the Isthmus of Tehuantepec in the south of Mexico the climate becomes more humid, particularly on the wetter Atlantic coast. The area supports several well-known plant families, including the legumes (Leguminosae), laurels (Lauraceae) and peppers (Piperaceae). The main genera of the peppers are Piper, with several hundred species of shrubs and small trees in the humid forests, and Peperomia with many small herbs growing on shaded rocks and trees.

Lowland deciduous forest, characterized by the red birch (Bursera simaruba), is widespread throughout the region. Where the rainfall is low but humidity nevertheless high, mahogany (Swietenia macrophylla) and other small- to medium-sized trees form seasonal evergreen

woodland. In dry areas, and where the soil is thin and poor, grasses and sedges predominate, but some woody plants are found, including chaparro (Curatella americana) and fat pork (Chrysobalanus icaco). Mangrove forests form along estuarine coasts; species include red mangrove (Rhizophora mangle), white mangrove (Laguncularia racemosa) and black mangrove (Avicennia germinans).

The Caribbean islands

The plant life on the islands is largely determined by the seasonal climate, with local variations in topography, soil and the underlying rock creating diverse ecological niches. In Cuba, for example, the locally endemic pino-hembra (Pinus tropicalis) colonizes the ridges of thin sandy soil on the Alturas de Pizzaras; in adjacent valleys with deeper sandy soil hardwood forest grows, dominated by

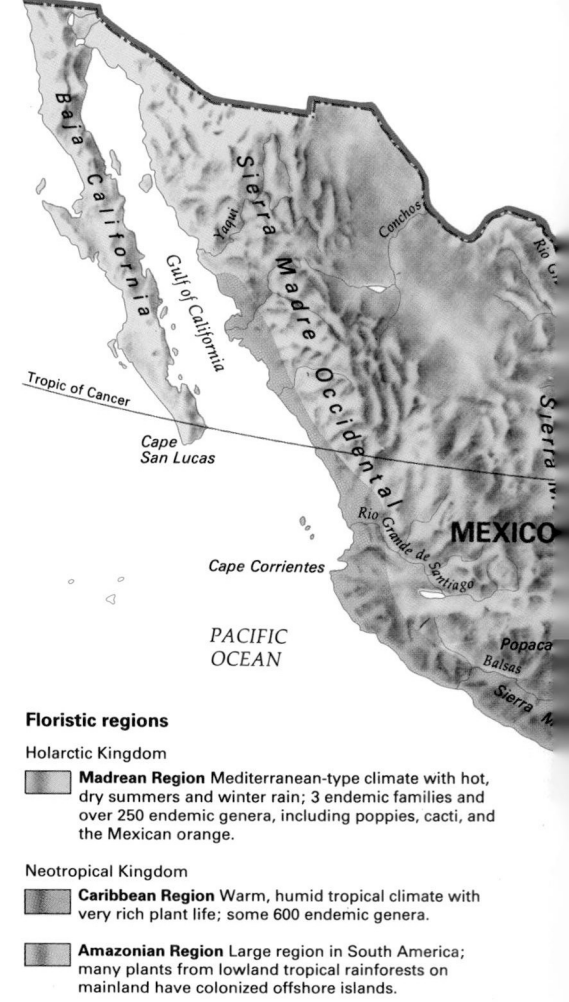

Floristic regions

Holarctic Kingdom

Madrean Region Mediterranean-type climate with hot, dry summers and winter rain; 3 endemic families and over 250 endemic genera, including poppies, cacti, and the Mexican orange.

Neotropical Kingdom

Caribbean Region Warm, humid tropical climate with very rich plant life; some 600 endemic genera.

Amazonian Region Large region in South America; many plants from lowland tropical rainforests on mainland have colonized offshore islands.

Tropical convolvulus The genus Ipomoea includes the perennial climbing morning glory, which scrambles up trees and over bushes in many parts of the region.

encina (*Quercus virginiana*). At the northern end of these heights, pure stands of Caribbean pine (*Pinus caribaea*) indicate magnesium-rich serpentine rock.

The smallest islands provide only a sandy or rocky shore, where the ubiquitous beach morning glory (*Ipomoea pes-caprae*) grows. Pedro Cays to the south of Jamaica has a strand woodland, dominated by buttonwood (*Conocarpus erectus*). Some low islands have a small element of endemism, particularly among cacti and agaves. Of the 580 species found on the Cayman Islands, for example, 21 are endemic, including the orchid *Schomburgkia thomsoniana*.

The larger islands, which have greater topographic diversity and higher rainfall, support more specialized plants: the cool, humid upland areas of Jamaica, Cuba and Hispaniola have a unique montane vegetation. On Hispaniola, at heights of between 2,650 and 3,000 m (8,700–9,850 ft), species of *Ilex, Lyonia* and *Garrya* grow in open woodland characterized by pino de guaba (*Pinus occidentalis*). The larger islands also support some plants found on the mainland. These include palms, the kapok tree (*Ceiba pentandra*), dogwood (*Piscidia piscipula*) and, in drier areas, scrubby legumes.

Desert scrub in northern Mexico Silvery brittlebush (*Encelia farinosa*) and spiny teddybear cholla (*Opuntia*) grow in arid regions, though *Opuntia* species are widespread. These jointed cacti may be treelike, creeping or with flattened segments.

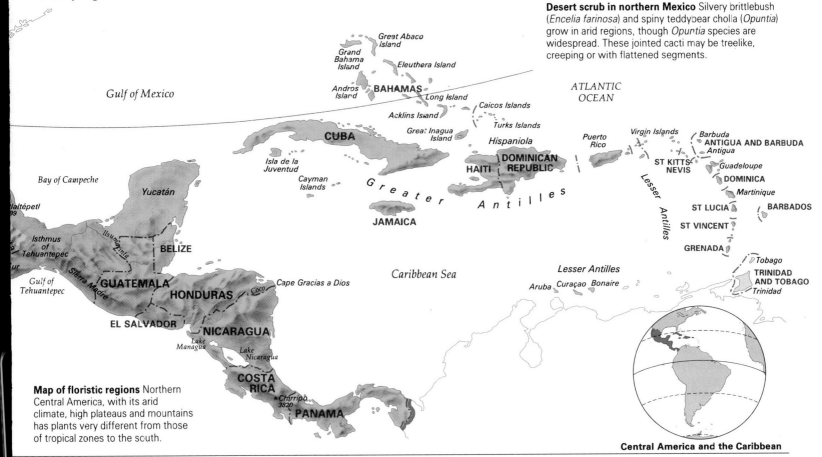

Map of floristic regions Northern Central America, with its arid climate, high plateaus and mountains has plants very different from those of tropical zones to the south.

Central America and the Caribbean

PLANTS AND POLLINATORS

In the forests of Central America the reproductive activity in plants is concentrated in the period when the fruits mature, between the end of the main dry season and the onset of the rains.

The flowering pattern of trees and shrubs usually conforms to seasonal changes, and in deciduous species may coincide with the cycle of leaf-fall or the flush of new leaves. Not all deciduous plants shed their leaves every year – whether they do or not depends on the length and severity of the dry season. However, in species of the tree *Tabebuia*, leaf-fall takes place every year and corresponds with flowering. In other species, the leaves fall when the fruits mature; this is particularly true of plants whose fruits are exposed, once the leaves have fallen, to the animals that feed on them and subsequently disperse the seeds. In others, such as the high climbers of the families Bignoniaceae and Malpighiaceae, leaf-fall coincides with the dispersal of seeds by the strong winds that blow. at this time.

Some species coordinate their flowering with the start of the rainy season. All the individuals open their flowers the same number of days after sufficient rain has fallen; these include the shrub *Hybanthus prunifolius* in Panama, a species of *Miconia* in Costa Rica, and cultivated coffee (*Coffea arabica*). The strategy is clearly coordinated with the availability

Shocking pink blooms (*above*) *Echinocereus pentalophus*, in common with other cacti, has become superbly well adapted to living in the hot, dry conditions of the desert zones. It has no leaves, but has developed a ridged stem that can expand to store water, and it also loses water very slowly.

Covered by flowers (*left*) The flowers almost obscure the rest of this prickly pear in their effort to attract a pollinator. Opuntias are very successful; even a tiny piece of stem can take root and grow into a new plant.

Tubular flowers of *Columnea* (*right*) Blossoms like these are typical of those pollinated by birds. This trailing epiphyte of the family Gesneriaceae inhabits the cloud forests of Costa Rica in the south of the region.

of appropriate pollinators. It results in mass fruiting, which ensures that some new seedlings will survive predation and mature into new plants. The fruits of *Miconia*, in particular, attract large numbers of birds that disperse the seeds.

Insect and bird pollinators

Although wind is important for the dispersal of some fruits and seeds, animals are generally more instrumental in the transfer of pollen from plant to plant. Some plants extend their flowering season by opening only a few flowers a day in what has been called a "trap-lining" pattern. It enables pollinating animals to visit a great many individual blossoms over a long period, spreads the burden of pollinators and reduces the competition for their services. Not all

---

PINES AT THEIR SOUTHERN LIMIT

On the fringes of any vegetation zone it is common to find widespread and versatile plants growing alongside species at the extremes of their natural range. *Pinus* is a genus of more than seventy species that range throughout the northern hemisphere. It reaches its most southerly limit in central Nicaragua, the Bahamas, Cuba and Hispaniola. Pino de guapa (*Pinus occidentalis*), known in Cuba as pino de la maestra, grows in a few small areas at high altitude in the eastern mountains of Cuba and in similar habitats on Hispaniola. It predominates in open conditions, or where soils are shallow. However, as the humus becomes enriched the pines are succeeded by hardwoods that flourish in the more favor-

able soil conditions. As the hardwoods reach maturity and die, the pines return again, and so the cyclic succession continues.

The wide range of sites and soils on Cuba has encouraged many endemic species to evolve into new species by diversifying from tropical plants that have reached their northern limit. This subtropical "interzone" consequently supports the anomaly of pines growing together with tree ferns on both Cuba and Hispaniola. However, the pines have retained their ability to grow in pure stands or in association with only a few other species in these areas, in contrast to the more usual multiplicity of different species that compete for dominance in most tropical forests.

plant species flower every year, which may protect their seeds from being eaten regularly by predators. Many tropical forest plants, such as orchids (Orchidaceae) and milkweeds (Asclepiadaceae) have sticky, spiny or clumped pollen, specially adapted to cling to the insects and birds that feed from the plant; the pollen is then deposited on the next flower that the animal visits. Some of the plant–animal relationships are quite haphazard, but where cross-pollination is essential, the plant is best served by an exclusive pollinator.

At least 3,000 species of orchids that grow in the New World tropical zones are pollinated by bees. Most of these orchids are epiphytes, growing on the bark of trees. They include the *Gongora* orchids, which are visited exclusively by the bright metallic blue or green male *Euglossa* bees. The flowers are extremely fragrant, yet they attract no other types of bees or other insects. It has been established that certain chemicals in the fragrance of the plant attract only one kind of bee, isolating closely related plants and avoiding the possibility of interbreeding between them. Other specialized flowers include *Aristolochia*, which has the appearance and odor of rotting meat, and attracts only carrion flies.

Hummingbirds and certain plants are thought to have coadapted; the depth of the tubular petals on some flowers are the same length as the bills of certain hummingbird species. In addition, the colors of the flowers, usually red or orange, attract the birds. However, some interdependent relationships between plants and animals are not as exclusive as they at first appear. The streamertails, or doctor birds, of Jamaica, for example, usually visit bell-shaped or tubular flowers such as *Gesneria* or *Lobelia*, but in arid places they use their long beaks to feed from the open, daisylike flowers of cacti such as prickly pear (*Opuntia*) and *Melocactus*. They are probably the only birds that can feed on *Melocactus* without being impaled on the needlelike spines that surround the flowers.

A number of alien or introduced plants are pollinated by native insects, and illegitimate crosses with different species of the same genus result in natural hybrids. For example, ginger lilies (*Hedychium*), introduced to Jamaica from Asia, have formed hybrids in the areas where they grow together.

A STOREHOUSE OF FOOD AND FLOWERS

Central America and the Caribbean are the home of many crop plants that now form staple foods around the world. In addition, several native species are used in industry, such as henequen (*Agave fourcroydes*), which provides fiber, and maguey (*Agave atrovirens*), a source of alcohol. Others have become popular garden plants, as well as providing substances with medicinal properties used as drugs today.

Edible tubers and fruits

The drier upland areas of Mexico and its southerly neighbors are the home of many useful plants that have long provided the staple foods of the indigenous peoples. Sweet potatoes (*Ipomoea batatas*), grain crops of the family Amaranthaceae,

Paradise flower Also known as pride of Barbados, *Caesalpinia pulcherrima* originates in the West Indies but has been introduced to many tropical countries. The flowers have very long stamens.

pulses and other legumes, gourds and squashes (*Cucurbita*), potatoes (*Solanum*) and maize (*Zea mays*) all originated from, and have been developed in, this region. Other members of the family Solanaceae, including tomatillos (*Physalis*), sweet peppers and chilli peppers (*Capsicum*), are now widely grown and add zest to the local cuisine.

In contrast, the rainforests of the lowland isthmus areas (from Panama to Nicaragua) and the larger islands have produced trees and forest climbers with edible fruits. The best known and most economically important of these are the cacao bean, from *Theobroma cacao*, and the avocado (*Persea americana*). Less familiar are the brown fruits of the mammee sapote (*Pouteria sapota*), a distinctive

evergreen tree with large leaves clustered at the ends of the twigs. Canistel belongs to the same genus and is similar, but with smooth, yellow-skinned fruits that resemble mangoes. Star apple (*Chrysophyllum cainito*) probably originated in the West Indies, but is widely cultivated and naturalized on the mainland. The robust climber chayote (*Sechium edule*), which is described as "a squash with prickles" by the Maya, is unusual in that the large seed germinates within the flesh of the fruit.

Some plants are not native to the region but now grow as freely as indigenous species. Pawpaw (*Carica papaya*), for example, is now a weed tree in the forests of Central America. Its milky sap is widely used for medicinal purposes and as a source of the enzyme papaine, used commercially to tenderize meat. This property has long been appreciated by the indigenous people, who use the leaves to make their meat more palatable.

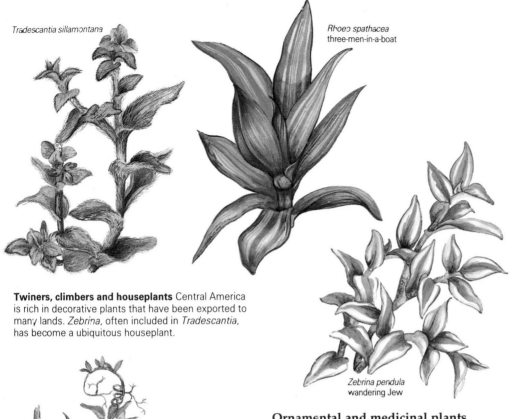

Tradescantia sillamontana

Rhoeo spathacea
three-men-in-a-boat

Twiners, climbers and houseplants Central America is rich in decorative plants that have been exported to many lands. *Zebrina*, often included in *Tradescantia*, has become a ubiquitous houseplant.

Cobaea scandens
cup-and-saucer vine

Zebrina pendula
wandering Jew

Solandra guttata
goldcup chalice vine

Solanum wendlandii
Costa Rican nightshade

treat warts. One example is nochebuena, or poinsettia (*Euphorbia pulcherrima*), a shrub that grows about 4 m (13.5 ft) high and produces spectacular red floral bracts in the dry season. A popular Christmas pot plant in Europe and garden plant in the tropics, it grows in deciduous tropical forests and some oakwoods of the Pacific side of Mexico.

Other plants with milky sap include flor de mayo or frangipani (*Plumeria*). *Plumeria obtusa* and *P. rubra* are open-branched shrubs or trees with large leaves and very fragrant white or red flowers. They grow wild in rocky thickets in warm areas, and are popular garden plants in the tropics. They are used medicinally against intestinal parasites and for skin and respiratory disorders.

Mexican breadfruit, known too as pinanona or monstera (*Monstera deliciosa*), also has a milky sap with irritant properties. This evergreen climber grows wild in forests in several parts of Mexico, but is probably better known as one of the most common houseplants worldwide. It can reach up to 15 m (50 ft) in height, and has large, deeply cut and perforated leaves. The long roots are used for weaving baskets and the fruit is edible.

One of the most famous and longest known medicinal plants is guayacan, or lignum vitae (*Guaiacum officinale* or *G. sanctum*). The gnarled tree grows to 4–8 m (13.5–26 ft). Its blue flowers are the national flower of Jamaica, and it produces orange-yellow fruit. An extract of the wood is used as a stimulant and to induce perspiration, the resin is used as a purgative, and other parts of the plant have been taken to treat rheumatism, skin ailments and venereal diseases.

Ornamental and medicinal plants

Mexico is particularly rich in ornamental plants, including dahlias, Mexican aster (*Cosmos*), *Zinnia*, marigolds (*Tagetes*), tree daisy (*Montanoa*), *Viguiera* and sunflower (*Helianthus*). Most are also aromatic, and some have medicinal properties, such as the margarita del mar or bay tansy (*Ambrosia hispida*). This annual, lowgrowing herb has numerous creeping branches and small lobed leaflets, with minute yellow-green flowers clustered in small heads. It occurs naturally on coastal dunes, and is used to control fever.

Some medicinal plants have a milky sap that has caustic properties. The sap of many euphorbias, for example, is used to

PRICKLY CUSTOMERS

It is a popular misconception that cacti grow only in the desert – in fact they are found in forests too. Some cacti, such as *Epiphyllum hookeri*, live in the forests as epiphytes clinging to other plants and drawing moisture from the air. Others are climbers, invading the crowns of trees or scrambling over rocks, putting out roots as they go. These include *Hylocereus undatus*, with its triangular stems, and *Cephalocereus swartzii*, with slender ribbed stems. Many of these cacti have huge, solitary, fragrant, nocturnal flowers, which have earned them names such as *reina de la noche* (queen of the night).

Typical cacti are barrel-shaped or columnar, with succulent, leafless, ridged, spiny stems. In the arid thickets of northern Mexico grows the largest cactus, *Pachycereus pringlei*, a massive columnar plant reaching 20 m (66 ft). Species of prickly pear also grow in thickets and thorn scrub, adopting creeping, shrubby or treelike habits. A spineless version is the smooth pear (*Nopalea cochinellifera*), which is often the host of the cochineal bug, the source of carmine dye. West Indian gooseberry (*Pereskia aculeata*) is also a fruit-bearing cactus, but unlike most cacti it has leaves at maturity.

The story of cocoa

For nearly 300 years after its discovery by Columbus in 1498, Trinidad was a sparsely inhabited Spanish colony. Initially the colonists spent their time growing tobacco for export to Europe and attempting to subjugate the indigenous Arawaks. Toward the end of the 17th century, missions of Capuchin friars set about employing the Arawaks as agricultural laborers, giving them protection from other Spanish settlers in return. They added maize and cotton to their crops, but put even greater emphasis on cacao trees; soon large plantations were supplying all of Spain's needs.

It is said that the Spaniards found cacao growing wild in Trinidad when they arrived, but this conflicts with other reports that they introduced the tree from Central America in 1625. The cacao they grew was of the Criollo variety, which has white or pale violet cotyledons (fleshy seed leaves). The chocolate flavor, which in this variety has no astringency, develops only after the seeds have been fermented and dried. Once processed, the beans produce highly flavored cocoa.

A world crop

In 1727 disease struck Trinidad's cacao crops, causing the young pods to wither on the trees. The economy collapsed completely and many settlers departed. The disease, referred to as "blast", was probably the condition now known as Cherelle Wilt, an exaggerated state of natural fruit-thinning. It was attributed by priests to the wickedness of the planters, and by the planters to "death winds" from the north or to drought. The response was a state of apathy that for the next fifty years reduced the social and economic life of Trinidad to a state of complete inactivity.

In about 1757 the Dutch introduced another variety, known as Amazonian Forastero, or Amelonado, probably from Venezuela. The plantations were soon reestablished, and the new variety was subsequently crossed with the white-beaned Criollo to create the Trinitario hybrids; the best selections of these have the vigor of Forastero combined with the quality of Criollo.

Long before this time, Spanish explorers had taken Amelonado cacao first to Brazil and then to the Spanish island of Fernando Póo, now called Bioko, off the coast of Equatorial Guinea in West Africa. There they established plantations worked by labor imported from eastern Nigeria and other parts of the Guinea coast. In 1879 one of these workers, Tetteh Quarshie, took cacao to what was then called the Gold Coast, farther west along the Gulf of Guinea. The forest soil, seasonal climate and peasant culture were

Cacao relatives (*below*) Cacao belongs to a genus of some 30 species of small trees that are found in the rainforests of both Central and South America. Other species are cultivated locally either for cocoa or for their edible fruits.

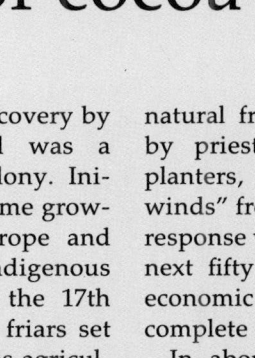

Cacao flowers (*above*) Cacao is a small evergreen tree with large thin leaves. The complex flowers have petal-like appendages, and are borne on the old wood of the trunk and the lower parts of the main branches.

Cacao pods (*left*) As the pods ripen they change from green to orange-yellow or red according to variety. They are large and heavy, and need the strength of the trunk or big branches to support their weight. The beans are rich in a fat known as cocoa butter, which is used in making chocolate.

ideal for its growth. In a few years the Gold Coast became the world's largest producer of cacao and, later, the center for scientific research on this plant. Most of the cacao trees in present-day Ghana are still of the same Amelonado variety.

Two-stage growth

The cacao sapling grows in two phases. First, upright shoots (called chupons) with spiral leaves develop; the terminal buds on these shoots produce three to five horizontal fan-branches (called jorquettes) with leaves in two rows. The leaves develop in series; the new leaves hang down and are often reddish at first. Eventually, with the change of season, chupon development resumes and the lower fan-branches wither and fall off, leaving a straight trunk.

The method of pollination of the small, scentless and nectarless flowers is unknown, but small midges probably perform this essential function. It is clear, though, that no one kind of insect is entirely responsible for pollination, in view of the success of this plant in a wide range of tropical areas.

DIVERSITY WITHOUT PARALLEL

TROPICAL TREES AND GRASSLANDS · WATER USE AND POLLINATION · STAPLE FOODS AND STIMULANTS

South America contains lush tropical rainforest, grasslands, arid deserts and high mountains. Such variety is reflected in the unparalleled richness and diversity of the plants, many of which are of great economic, medicinal or decorative value. The sheer size of the rainforests gives them an added significance in terms of global ecology. The total number of plants in the region is unknown, and many have yet to be investigated. Many plants have also migrated from North America – the Andes provide a route through the tropics, and seeds may be carried by birds. A few primitive plants, which are related to species found in Australasia, are ancient remnants from the time when South America was part of the giant landmass called Gondwanaland, from which it broke away some 64 million years ago.

COUNTRIES IN THE REGION

Argentina, Bolivia, Brazil, Chile, Colombia, Ecuador, Guyana, Paraguay, Peru, Surinam, Uruguay, Venezuela

EXAMPLES OF DIVERSITY

	Number of species	Endemism
South America	70,000–80,000	very high
Galapagos Islands	543	55%
South Atlantic islands	170	10%

PLANTS IN DANGER

	Threatened	Endangered	Extinct
South America		Very many	
Galapagos Islands	122	9	–
South Atlantic islands	4	1	1

Examples Aechmea dichlamydea; Amaryllis traubii; Dalbergia nigra; Dicliptera dodsonii; Glomeropitcairnia erectiflora; Legrandia concinna; Mimosa lanuginosa (snow mimosa); Mutisia retrorsa; Persea theobromifolia (Rio Palenue mahogany); Spergularia congestifolia

USEFUL AND DANGEROUS NATIVE PLANTS

Crop plants Anacardium occidentale (cashew nut); Ananas comosus (pineapple); Arachis hypogaea (peanut); Bertholletia excelsa (brazil nut); Capsicum annuum (chilli peppers); Carica papaya (papaya); Hevea brasiliensis (rubber); Lycopersicon esculentum (tomato); Nicotiana tabacum (tobacco); Passiflora edulis (passion fruit); Phaseolus vulgaris (kidney beans) Solanum tuberosum (potato); Theobroma cacao (cocoa);
Garden plants Begonia species; Berberis linearifolia; Bougainvillea species; Buddleia globosa; Calceolaria polyrrhiza; Fuchsia corymbiflora; Gunnera manicata; Jacaranda mimosifolia; Tradescantia albiflora
Poisonous plants Datura species; Dieffenbachia species (dumbcane); Lantana camara (yellow sage); Strychnos toxifera; Thevetia peruviana

BOTANIC GARDENS

Department of Botany and Agriculture, Castelar (5,000 taxa); José C Matis, Bogotá; Rio de Janeiro (7,000 taxa); São Paulo (1,000 taxa); University of Southern Chile, Valdivia (1,200 taxa)

TROPICAL TREES AND GRASSLANDS

Some South American plants are remnants of the plant life of the ancient supercontinent of Gondwanaland. Examples of these are certain primitive genera of trees, including monkey puzzles (*Araucaria*) in Brazil and Chile, and the podocarps (*Podocarpus*), which also occur in Australasia. *Trigonobalanus excelsa*, a Colombian tree, is the only South American representative of a genus of three species; it forms an evolutionary link between the oaks (*Quercus*) and beeches (*Fagus*) of the northern hemisphere and the beeches (*Nothofagus*) of the southern hemisphere.

The tropical zones of northeastern South America are extremely rich in plants. The principal families are the orchids (Orchidaceae), daisies (Compositae), Bignoniaceae and Melastomaceae, with 400 genera endemic to the region. The 30 species of fern-tree (*Jacaranda*) are centered in the area but not restricted to it. Among the many lianas that contribute to the structure of the Amazonian forests are spiny *Bougainvillea*, *Cobaea*, golden trumpet (*Allamanda*) and passion flowers (*Passiflora*). Many plants in the rainforest – particularly orchids and bromeliads (Bromeliaceae) – are epiphytes, relying on a host tree for anchorage and support.

The spine of the continent

The Andes dominate South America. Their altitude (maximum 6,960 m/22,800 ft) means that they provide a cool temperate route for southern plants such as the Antarctic pearlwort (*Colobanthus quitensis*) to migrate north to Peru and beyond, and for North American groups, such as *Oreomyrrhis*, to move south into Patagonia. The Andes also obstruct the rain-bearing winds, giving rise to the Atacama Desert in northern Chile and southern Peru, and the arid Patagonian steppe of Argentina.

The Atacama Desert is largely devoid of plant life; a few species grow locally, and coastal fogs support the cacti *Eulychnia* and *Copiapoa* and the bromeliad *Puya berteroana* – other species of *Puya* occur at higher altitudes in the Andes. Farther south, central Chile enjoys a Mediterranean-type climate in which aromatic trees such as the boldo (*Peumus boldus*), raran (*Myreugenia obtusa*) and huacan (*Myrica pavonis*) flourish. The annuals

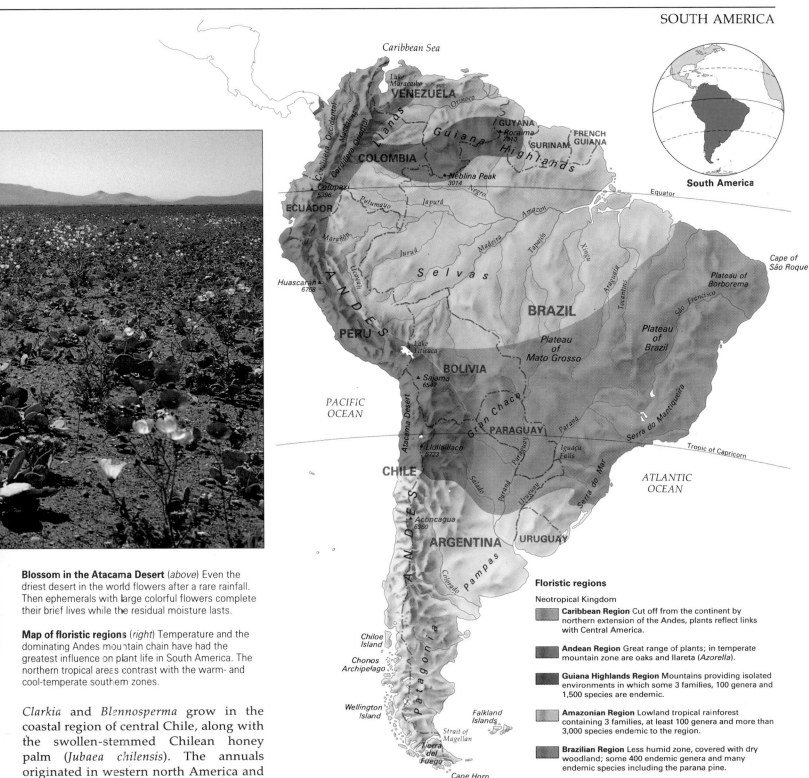

South America

Equator

Tropic of Capricorn

Floristic regions

Neotropical Kingdom

Caribbean Region Cut off from the continent by northern extension of the Andes, plants reflect links with Central America.

Andean Region Great range of plants; in temperate mountain zone are oaks and llareta (*Azorella*).

Guiana Highlands Region Mountains providing isolated environments in which some 3 families, 100 genera and 1,500 species are endemic.

Amazonian Region Lowland tropical rainforest containing 3 families, at least 100 genera and more than 3,000 species endemic to the region.

Brazilian Region Less humid zone, covered with dry woodland; some 400 endemic genera and many endemic species including the parana pine.

Holantarctic Kingdom

Chile-Patagonian Region Cool southern zone, with 8 endemic families. Chile contains greatest diversity, but here plants are linked across the Andes.

Blossom in the Atacama Desert (*above*) Even the driest desert in the world flowers after a rare rainfall. Then ephemerals with large colorful flowers complete their brief lives while the residual moisture lasts.

Map of floristic regions (*right*) Temperature and the dominating Andes mountain chain have had the greatest influence on plant life in South America. The northern tropical areas contrast with the warm- and cool-temperate southern zones.

Clarkia and *Blennosperma* grow in the coastal region of central Chile, along with the swollen-stemmed Chilean honey palm (*Jubaea chilensis*). The annuals originated in western north America and were probably dispersed by migrating birds. Birds are also responsible for introducing many plants to the Galapagos Islands, which lie about 1,000 km (600 mi) off the coast of Ecuador. Many plants on these islands show similarities to species found in the Andes.

Southern grasslands

Moving southward across Argentina, the subtropical forests give way to open grassland – the pampas. This habitat is dominated by many species of feather grass (*Stipa*) and, especially in rocky areas, by the enormous silvery plumes of pampas grass (*Cortaderia selloana*), which grows 2–3 m (6.5–10 ft) tall. In southeast

Argentina the arid Patagonian meseta is a sparse grassland of drought-resistant species of feather grasses, tough, narrow-leaved fescues (*Festuca*) and piojillos (*Poa*). Some areas are dominated by shrubs that have adapted to the dry conditions (xerophytes), such as mata negra (*Verbena tridens*) and the spiny *Chuquiraga* and *Mulinum*; there are also many spiky-leaved, lowgrowing plants such as the red-flowered mata guanaco (*Anarthrophyllum rigidum*).

At the southern tip of the continent lies

Tierra del Fuego. There are fewer northern plants here; southern groups such as the abrojos (*Acaena*) and species such as *Oxalis magellanica* and the dandelion *Taraxacum gilliesii* are widespread.

The Falkland Islands (Islas Malvinas) off the east coast are a botanical extension of Patagonia, though with fewer species. Thirteen species are unique to the islands, but most of these show strong similarities to mainland plants – only the cabbage-relative *Phlebolobium macloviana* is truly distinctive.

WATER USE AND POLLINATION

The wide range of environmental conditions in South America have resulted in a great variety of plant adaptations. In the equatorial forests of Amazonia a large number of trees, such as *Martiodendron parviflorum* and mahogany (*Swietenia macrophylla*), have buttress roots at the base of their trunks. These structures resemble the supports found on the external walls of cathedrals, and are thought to perform a similar function.

The high humidity of these forests means that water tends to settle on leaves. If the water were to remain on the surface of the leaf, algae would develop and impair photosynthesis by blocking out the light. In order to avoid this, many unrelated species of forest plants have leaves with long, pointed drip-tips (apices), which encourages water to drip from the leaves.

Lianas flourish in the tropical forests. Some species, such as *Cobaea trianaei* in Colombia, are pollinated by bats, while others, including the Chilean copihue (*Lapageria rosea*) and its southern relative the coicopihue (*Philesia magellanica*), are pollinated by hummingbirds. The southern forests also support mistletoes (*Misodendrum*), that are found nowhere else in the world. These green-leaved plants are hemiparasites – they are able to photosynthesize, and therefore manufacture a certain amount of food, but nevertheless derive most of their nutrients from the trees, usually southern beeches (*Nothofagus*), on which they grow.

Hostile environments

The highlands of Venezuela (paramos) are as much as 2,400 m (7,800 ft) above sea level. The plants that live there have to be specially adapted to deal with high temperatures during the day and low temperatures at night – conditions typical of any high-altitude tropical habitat. The frailejon (*Espeletia*) have adaptations typical of this environment. They have leaves arranged spirally in dense rosettes that close up at night to insulate the growing tissues from the cold. The leaves are also covered with fine hairs that help reduce water loss during the day.

In Patagonia the flanks of the Andes are clothed with temperate forests, but farther east the conditions become increasingly arid and the tree cover soon

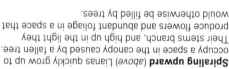

Spiralling upward (*above*) Lianas quickly grow up to occupy a space in the canopy caused by a fallen tree. Their stems branch, and high up in the light they produce flowers and abundant foliage in a space that would otherwise be filled by trees.

High in the Andes (*below*) Many plants have thick leaves felted with silky hairs to enable them to cope with the frosty nights, high daytime temperatures and the intense ultraviolet radiation found at altitudes of 4,800 m (15,800 ft).

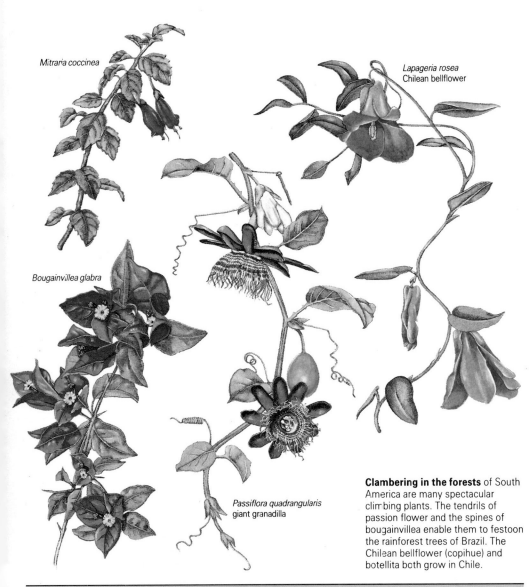

Mitraria coccinea

Lapageria rosea
Chilean bellflower

Bougainvillea glabra

Passiflora quadrangularis
giant granadilla

Clambering in the forests of South America are many spectacular climbing plants. The tendrils of passion flower and the spines of bougainvillea enable them to festoon the rainforest trees of Brazil. The Chilean bellflower (copihue) and botellita both grow in Chile.

disappears. The plants have developed many adaptations to the dry, windy conditions. Shrubs generally have leaves that are reduced to narrow, hard, often spiny structures; many species from quite unrelated groups show similar adaptations to the lack of moisture. The southernmost cactus in world, *Maihuenia patagonica*, also lives here.

A study of the leaf structure of 284 Patagonian species showed that, as one moves eastward into the more arid areas, the plants increasingly exhibit features that help them to combat water stress. These include, for example, leaves with a thick cuticle and an extra outer layer of epidermal cells, a covering of hair on the lower leaf surface, a high level of sticky secretions from the leaves and more abundant tissues for storing water (xylem): all these features help to reduce the amount of water lost.

In arid areas of Patagonia many plants have a cushion habit of growth. They include the llareta or balsam bog (*Bolax gummifera*) and *Azorella trifurcata*, both in the carrot family (Umbelliferae), and the white-flowered *Benthamiella nordenskjoldii* (Solanaceae), a relative of the potato. These cushion plants provide their own microclimate to cope with the dry, windy environment. The small flowers of these species are grouped into a mass that is conspicuous to pollinators. Cushion plants grow in similar conditions at very different altitudes: the yellow-flowered *Oreopolus glacialis* occurs at only 50 m (160 ft) above sea level in the cool conditions of the extreme south in Tierra del Fuego, and at 4,000 m (13,000 ft) in the high Andes above Santiago, some 20 degrees farther north.

Some plants have developed long taproots that enable them to reach water and also provide secure anchorage in the unstable, often volcanic soils. To counter the great difference between day- and nighttime temperatures typical of arid and high-altitude environments, the leaves of many plants are arranged in tight rosettes; unusual examples are a violet (*Viola cotyledon*) and the whitish yellow flowered *Menonvillea nordenskjoldii*.

Other plants of the high Andes include natural migrants and colonizers, such as the Antarctic pearlwort. With male and female flowers on the same plant, it canself-pollinate; it produces numerous seeds that germinate readily, a necessary mechanism for a colonizing plant.

PLANTS OF THE GALAPAGOS

The Galapagos Islands have never been connected to the South American mainland; they rose from the ocean floor as a result of volcanic activity that began several million years ago. The plants that colonized the islands were almost certainly carried from the mainland by birds or on the ocean current; they all have related species in the Andes. Seven genera have since evolved; four of these, *Darwiniothamnus*, *Lecocarpus*, *Macraea* and *Scalesia*, belong to the daisy family (Compositae), there are two genera of cacti, *Brachycereus* and *Jasminocereus*, and the last genus is a cucurbit (*Sicyocaulis*).

Of all the species that grow on the Galapagos, 43 percent are found nowhere else in the world. The level of endemism is much lower than that of more isolated Pacific islands such as Hawaii, where the figure is 95 percent. This is because of their relative youth and the proximity of the Galapagos to the mainland. Most of the endemics (67 percent) are in the arid lowlands, the oldest part of the islands, while 29 percent are found in the rather moister and newer uplands. Only 4 percent of the endemic species grow in coastal areas; the ocean brings in many immigrants but also causes a high rate of extinction as a result of wave-action.

An additional 181 plant species are aliens, brought deliberately or by accident to the islands by humans. People and animals are also responsible for bringing many of the native plants, including *Borreria linearifolia* and a scrambling relative of the cucumber, *Sicyos villosa*, to the verge of extinction; several species have already been lost.

STAPLE FOODS AND STIMULANTS

South America is the source of many plants that are now grown commercially in other parts of the world. Conversely, some species have been imported into the continent and successfully exploited. It is difficult to assess the extent to which food plants were transported between Central and South America prior to the arrival of European settlers in the 16th century, but there is substantial evidence that many popular and widely cultivated crops originated in the south.

Cash crops

South America is the original source of rubber (*Hevea brasiliensis*). The Brazilian monopoly of this crop was destroyed when wild seeds were taken from the Amazon forest and used to establish plantations in Malaysia.

Cocoa (*Theobroma cacao*), from which chocolate is made, also originates in the region. This tree, which produces seeds rich in stimulating alkaloids, probably originated in the upper Amazon, but it is now widely cultivated in both western Africa and Southeast Asia, while coffee (*Coffea arabica*), popularly associated with tropical South America, was introduced from Africa. Less widely known beverages include the Argentinian drink yerba maté, which is made from the dried leaves of tea (*Ilex paraguayensis*). In Chile beverages called aguitas are prepared by infusing the leaves of certain aromatic species, notably boldo (*Peumus boldus*). This shrub contains certain essential oils, such as terpinol and eugenol, and the drink is taken after meals as a digestive.

Foods and flowers

Many widely cultivated crops originated in South America. The peanut (*Arachis hypogea*) has close relatives in northwest Argentina, Paraguay and Uruguay, and is known from pollen evidence to have grown on the coast of Peru 3,800 years ago. The potato (*Solanum*) comes from the temperate areas flanking the Andes. The best-known species, *S. tuberosum*, probably originated from cultivation in the area spanning Bolivia to Colombia, but other wild species have been used subsequently in breeding programs.

The tomato (*Lycopersicon*) comes from the northern Andes and the Galapagos Islands. The cultivated chilli peppers (*Capsicum*) seem to have developed from *C. baccatum*, which can be found from southern Peru through Bolivia and Paraguay to southwest Brazil. Papaya or pawpaw (*Carica papaya*) probably originated in Peru. It is now a widespread weed tree of tropical forests, and its fruits are of increasing economic value. Plants of the genus *Oxalis* are not usually considered as a source of food, but the oca (*O. tuberosa*) has long been cultivated in the Peruvian Andes as a root vegetable.

South America has produced many plants that decorate gardens throughout the world. Notable examples are the spiny barberry, African marigolds (*Tagetes*), fuchsia (which has been hybridized

Nothofagus obliqua

Nothofagus pumilio

Nothofagus procera

Podocarpus andinus
plum-fruited yew

Araucaria araucana
monkey puzzle

Relict trees These primitive trees are all remnants of the ancient vegetation of the supercontinent of Gondwanaland. The timber of the southern beeches (*Nothofagus*) has many uses. The podocarp is used ornamentally as a hedge; the monkey puzzle is used for shelving and planted as a curiosity.

MANIHOT: POISONOUS BUT EDIBLE

When European explorers discovered South America, the shrubby tree, *Manihot esculenta* (locally called cassava, manioc, aypi, yuca and tapioca) was widely cultivated in the tropical lowlands, where it probably originated. The tree grows well on poor soils. Its large, tuberous roots are an excellent source of starch, though they contain little protein. The major problem with using manihot as a food source is that the plant contains high levels of the poison cyanide. Moreover, the varieties that yield the most starch also contain the most cyanide. The poison is removed by squeezing the pulverized tubers in water and then laying out the meal to dry.

Manihot or cassava has been widely introduced into other tropical regions, where it has become an important staple food crop. Intensive research is currently being directed toward developing cultivated forms that combine low levels of cyanide with a high yield of starch. These efforts have been hampered by the slow rate of reproduction from seeds, so vegetative reproduction and modern tissue-culture techniques are being used to accelerate these breeding programs.

to produce a great variety of ornamental cultivars), petunia (*Petunia*) and godetia (*Clarkia tenella*).

Medicines and hallucinogens

South American plants are the source of a variety of drugs. Quinine is an anti-malarial alkaloid derived from the Jesuit's or Peruvian bark species, *Cinchona*. Coca (*Erythroxylum coca*) has been used for at least 5,000 years by Andean Indians as a stimulant. The leaves are mixed with lime to increase the stimulant effect and then chewed, but not swallowed. Used in this way they help to combat dietary deficiency by maintaining blood glucose levels and reducing hunger pangs. Coca is also the source of cocaine, long used as an anesthetic but now notorious for its role in drug abuse.

Plants containing chemicals that induce hallucinations are, or were, important in many South American cultures. In the west of the region, the hierba loca or huedhued (*Pernettya furens*) of Chile and the taglii (*P. parvifolia*) of Ecuador – both members of the heath family (Ericaceae) – contain compounds that induce psychic alterations. The arbol de los brujos (*Latua pubiflora*) is used by the Mapuche Indians, who live in southern central Chile, for similar purposes.

Plants of the genus *Datura*, which includes the jimson weed or thorn apple, are rich in alkaloids such as hyoscyamine and hyoscine – substances widely used in the treatment of asthma. The tree *D. arborea*, found growing from Colombia to Peru, is valued for its hallucinogenic properties. Farther south, the Mapuche people use *D. candida* and *D. sanguinea* to correct the behavior of unruly children: they believe that the spirits of their ancestors chastise the children during their hallucinations.

In Patagonia and Tierra del Fuego the branches of the barberry or humerzamaim (*Berberis buxifolia*) were greatly prized by the Yahgan Indians for making arrows. Not only does the species provide straight, true shafts, but when the bark is peeled off the wood underneath is yellow – a color that was of mystical significance to the Indians.

A marvel of Peru *Mirabilis jalapa* is also known as the four o'clock plant because its fragrant flowers open late in the afternoon. The specific name is an allusion to the fact that the plant was thought at one time to be a source of the purgative drug jalap. It is often grown in gardens as an annual.

Epiphytes in the rainforest

Epiphytes are plants that grow on other plants without – unlike parasites – deriving any nutrients from their hosts. In the rainforest, epiphytes take advantage of the trees, which provide them with an elevated position away from the gloom of the forest floor and closer to the light. The angles between the branches and trunks, and the crevices in the bark – places where moisture and organic debris accumulate – are favored by many species. Epiphytic ferns and mosses survive on water trickling down the tree trunks.

Prominent among the rainforest epiphytes are numerous species of orchid (Orchidaceae). Many produce long roots that dangle in the humid air and absorb moisture. In many instances the roots are green and able to photosynthesize. Palms such as the Brazilian jara (*Leopoldinia pulchra*) often support a wide range of epiphytes, as their fibrous stems are porous and retain water.

Another group of epiphytes, the bromeliads (Bromeliaceae), have water-absorbing hairs on their leaves and stems. Many bromeliads, such as *Neoregelia* and *Guzmania*, have tight spiral clusters of leaves arranged as inverted cones, which retain rainwater. The small pools of water provide a habitat for numerous insects, species of the aquatic plant bladderwort (*Utricularia*), and even frogs (genus *Hyla*).

Epiphytes under attack

The Amazon rainforest is incredibly rich in epiphytes, with many trees supporting hundreds of plants; the species come from several plant families. Orchids and bromeliads predominate, and have the most colorful and complex flowers. The exotic nature of their blooms has led to many species, especially orchids such as *Cattleya*, *Laelia* and *Epidendrum*, being collected to the point of extinction. These plants are highly prized not only as specimens but also for use in horticultural

Spider orchid (*above*) These extraordinary flowers form part of a long inflorescence typical of the genus *Brassia*. In some species it may measure 60 cm (2 ft) or more. These orchids form large plants and produce many roots.

Bromeliads perch on a rough-barked tree (*right*) Epiphytes can gain a better foothold on host trees that have rough bark than on those whose bark is smooth. Sometimes a branch may support many bromeliads, and when their tanks are full of water may break under their combined weight.

Flamboyant purple orchid (*below*) Many orchids have developed swollen stems known as pseudobulbs for storing water. In *Cattleya skinneri* these are green and capable of photosynthesis, and like a small cucumber in appearance.

hybridization programs. More destructive than the actions of plant collectors, however, has been the removal of the rainforests – the felling of one tree results in the destruction of numerous epiphytes that depend on it.

Other families of plants besides the orchids and bromeliads contain epiphytic species. The aroids (Araceae) are represented by the genus *Anthurium*. The popular houseplant known as the flamingo flower (*A. andraeanum*) has a conspicuous 12 cm (4 in) red leaf bract underlying the pencil-like flowerhead. The gesneriads (Gesneriaceae) also include prominent epiphytes such as *Columnea* and *Cedonanthe*, which are pollinated by birds. The latter species normally grows in association with ants' nests, and its seeds, which are the same size as ant eggs, are distributed by these insects.

It is not known what creature pollinates the Amazon moonflower (*Strophocactus wittii*); this white-flowered epiphytic cactus opens its strongly perfumed blooms for only a few hours on a single night. It might seem remarkable to find a cactus in a rainforest, but survival in this habitat – as in the desert – requires the ability to cope with water stress.

Giant herbs of the banana family

The lobster claws (*Heliconia*) inhabit both the moist humid, semideciduous, monsoon forests of the lowlands and the rainforests of Central and South America. These two types of forest have many species in common. The monsoon forest tends to have fewer species, in particular fewer lianas and epiphytes, but a better developed ground layer of which the heliconias form a part. In the monsoon forests they grow with tall bamboos, palms and *Calathea insignis*, a cormous plant with thin variegated foliage widely sold as a houseplant.

The large leaves and fantastic inflorescence of *Heliconia* are shared by other members of the banana family (Musaceae). The large, colorful, boat-shaped bracts conceal inconspicuous flowers. Copious amounts of nectar ooze into the bracts; the nectar is diluted with rainwater to produce a sweet drink for bird pollinators. The bird of paradise flowers (*Strelitzia*), South African relatives of the lobster claws, also have large sturdy flowerheads composed of brilliant-colored bracts to attract birds. By far the best-known species is *S. reginae*, with its spiky inflorescence of orange and purple.

One or two species of *Strelitzia* develop trunks, but the only woody genus of the banana family is *Ravenala*, represented in Madagascar by the traveler's palm (*R. madagascariensis*) and in Brazil and Guyana by *R. guianensis*. These both have a fan of paddle-shaped leaves at the top of their trunks.

Large, broad leaves are characteristic of all the Musaceae. They tend to tear easily along the parallel veins that run at right angles to the midrib. Another common feature is the way in which the leaf sheaths are rolled to form a stem. This false stem may grow to a height of more than 10 m (33 ft) in a banana "tree".

Lobster claws in the rainforest Three species of *Heliconia* growing together. Although these bizarre plants are found only in tropical Central and South America, they take their generic name from Helicon, the mountain supposed to be the home of the Muses in central Greece.

PLANTS OF FOREST AND SNOW

EFFECTS OF THE ICE AGE · PLANTS OF THE MOUNTAIN SNOWS · TREES – THE MAIN RESOURCE

In the Nordic countries, as elsewhere, the plant life is governed by changes in climate, topography, soil type and disturbances of the land. Here the greatest influence is still the cold of high latitudes; during the last ice age the entire region was covered by the ice sheet. It has left a legacy in the relative paucity of plant species to be found. In the far north tundra plants and peat bogs predominate, giving way farther south to coniferous forests of pine, spruce and birch in the rugged mountains of Norway and around the lowlying lakes of Finland. In the south the climate is milder, and the fertile plains are intensely cultivated. Far out in the Atlantic Ocean, some 910 km (570 mi) west of Norway, lies the volcanic island of Iceland, grassy, windswept and treeless, and the new island of Surtsey, already being colonized by plants.

Fall colors in a rugged northern landscape Lichens cover the rocks and boulders, and a stunted silver birch tree prepares to drop its leaves for winter. Trees are rare in this harsh evironment; the few plants that are tough enough to survive in the northern tundra zones cling to the ground.

COUNTRIES IN THE REGION

Denmark, Finland, Iceland, Norway, Sweden

DIVERSITY

	Number of species	Endemism
Denmark	1,400	1
Finland	1,100	very low
Iceland	470	1
Norway	1,700	1
Sweden	1,700	1

PLANTS IN DANGER

	Threatened	Endangered	Extinct
	16	10	3

Examples *Braya linearis; Cephalanthera rubra; Gentianella uliginosa; Liparis loeselii; Najas flexilis; Oxytropis deflexa* subsp. *norvegica; Papaver lapponicum; Platanthera obtusata* subsp. *oligantha; Polemonium boreale; Potamogeton rutilus*

USEFUL AND DANGEROUS NATIVE PLANTS

Crop plants *Picea abies* (Norway spruce)
Garden plants *Betula pendula* (silver birch); *Papaver nudicaulis; Saxifraga oppositifolia* (purple saxifrage); *Sorbus aucuparia* (mountain ash); *Trollius europaeus* (globe flower)
Poisonous plants *Aconitum napellus; Taxus baccata* (yew)

BOTANIC GARDENS

Copenhagen University (25,000 taxa); Gothenburg (12,000 taxa); Oslo University (8,000 taxa); Reykjavik (3,000 taxa); Uppsala (10,000 taxa)

EFFECTS OF THE ICE AGE

Much of northern Scandinavia lies beyond the Arctic Circle, where even at low altitudes conditions are extremely harsh. These are also found in the higher altitudes of the mountains that run down the western coast. During the last ice age the polar ice cap extended southward and covered the whole of the region. Many plant species migrated south ahead of the advancing ice, but after the ice had retreated with the onset of warmer climatic conditions some 10,000 years ago some species did not move north again to reoccupy their former areas.

The effects of the ice age are one reason why the northern areas have relatively few species of plants compared with those of the south. The poor quality of the soil, which is often low in minerals and other essential plant nutrients, is another contributory factor. An extreme example of the poor diversity is found on Spitz-bergen, the principal island of the Svalbard group north of Norway: it has only 110 recorded species of flowering plants.

In the north the growing season – regulated by sunlight and temperature – may be only 100 days long, compared with 200 days a year in the south of the region. The plants have had to adapt to periods of continuous daylight, still with relatively low temperatures, in summer, and in winter to long stretches of darkness and prolonged severe frost.

In the far north of the region, the undulating plains of Lapland support Arctic tundra; mosses, lichens and grasses are prominent, along with hardy dwarf shrubs such as ling or heather (*Calluna vulgaris*), crowberry (*Empetrum hermaphroditum*), bilberry (*Vaccinium*) and small cushions of *Diapensia lapponica*. In the southern limit of the tundra belt grow birches (*Betula pubescens* and *B. tortuosa*). Few tree species can survive this far north, though there are some isolated stands of Scots pine (*Pinus sylvestris*) or birch growing amid areas of barren rock and subarctic tundra.

Coniferous forest

Farther south the tundra gives way to coniferous forest known as taiga, which covers much of northern and central Sweden and Finland. Here, Norway spruce (*Picea abies*) and Scots pine predominate. Scots pine grows farther north than Norway spruce and is also found at higher altitudes. The forest forms a mosaic with the extensive wetlands of the region; the bogs, fens and mires are often rich in plant species, particularly sedges and mosses. It also supports a small number of broadleaf trees such as birch, aspen and alder. At the taiga's southern limit these give way to mixed conifer and deciduous forest where spruce and pine mingle with hazel, beech (*Fagus*), mountain ash (*Sorbus aucuparia*), willow (*Salix*), ash (*Fraxinus*) oak and even the small-leaved lime, usually found in more temperate areas.

Farther south still there are pure deciduous forests – though many of the beechwoods are in fact plantations. Although the trees themselves are often magnificent, most of the shrubs and herbs are poor and not at all like those of the original wildwood, which was destroyed by early settlers.

In the mountains, as altitude increases, the forests give way to low willow scrub. At the treeline tall herbs grow, with globe flower (*Trollius europaeus*), wood cranesbill (*Geranium sylvaticum*), northern wolfsbane (*Aconitum septentrionale*) and alpine sowthistle (*Lactuca alpina*) prominent. Of these only globe flower is found much farther up the mountain slopes. Then follows a species-rich moist zone with sedges, cotton-grass and cloudberry (*Rubus chamaemorus*). This gives way to the plants of the open *fjäll*; and close to the summit grows the glacier buttercup (*Ranunculus glacialis*), often at a greater altitude than any other flowering plant. In the mountains there are many small streams, and attractive herbs such as yellow saxifrage (*Saxifraga aizoides*) – some of whose flowers are actually orange – grow along their margins.

The islands of the north

On the Faeroe Islands, which lie midway between Denmark and Iceland, mosses and sedges grow on wet ground, while heathers and grasses thrive in the relatively dry areas. On Iceland the diversity of species is very poor; in the lowlands the grassland is extensive, while mosses, rushes, sedges and cotton-grass are found in wet areas. Subarctic tundra occurs above 200 m (650 ft), but large areas are so cold or dry, or the soil so lacking in plant nutrients, that nothing can grow.

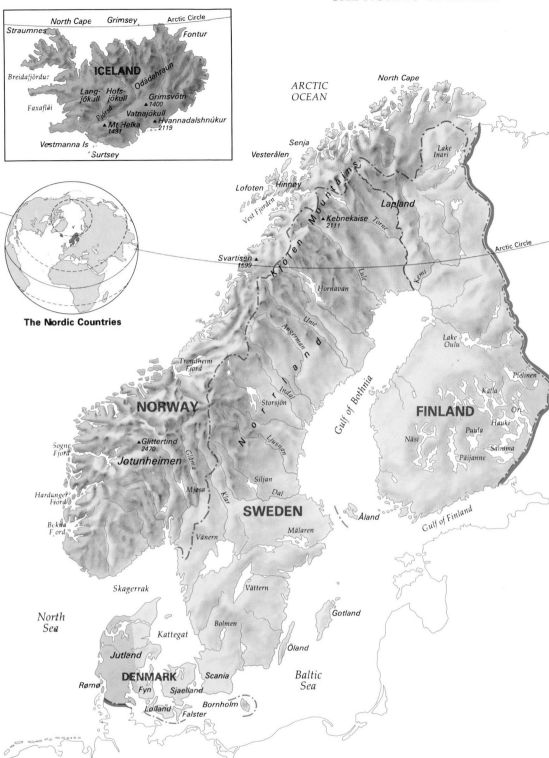

The Nordic Countries

Floristic provinces

Holarctic Kingdom/Circumboreal Region

Arctic Province Long, cold winter and brief summer; characterized by dwarf willows, small birches, dwarf bilberry and bearberry and the mountain avens.

Atlantic-European Province Climate moderated by warm North Atlantic Drift; characterized by ivy, heather and the primrose *Primula scandinavica*.

Northern European Province Relatively new range of plants so few endemic species. Characteristic trees are conifers, e.g. pines and spruces.

Central European Province Continental European climate; plants of mountains farther south (Alps and Carpathians) found at this northern limit of the province.

Map of floristic provinces The Nordic countries are all included in the Circumboreal Region, the Earth's cool northern temperate zone. Much of the plant life reflects the harsh climate of these northern lands.

PLANTS OF THE MOUNTAIN SNOWS

Arctic alpine plants grow at high altitudes and high latitudes, both of which are very cold in winter. The mountain avens (*Dryas octopetala*), for example, is only found high up in the mountains of southern Norway, but will grow almost at sea level in the north of the country. Arctic alpines are well adjusted to snow, which often arrives in the northern mountains in September and lies until the following spring or summer. The icy wind whirls the snow into uneven drifts, from a few centimeters deep to 3 m (10 ft) or more. In sites where the snow has been blown away completely, plants such as trailing azalea (*Loiseleuria procumbens*) are exposed to bitterly cold, drying gusts of wind and are scoured by wind-driven ice crystals that sweep close to the ground. A few meters away, neighboring plants may lie protected beneath a thick, insulating blanket of snow, but even they are vulnerable to grazing by lemmings.

In spring the sun soon melts what little snow and ice remains on plants hardy enough to survive on the exposed ridges, whereas those that are protected by deep snow have to wait several more weeks before the blanket melts and they too can take advantage of the precious sunlight to photosynthesize. The active lives of such plants are telescoped into the few brief weeks of summer, during which time they must grow, flower, fruit and lay down reserves for the following year.

High on the mountains the snow encourages unusual growth forms in isolated conifers. Young spruce trees develop as low cushions that are covered by snow in the winter. The trees reproduce by layering; the long side branches form adventitious roots that penetrate the soil. Upward-growing shoots that survive above the level of the snow develop lateral shoots only on the relatively well protected side of the trunk that faces away from the prevailing wind. In the extreme cold immediately above the winter snowline, the stunted trunks of spruce trees often develop no branches at all.

Pink-flowered trailing azalea (*above*) This prostrate shrub forms a dense evergreen mat, a growth form that not only helps it to survive the rigors of an Arctic winter but also combats the summer drought experienced in a freely draining habitat.

Creeping shrubs of the Arctic (*right*) These lowgrowing plants spread slowly. The bearberry is pollinated by insects, but the dwarf willows are wind pollinated. Male and female catkins grow on different plants, and natural hybrids are common.

Arctostaphylos uva-ursi
bearberry

Salix reticulata
reticulate willow

Salix arctica
Arctic willow

A mossy forest The storm gaps of Fiby Urskog are frequently invaded by birches and aspen (*Populus tremula*) before the Norway spruce regains its dominance. The forest is carpeted with many mosses and lichens, which flourish in the moist shelter.

THE PRIMITIVE FOREST OF FIBY URSKOG

In Europe there are very few primitive forests – plant communities that have developed naturally without any human disturbance. The storm gap structure of Fiby Urskog, 16 km (10 mi) west of Uppsala in central Sweden, is one of the earliest examples of a cycle of forest growth to be discovered.

In central Europe the Norway spruce is known to live for 400 years, yet at Fiby Urskog its lifespan seldom seemed to exceed 250 years, and many of the trees were very small. In the 1930s the forest was under threat of being cut down because it was assumed the trees had been felled previously, but they proved to be much older than their height and growth suggested. Tree ring samples taken from as near the ground as possible showed that the ages recorded were considerably greater than those noted at the conventional sampling height of 1.3 m (4.3 ft). This stunted growth occurred because the spruce trees growing in the shelter of large mature trees developed as dwarf trees; they grew well-formed lateral branches but very poor leading shoots. These dwarf trees, which may take as long as 40 years to grow to a height of 1.3 m (4.3 ft), can survive for many years. Mature trees seldom reach old age as they are often felled in groups by the wind on the unstable soils. Dwarf trees then take the opportunity to grow rapidly through the resulting gaps in the canopy.

Indicator plants

Above the treeline there is less drifting, and the snow cover is more uniform in depth. On the bare, exposed ridges the sparse plant life is often dominated by dry lichens that form low, dense piles interspersed with grasses and isolated, stunted, cold- and drought-resistant shrubs. The most favorable position for plants is on the flanks of the ridges, just a meter or two beneath the summit. Here the plants lie midway between an upper zone that lacks protection from the cruel winds of the summit, and a lower zone where snow lie is prolonged. In the lower alpine belt this zone is often occupied by juniper or dwarf birch, with bilberry taking over at higher altitudes.

Lichens are excellent indicators of the environmental conditions in the mountains. *Haematomma ventosum* is known as the wind lichen because it grows in exposed, windswept places in the alpine belt, as does the three-leaved rush (*Juncus trifidus*). *Parmelia olivacea* grows on birches in positions that mark the general level of the winter snowline. Lichens can survive without water for long periods, but they dislike snow cover.

Liverworts, in contrast, are found beneath snow beds that melt for such a brief period each summer (and sometimes not at all) that even the most specialized annual flowering plants cannot complete their growing cycle. Carpets of tiny, dark liverworts, interspersed with mosses, absorb what little radiation there is, even from beneath the snow. They are well suited to this strange environment, where they lack competition and receive a plentiful supply of water. Independent of insect pollination, they begin to produce spores as soon as the snow melts, and simply suspend activity when the next snow falls.

Plants of the snow beds

The plants of the late-melting snow beds, such as mountain sorrel (*Oxyria digyna*), starry saxifrage (*Saxifraga stellaris*) and pygmy buttercup (*Ranunculus pygmaeus*), have very short summer lives and must spring into growth as soon as the snow melts. Some of these specialists have next year's leaves already present in tiny buds concealed in the axils of the leaves. Some snowbed plants are almost full size as soon as they unfold their leaves; others first produce flowers on stalks only a few centimeters above the ground, with the main leaves following later.

TREES – THE MAIN RESOURCE

Left to nature, most of the region would consist of mountain, lake or forest. Even today more than half of Sweden is covered with trees. Although the forests of Sweden and Finland are extensive, they have been greatly altered by foresters. Large areas now consist of conifer plantations managed by the method of clear-felling, sometimes with burning of the woody waste to assist in recycling precious nutrients before new trees are planted. Forests are one of the principal resources of both these countries, whose skills in forestry and technological ability in timber processing and paper manufacture are renowned.

Although the forested areas are large, the rotation times of plantations in the far north are very long; Scots pine takes 130 years to reach maturity in the lowland forests near the Arctic Circle, and much longer in the mountain forests. Grazing by herbivores, such as elk and other deer, also causes regeneration problems. In Norway and Denmark forestry is a relatively minor industry: when the original wildwood of Denmark was cleared, it was replaced by heathland.

Iceland has not always been so nearly treeless. Ten million years ago, when the climate was much like that of present-day Florida in the United States, it supported forests of giant conifers. Even 5,000 years ago it had extensive forests, and when the first permanent settlers arrived in AD 874 there were still large stands of birch and mountain ash the meadows. Almost all these woods have been destroyed by felling and sheep grazing.

Change and conservation
In the north of the region farming is in decline, though until well into the 20th century local farmers made every effort to use all the grazing land available. In Norway particularly, cattle were taken up into the mountains for summer grazing, and hay was gathered from tiny fields. Many meadows in central and northern Sweden have now been abandoned and are gradually reverting to woodland.

Forestry in Sweden is regulated by law, and conservation interests have to be taken into account when groups of trees are felled or clearings replanted. Trees in which endangered birds nest must be left standing. Old deciduous trees such as

JOURNEY THROUGH LAPLAND

Carl Linnaeus (1732–78) is perhaps the world's most famous botanist. He devised the dual name (binomial) system used to provide scientific names for plants and animals. In 1732 the Swedish Royal Society of Science financed Linnaeus on a journey to the north so that he could make the first scientific survey of the plants, animals and geology of Lapland. North of Gävle on the central east coast of Sweden, he came upon the twinflower. It had already been given a scientific name by another botanist, but was later renamed after Linnaeus himself; it is still known as *Linnaea borealis*.

On his return journey he stopped at Tornio on the northern coast of the Gulf of Bothnia, where he discovered that the deaths of local cattle were caused by the animals eating a highly poisonous water hemlock. His drawings of northern species such as mountain avens, *Cassiope tetragona*, starry saxifrage and Arctic rhododendron (*Rhododendron lapponicum*) still survive, as does his *Flora Lapponicum*, published in Amsterdam in 1737.

In 1988 the Linnean Societies of London and Sweden made a journey to Lapland that covered much of the same ground as Linnaeus. Their plant lists and observations show that most of Lapland's natural communities are still flourishing, despite the pressures of the 20th century.

The twinflower (*left*) *Linnaea borealis*, named after the Swedish botanist Carl Linnaeus, is a woodland plant of the taiga, and the only species in the genus. A dainty shade-lover planted in rock gardens and moist peaty places, it flowers from May to July.

aspen (*Populus tremula*), which are valuable habitats for birds and insects, must also be left untouched, as must zones of slow-growing trees around mires. In southern Sweden it is forbidden to replace deciduous forest with conifers after clearing if the area previously supported 70 percent or more deciduous trees. This law aims to prevent the loss of deciduous trees, which have proved less profitable to grow than conifers. Until the 1980s deciduous trees (particularly birch) growing in conifer plantations were frequently killed by chemicals sprayed from the air, a practice that is now forbidden.

The excessive planting of exotic trees is another cause for concern; the most common species is the North American lodgepole pine (*Pinus contorta*), which is frequently cultivated in the north, but not permitted in southern Sweden. Recent infestations of this tree have shown the dangers inherent in huge monocultures of exotic species.

Forests of the treeline

Today the most vigorous debate in forestry concerns the future of the long-established forests close to the treeline. Because the Swedish mountain chain is so long, the area occupied by these trees extends for several hundred kilometers. As it takes over 200 years for a tree to grow to harvestable size here, forestry is of little or no economic value in the long run, and problems with the regeneration of new trees may cause the treeline to drop even lower in some areas. This creates problems for the Lapps, who live in the north, as their reindeer have traditionally grazed the lichens that grow on the trees of this zone when the ground is covered in snow.

Under pressure from conservationists, complex new legislation has been passed to protect the existing forests from excessive felling, though there is still much debate. The Swedish nuclear energy program is being run down, and some alternative source of natural fuel will be required to solve the energy problem. This may eventually result in the establishment of immense, monotonous plantations of willow to supply timber that can be burned in power stations.

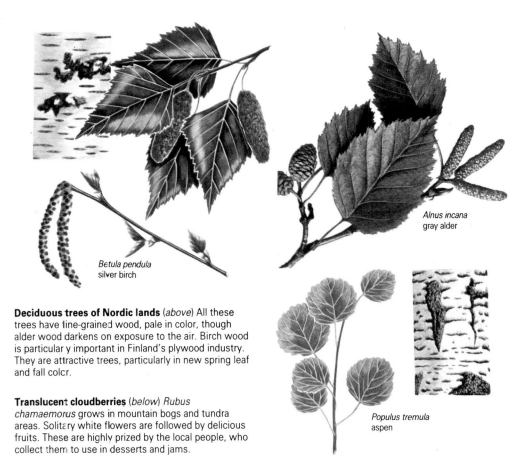

Betula pendula
silver birch

Alnus incana
gray alder

Populus tremula
aspen

Deciduous trees of Nordic lands (*above*) All these trees have fine-grained wood, pale in color, though alder wood darkens on exposure to the air. Birch wood is particularly important in Finland's plywood industry. They are attractive trees, particularly in new spring leaf and fall color.

Translucent cloudberries (*below*) *Rubus chamaemorus* grows in mountain bogs and tundra areas. Solitary white flowers are followed by delicious fruits. These are highly prized by the local people, who collect them to use in desserts and jams.

Surtsey: colonizing a new island

The island of Surtsey emerged in 1963 as the result of a spectacular series of eruptions from a submerged volcano some 40 km (25 mi) south of Iceland. By 1967 an island with an area of almost 3 sq km (1 sq mi) had been formed, some of which has already been eroded. Biologists immediately recognized the unique value of Surtsey as a natural laboratory for the study of succession – the way in which an area comes to be colonized by a series of plant and animal species that continue to interact and develop until a relatively stable community has evolved.

The development of plant life on Surtsey has been fairly slow, hampered by severe winters with low temperatures, high winds and scouring sand, while waves often sweep coastal plants away. Seeds and plant fragments frequently arrived by sea, swept from islands such as Geirfuglasker (5 km/3 mi away) or the larger Heimaey, one of the Vestmanna Islands (20 km/12 mi away). The spores of mosses and ferns were carried in on the wind, and snow buntings visiting the island deposited seeds there.

In the first few years colonization was so slow that regular counts could be made of the individual flowering plants. In 1965 all 30 small plants recorded were species of annual sea rocket (*Cakile maritima*); none flowered, and all of them were later killed by volcanic ash.

The early stages of succession

Virgin land is usually first colonized by the lower plants (algae, lichens, mosses and ferns). Surtsey is unusual because the flowering plants were the first to become established. In the early years the number of flowering plants fluctuated considerably, with a sudden jump from a total of 199 in 1972 to 1,273 in 1973. The latter included 33 sea rocket plants and 586 plants of common scurvy grass (*Cochlearia officinalis*). There were 66 plants of lyme grass (*Leymus arenarius*), which is important in the formation of coastal dunes, and just one of red fescue grass (*Festuca rubra*), which is able to grow on both dunes and cliffs. Sea sandwort (*Honkenya peploides*, 548 plants), with its extensive underground rhizomes, was doing well, and there was just one chickweed plant (*Stellaria media*). The young dunes attracted nesting seabirds that enriched the developing soil with their feces.

Although mosses arrived later than the mainly coastal flowering plants, by 1969 they had colonized lava surfaces in the interior of the island. By 1972, 85 species were recorded, often growing in warm sites close to volcanic vents. The first lichens were observed in 1970. The delicate, brittle bladder-fern (*Cystopteris fragilis*) had also arrived.

Plants are now growing well on the sand dunes, and Surtsey has a rudimentary vegetation on its coastal cliffs. There are as yet only a few species of flowering plants. After nearly three decades the plant life is still in an early stage of succession, and competition between species has yet to begin.

Spreading sea sandwort (*left*) This succulent herb is a widespread pioneer plant. It grows on beaches, where it lives on shifting sand and loose shingle.

Mosses growing in the crater (*right*) Normally mosses are among the first plants to colonize a new habitat. It took three years for them to reach Surtsey from Iceland, Heimaey and other small adjacent islands. When they die their remains, together with those of other plants, are changed into humus, which contains valuable plant nutrients and improves the waterholding capacity of the developing soil.

Blue-green oyster plant (*below*) *Mertensia maritima* is a plant of northern seashores. Together with sea sandwort, it was carried by the sea from Heimaey and other small islands to the northwest.

The Arctic bearberry

There are two species of *Arctostaphylos* found in the Arctic regions of the northern hemisphere, and both are creeping shrubs. The evergreen bearberry (*A. uva-ursi*) grows on gravelly slopes, rocks or sandy heaths, its flexible branches rooting at intervals. Young branches are often covered with sticky hairs, which probably give some protection in the harsh conditions before the peeling bark of maturity develops. It has pink-tinged flowers in spring and glossy red berries late in summer. The leaves have medicinal value – the plant is cultivated for use in medicines for cystitis and disorders of the urinary tract. The bearberry's Arctic relative, the deciduous *A. alpina*, is rare in Europe. Its branches are brittle, and carry pure white flowers and black fruits.

These plants of the Arctic, like those of alpine regions, have had to adapt to a variety of difficult conditions, one of which is snow. Although snow insulates, its weight could crush plants unless they are prostrate. Snow water comes with the thaw at a time when the plant is dormant. If the plant is on a slope the water runs off, so for most of their growing season these plants have to cope with drought. In addition, exposed habitats make plants vulnerable to desiccating winds, which further increases their water stress. The small, leathery leaves and low, compact growth form help the plants to survive these conditions.

Fall colors of the alpine bearberry As winter approaches the green chlorophyll in the leathery leaves breaks down, and red pigments (carotins) display their brilliant presence.

HEATH, MARSH AND WOODLAND PLANTS

HIGHLAND HEATHS, LOWLAND WOODS · LIMESTONE REFUGES · MAKING WAY FOR AGRICULTURE

The British Isles have a diverse plant life, its distribution being determined largely by the temperate climate and abundant rainfall. On the west coasts of Britain and Ireland species normally associated with warmer regions are able to thrive in the mild conditions resulting from the influence of the Gulf Stream. Highland Britain is dominated by tracts of open moorland, where only a limited range of plants can grow because of the poor soil. The lowlands, in contrast, are fertile areas of undulating land; once densely forested, the natural plant life here is now scrubland, developing into broadleaf forest. The large human population places great pressure on the native plant life. Few areas have been left untouched and many habitats – notably marshes and meadows – are now much reduced in extent.

COUNTRIES IN THE REGION

Ireland, United Kingdom

DIVERSITY

	Number of species	Endemism
Ireland	1,000	1
United Kingdom	1,800	16

PLANTS IN DANGER

	Threatened	Endangered	Extinct
	11	2	1

Examples *Apium repens* (creeping marshwort); *Cypripedium calceolus* (lady's slipper orchid); *Damasonium alisma* (starfruit); *Dianthus gratianopolitanus* (Cheddar pink); *Epipogium aphyllum* (ghost orchid); *Liparis loeselii* (fen orchid); *Ranunculus ophioglossifolius* (adder's tongue spearwort); *Rumex rupestris* (shore dock); *Saxifraga hirculus* (marsh saxifrage); *Trichomanes speciosum* (Killarney fern)

USEFUL AND DANGEROUS NATIVE PLANTS

Crop plants *Daucus carota* (carrot); *Pastinaca sativa* (parsnip); *Porphyra umbilicalis* (laver); *Rubus idaeus* (raspberry)
Garden plants *Betula pendula* (silver birch); *Calluna vulgaris* (ling or heather); *Caltha palustris* (marsh marigold); *Centaurea cyanus* (cornflower)
Poisonous plants *Aconitum napellus* (monkshood); *Arum maculatum* (lords and ladies); *Atropa belladonna* (deadly nightshade); *Ligustrum vulgaris* (privet); *Tamus communis* (black bryony)

BOTANIC GARDENS

National Botanic Gardens, Dublin (25,000 taxa); Oxford (10,000 taxa); Royal Botanic Gardens, Edinburgh (12,000 taxa); Royal Botanic Gardens, Kew (30,000 taxa)

HIGHLAND HEATHS, LOWLAND WOODS

Highland Britain is typified by large tracts of grassland dominated by common bent (*Agrostis capillaris*), sheep's fescue (*Festuca ovina*) and bracken (*Pteridium aquilinum*). Limestone areas – such as the rocky grasslands of the Burren in western Ireland – support a more varied range of plants. In many such places heavy rainfall washes the nutrients from the soil, creating acid moorland dominated by the purple-flowered ling or heather (*Calluna vulgaris*) and other members of the heath family (*Ericaceae*). The extensive forests of Scots pine (*Pinus sylvestris*) and birch (*Betula*) that used to grow in these areas have long since been cut down to create grazing for sheep. However, coniferous trees have been planted extensively in recent years for timber production.

Some highland plants, such as twinflower (*Linnaea borealis*), bog rosemary (*Andromeda polifolia*) and bilberry (*Vaccinum myrtillus*), are related to species of northern and central Europe. Arctic and alpine plants, such as moss campion (*Silene acaulis*), grow above the treeline in Scotland. Many of these northern plants are relicts, survivors of the last ice age, which ended some 10,000 years ago.

Fertile lowlands

Although the soils of lowland Britain are deeper and richer than those of the highlands, the natural vegetation is often restricted to steeper escarpments and heaths. The dominant species here are ling, bell heather (*Erica cinerea*) and gorse (*Ulex*), along with petty whin (*Genista anglica*) and, in the southwest, pale dog-violet (*Viola lactea*). Birch or Scots pine will gradually invade these areas unless prevented from doing so by fire, grazing or conservation measures.

Where the heathland becomes permanently waterlogged, species such as cross-leaved heath (*Erica tetralix*), purple moor-grass (*Molinia caerulea*), carnivorous butterworts (*Pinguicula*), sundews (*Drosera*) and a variety of sphagnum mosses abound. These heath and bog plants show many similarities to plants of the Atlantic coast of continental Europe.

In marshy areas reeds (*Phragmites australis*) and reedmace (*Typha*) grow in open water, while willows (*Salix*) and alders (*Alnus*) are the dominant trees lining the

banks. Many areas of marshland, including the fens of eastern Britain, have been drained so that the land can be used for agriculture.

The chalklands of southern Britain have well-drained soil and receive low summer rainfall – conditions that are ideal for a distinctive and varied community of flowering plants, many of which are found in southern Europe. The chalklands are maintained by sheep grazing; when this ceases scrub vegetation, including hawthorn (*Crataegus monogyna*), encroaches and ash trees (*Fraxinus excelsior*) form an open-canopied woodland with various species growing underneath. Beech (*Fagus sylvatica*) eventually takes over on the steep slopes with shallow soils. The dense shade of the beech canopy and the thick layer of leaf litter on

A woodland floor in spring An open patch of woodland before the leaves unfurl to cast their shade. It is carpeted with ferns, bluebells and red campion (*Silene dioica*)

Map of floristic provinces The plant life of the British Isles has developed only since the last ice age. Not surprisingly, it has strong links with the plants of much of continental Europe.

the forest floor restricts the growth of other plants.

Oaks (*Quercus robur* and *Q. petraea*) and hazel (*Corylus avellana*) form the natural vegetation on more favorable soils. A wide variety of spring flowers, including lords and ladies (*Arum maculatum*), lesser celandine (*Ranunculus ficaria*), primrose (*Primula vulgaris*) and bluebell (*Hyacinthoides non-scriptus*), live in these woods, taking advantage of the light before it is filtered out by the growth of the tree canopy. In early summer, when the leaves of the trees shut out most of the light, shade-dwelling plants such as bugle (*Ajuga reptans*) come into flower. Few areas of native forest remain today, though some examples are now being protected and maintained under the auspices of various conservation groups.

Coastal plants

Plants that thrive in salty conditions (halophytes) are restricted to coastal areas that are flooded by the sea; the abundant rainfall of the region prevents the formation of inland salt marshes. The succulent glassworts (*Salicornia*) grow in places that are regularly inundated, but thrift (*Armeria maritima*) and red fescue grass (*Festuca rubra*) grow farther inland and are covered only by high spring tides.

On the coastal sand dunes a community of salt-tolerant species that includes sea rocket (*Cakile maritima*), prickly saltwort (*Salsola kali*) and oraches (*Atriplex*) develops at or just above the highwater

mark. Farther inland, marram (*Ammophilia arenaria*), lyme grass (*Leymus arenarius*) and sea bindweed (*Calystegia soldanella*) have adapted to survive burial by the encroaching sand. A variety of species cling to the sea cliffs, including some woodland plants, together with salt-tolerant succulents such as roseroot (*Sedum rosea*), cushions of thrift and carpets of plantain (*Plantago maritima*).

Floristic Provinces

Holarctic Kingdom/Circumboreal Region

Atlantic-European Province Covers much of western Europe; influenced by maritime climate, and characterized by ivy, holly, heather and gorse species. Only 16 species are endemic.

LIMESTONE REFUGES

Chalk and other limestone grasslands provide refuges for plants that have adapted to cope with a soil that often lacks trace elements and tends to dry out in summer. Specialized communities of plants thrive in these open habitats, which have traditionally been maintained by the grazing of sheep. Since late medieval times rabbits, which had been introduced from Europe, have also kept the grass short and encroaching scrub at bay. Although much of the old chalk grasslands have been plowed up, some extensive areas still survive.

Adaptations to grazing

Grassland plants have been able to withstand sustained grazing. Red fescue grass, which is typical of the low tufted grasses of these areas, survives because its lower parts are green. The leaves grow from the base rather than the apex, so they can continue to grow even when the tops have been bitten off. Plants such as plantains (*Plantago*) and daisies (*Bellis perennis*) survive the trampling feet of animals by forming rosettes of leaves that lie flat on the ground. The stemless thistle (*Cirsium acaulon*) not only forms a rosette but has leaves armed with prickles as a further deterrent to feeding animals.

Other ground-hugging plants, such as wild thyme (*Thymus praecox*), produce long spreading branches that form a dense mat. Some members of the pea family, including birdsfoot trefoil (*Lotus corniculatus*) and horseshoe vetch (*Hippocrepis comosa*), form an even stronger mat by producing roots at intervals along their stems and anchoring themselves firmly to the ground. This makes it difficult for animals to eat the plants.

The thin, well-drained chalk soils of the downlands dry out frequently in the summer, and many plants in these areas produce long roots able to reach moisture that is inaccessible to shallow-rooting species. For example, salad burnet (*Poterium sanguisorba*) may produce a tap root of 1 m (3 ft) or more. Plants such as the crimson pyramidal orchid (*Anacamptis pyramidalis*) and the frog orchid (*Coeloglossum viride*) have tuberous roots that store water to tide them over the dry season. The ground-hugging plants are also well adapted to the climatic conditions: their water loss is reduced because

PLANTS OF THE LIZARD

The Lizard peninsula, in the extreme southwest of mainland Britain, has a plant life unique to the region – a result of the mild, sunny climate, varied geology, and proximity to the sea. The plateau, which lies about 100 m (330 ft) above sea level, is surrounded by ancient rocks that form steep cliffs along the coast. Much of the upland area is composed of magnesium-rich (serpentine) rock, poor in nutrients and overlain by wind-deposited soil. The poor soil and strong prevailing westerly winds are unsuitable conditions for agriculture, and they also prohibit the growth of forest.

A number of clovers, notably the long-headed clover (*Trifolium molineri*), twin-flowered clover (*T. bocconei*) and upright clover (*T. strictum*), represent the northern limit of a range of plants more characteristic of the Mediterranean and southwest Europe. Several other plants of Mediterranean extraction also grow on the Lizard: two tiny rushes, dwarf rush (*Juncus capitatus*) and pygmy rush (*J. pygmaeus*) flourish in winter-wet hollows; in spring early meadow grass (*Poa infirma*), which thrives along paths and in open ground, is very common.

The heathlands that have suffered soil erosion are dominated by Cornish heath (*Erica vagans*). Spring sandwort (*Minuartia verna*) and a dwarf variant of juniper (*Juniperus communis*) – two plants more commonly found in alpine areas – grow on steep slopes. The mild climate has also made it possible for several exotic garden species, such as iris and gladiolus, to spread throughout the peninsula.

In copse and wood Moisture-loving herbs like lords and ladies and the lesser celandine flower early in these habitats, and the rare, tall lady orchid may occasionally be found on chalky soils in the south. The field maple forms a small tree also found in many hedgerows.

Orchis purpurea
lady orchid

Acer campestre
field maple

Arum maculatum
lords and ladies

Ranunculus ficaria
lesser celandine

Clifftop plants A drystone wall provides a niche for sea campion (*Silene maritima*) and thrift (*Armeria maritima*). Their cushion growth form enables them to survive the constant buffeting from sea winds.

the lower surface of their leaves is pressed close to the ground.

Where grazing has been discontinued, chalk grassland gradually reverts to woodland – the natural vegetation of the downs before they were cleared for agriculture. The first phase is the invasion of the turf by coarse grasses and scrub dominated by the spiny shrub hawthorn (*Crataegus monogyna*). These shade out the finer species, and also those that form rosettes and mats. Other shrubby species that move in include dogwood (*Cornus sanguinea*), the wayfaring tree (*Viburnum lantana*) and privet (*Ligustrum vulgare*), and sometimes the spiny coniferous shrub juniper (*Juniperus communis*).

Tolerating metallic soils

Highly specialized communities of plants live in areas where the soil has been contaminated by metals. In some places this happens naturally – such as the Shetland Islands in the far north and the coastal heaths on the southwestern tip of Britain. Here magnesium-rich (serpentine) rock underlies the soil. In areas where metals are mined there may also be abnormal surface concentrations of metal in the earth and rocks that are left behind. Here the plants might have to tolerate well-drained soil that is liable to dry out, and is deficient in some trace nutrients. The toxicity of the metal ensures that relatively few plants are able to become established.

Several species of grass, such as sheep's fescue and sweet vernal grass (*Anthoxanthum odcratum*), and also ling, bladder campion (*Silene vulgaris*) and other species, have variants that can tolerate high levels of metals in the soil. Plants that colonize such areas are widely spaced at first, but in time may form a more or less continuous cover. Birch trees are also able to survive: the fungal system (mycorrhiza) that facilitates the uptake of nutrients from the soil is tolerant of metals.

The combination of chemical stress, drought and unstable soil ensures that the vegetation never develops into full woodland. Where calcium is present, the effect of the metals is more tolerable. As a result a closer turf develops, which in summer is bright with flowers. Alpine penny-cress (*Thlaspi caerulescens*) and spring sandwort (*Minuartia verna*), with clusters of white flowers, and the yellow or purple flowers of mountain pansy (*Viola lutea*) are typical plants of metal-bearing soils.

MAKING WAY FOR AGRICULTURE

The landscapes of Britain and Ireland have nearly all been fashioned by human activity. The native plant life has been profoundly affected by the felling or planting of trees, by agricultural practices and the expansion of urban and suburban settlement; in addition, the number of plants has increased considerably through the introduction, both deliberate and accidental, of alien species. Relatively few plant species have become extinct in recent years (though many have become rarer or more localized). Those that have disappeared are mostly cornfield weeds that have been unable to tolerate modern herbicides and fertilizers. Thorow-wax (*Bupleurum rotundifolium*) and lamb's-succory (*Arnoseris minima*) are now extinct, and corn-cockle (*Agrostemma githago*) and pheasant's eye (*Adonis annua*) are very rare indeed.

Destruction of natural habitats

Modern trends in agricultural management have also damaged the chalk and lowland limestone grasslands. These were extensively plowed at the beginning of the 19th century to increase arable cultivation, after centuries of use as sheep grazing, and World War II and its aftermath saw an expansion of the process. Even if farmers were to stop cultivating areas of land now, it would take many centuries for the natural plant life of the chalk downlands to become reestablished, especially as the soil has been treated with fertilizers that favor coarse, dense grasses. Species that were characteristic of old grasslands, like pasque flower (*Pulsatilla vulgaris*), with its violet-blue flowers, and burnt-tip orchid (*Orchis ustulata*), with spikes of pinkish flowers, are in decline and in need of protection.

The meadow, which was traditionally managed to produce hay and grazing land, provides a lowland habitat for a rich and beautiful array of plants. Old meadows display a wealth of color in spring and early summer, the grasses interspersed with cowslip (*Primula veris*), green-winged orchid (*Orchis morio*), tiny adderstongue (*Ophioglossum vulgatum*)

A flowery hay field Such a profusion of buttercups, clover and other wild flowers growing among meadow grasses is becoming a rare sight. Traditionally, grazing animals would be moved in after the hay was cut.

the hay-rattle (*Rhinanthus minor*), and several species of clover and vetch. However, meadows are now much reduced in extent: since 1945 some 95 percent have been destroyed or irretrievably damaged. Pockets remain, but they are unlikely to survive without conservation measures. Similar plant communities are found both in upland parts of Britain, and in the areas of blown sand on the coasts of western Ireland and western and northern Scotland known as "machair". These areas are also vulnerable, as they are increasingly under threat from changing agricultural practices.

Woodland is another habitat with a history of intense human management. Little untouched woodland survives, except in wetland pockets such as the Gearagh, an area of oak swamp woodland in the southeast of Ireland, or the native pines of the Black Wood of Rannoch and elsewhere in the Highlands of Scotland. Managed and planted woodlands were, however, often established on sites of ancient woods, recognizable by the presence of bulbous plants and trees such as small-leaved lime (*Tilia cordata*). By about a thousand years ago much of the woodland had already been cleared, and since then woodland areas have been extensively managed for hunting, timber extraction and grazing. A typical plan of management was the growth of tall "standard" oak (*Quercus robur*) interspersed with coppiced hazel or ash, the young growth being used for poles and to make baskets and furniture.

Some changes that have been imposed on the landscape have been exploited by certain plants. The Oxford ragwort (*Senecio jacobaea*) was brought from the lava slopes of Etna in Sicily and grown in the Oxford Botanic Garden during the 18th century; by 1800 it was established on walls throughout the city. During the 19th century it underwent a spectacular expansion to many parts of England, and

Poisonous henbane (*above*) In the Middle Ages this tall, evil-smelling annual was used in magic ritual and in love potions. Now it grows on wasteland. The dried leaves and seeds are used in the preparation of sedative drugs.

Introduction to the railroad (*below*) Oxford ragwort has established itself on railroad ballast, walls and wasteland. It is closely related to the native ragworts and groundsels of Britain, which have similar yellow daisy flowers.

in the early years of the 20th century spread as far as Wales and Scotland. Its preferred habitat was railroad ballast, a substrate not unlike the dry sulfurous lavas of Sicily, and the rapidly expanding Victorian rail network enabled it to spread throughout Britain.

Traditional uses of plants

A whole range of plants, many of them weeds, have been grown for thousands of years, and used for many different purposes. Most common wayside plants, such as the stinging nettle (*Urtica dioica*), have some history of use – nettles are still gathered and eaten in spring and were once used as a source of fiber. In the 16th century starch from the tubers of lords and ladies (*Arum maculatum*) was used to stiffen the ruffs that were fashionable at the time. The celandine was erroneously thought to cure hemorrhoids; the greater celandine (*Chelidonium majus*), however, is a genuinely effective remedy for the removal of warts.

Some plants are reminders of ancient cultivation: woad (*Isatis tinctoria*) was once grown from western Turkey to the British Isles as a source of blue dye. The plants of medieval herb and kitchen gardens have spread beyond their original confines into the countryside beyond. This is very noticeable in Ireland – as there are rather few native species, such exotics are all the more conspicuous. Several Irish castles have parsley (*Petroselimum crispum*) growing on their walls and banks, together with henbane (*Hyoscyamus niger*), a poisonous nightshade that was used as a sleeping draught or primitive type of anesthetic.

FRITILLARY OF THE WATER MEADOWS

The fritillary or snakeshead (*Fritillaria meleagris*) is a handsome bulbous member of the lily family, with nodding, bell-shaped, purple or white flowers. It can be found in parts of southern and central England. Its characteristic habitat is the flood plains of lowland rivers, and in particular water meadows. These meadows, mostly established between 1600 and 1800, were traditionally managed by controlled flooding during the early part of the year. On clear nights, water would be let on to the meadow through a system of dykes and sluices to prevent the formation of ground frost, thus encouraging an early growth of grass. The

meadow was grazed from March to mid-April, after which the grasses were left to grow so that a crop of hay could be cut in late July or early August.

The fritillary was able to exploit this scheme of management. The plants flower in late April to early May, and by the time the meadow was mown for hay they would have set seed and died down. Fritillaries are unpalatable to animals, and consequently would have survived the grazing period during the early spring. A few water meadows do still exist, but as the traditional pattern of management has almost completely ceased the numbers of fritillaries have been greatly reduced.

Plants of the peat bogs

Large areas of the western and northern parts of the British Isles are covered by peat – an accumulation of partly decomposed plant remains in which decay is inhibited by waterlogging (as there is a lack of oxygen for the microorganisms that promote decay). If such conditions persist for any length of time, a considerable mass of peat can accumulate. Over millions of years this mass may become compacted to form lignite or brown coal, and finally coal itself. Peat is characterized by a relatively high level of carbon and a very low level of nitrogen, as well as of other nutrients such as phosphorus and potassium.

Raised bog develops when an area of wetland – a fen or a lake – becomes densely covered with vegetation. The rotting remains of this build up over time into a characteristic dome of peat. Blanket bog forms a general covering of the landscape in areas such as western Ireland, Dartmoor in southwest England and the mountains of Scotland in the north, where rain falls regularly throughout the year and thus exceeds evaporation and plant transpiration. Valley bog, a variant of raised bog, develops in hollows or valleys where poor drainage allows localized bogs to build up.

The formation of peat bog is largely the result of the growth of species of bog moss (*Sphagnum*), which first form hummocks and then a more extensive sponge-like structure. The bog mosses retain water and create their own self-contained habitat. Nutrients can enter the system only through rain and wind, and the bog therefore becomes progressively depleted of minerals. Some recycling of nutrients may occur, but it is restricted because of the limited level of decay. Nutrients are generally leached out by water percolating downward through the soil.

Huge areas of Ireland are covered by bogs, but many of these – principally the raised bogs of the midlands, such as the great bog of Allen – have gradually been cut away for domestic and industrial fuel. The conservation of these raised bogs is of great importance, as few extensive areas remain.

Carnivorous plants

The bogs are the home of several species of carnivorous plant, which have adapted to their mineral-deficient habitat by trapping and digesting insects and other creatures in order to obtain nutrients.

Sundews (*Drosera*) and butterworts (*Pinguicula*) trap and digest insects on their sticky leaves. The pitcher plant (*Sarracenia purpurea*), a native of North America, was introduced into Ireland in 1909 and now forms large populations on several bogs. As its name suggests, the leaves form flasklike structures filled with fluid. Insects are attracted by the bright red or purplish coloration of the leaf; they slip on the walls, tumble into the fluid and are slowly digested.

In acid pools in bogs grow species of bladderwort (*Utricularia*). These plants trap small underwater invertebrates in tiny sacs. Trigger hairs on the "door" of a sac detect the presence of an animal; the door is flicked open and the subsequent inrush of water sweeps the prey in. Enzymes secreted into the bladder slowly digest the victim.

Carnivorous sundew (*above*) Sundews supplement their diet by "eating" visiting insects, which they capture by means of their sticky tentacles. Glands in the tentacles secrete digestive fluids to break down the soft insect body.

Creeping bogbean (*right*) This widespread aquatic plant grows in the wettest parts of a bog, where it spreads by rhizomes. The striking fringed flowers are pink in bud.

Aquatic insect-eater (*below*) Bladderwort has no roots, but captures prey in sacs on the finely cut leaves. Glands in the bladder remove the water and the decomposed animal, the sac flattens and the trap is set once more.

Bladder

Sensitive trigger hairs open trap door

Tiny crustacean activates trigger hair

Flowers open above water

Leaf

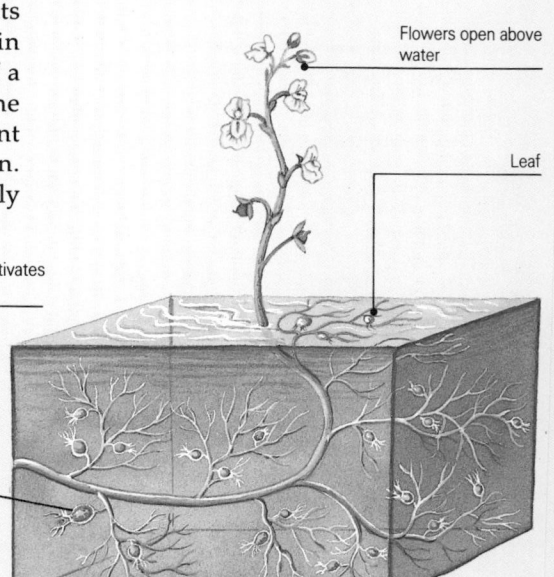

Nitrogen fixing legumes

Lupinus nootkatensis and another lovely lupin, the shrubby yellow tree lupin (*L. arboreus*), naturalized along coasts of southern Britain, originated in western North America. They were introduced to Britain at the end of the 18th century.

The roots of legumes (in common with a few other plants) have a remarkable association with a nitrogen-fixing bacterium. The bacteria enter the roots, where they cause the cells to divide to form the globular nodules where nitrogen fixation takes place.

The bacteria are able to split atmospheric nitrogen and incorporate it into amino acids (protein building blocks) that are used by the plant. This allows leguminous plants to thrive in stony or sandy soils low in nitrogen. When they die the nitrogen is released into the soil. This property is exploited in crop rotation systems, when peas or beans are planted every third year to replace some of the nitrogen removed by other crops. Lupins or other fastgrowing nitrogen-fixers are also grown as a green manure to be plowed back into the soil before the main crop is sown.

Lupins, like other members of the pea family (Leguminosae), have a distinctive keeled flower and a characteristic pod. Some legumes have seeds rich in protein, and a variety of peas, beans and lentils are important food crops; others, such as rape and clover, are used to feed animals.

Alien lupin *Lupinus nootkatensis* is an alien species that grows on the gravelly soils by rivers in Scotland, in the north of the British Isles. Like much of the plant life that has been introduced, it has found a suitable niche in the wild.

A COLORFUL, RICHLY SCENTED LAND

OAK FOREST AND SCENTED SCRUBLAND · COPING WITH EXPOSURE · FRUITS, PERFUMES AND HERBAL REMEDIES

France has a particularly rich temperate plant life, strongly influenced by the diverse geology and climate of the region. The fertile soils support some 5,000 native species. Gorse, pine trees, and scented thyme and lavender grow along the Atlantic coastline, where they are influenced by the maritime climate, while in the Mediterranean area to the south aromatic shrubs and rock roses have adapted to the hot, dry summers and warm wet winters. On the snowbound ridges of the Alps and Jura Mountains, and in the Pyrenees, small alpine and cushion plants have adapted to cope with the thin, poor soils and harsh conditions. The cliffs of the mountains and river gorges have evolved a unique plant life, while Corsica and some of the smaller islands have become refuges for many rare and endemic species.

COUNTRIES IN THE REGION

Andorra, France, Monaco

DIVERSITY (France)

Number of families	about 160
Number of genera	about 1,000
Number of species	4,300–4,450
Endemics	73

PLANTS IN DANGER

Number of threatened species	116
Number of endangered species	10
Number of extinct species	4

Examples *Aldrovanda vesiculosa; Angelica heterocarpa* (Atlantic angelica); *Caldesia parnassifolia; Hammarbya paludosa* (bog orchid); *Hormatophyllum Pyrenaica* (Pyrenean alyssum); *Leucojum nicaeense; Lythrum thesioides; Primula allionii; Saxifraga florulenta; Viola hispida* (raven violet)

USEFUL AND DANGEROUS NATIVE PLANTS

Crop plants *Apium graveolens* (celery); *Beta vulgaris* (sugar beet); *Coriandrum sativum* (coriander); *Pyrus communis* (pear)
Garden plants *Campanula allionii; Cistus albidus; Corydalis lutea; Galanthus nivalis* (snowdrop); *Gentiana lutea* (yellow gentian); *Geranium pratensis* (meadow cranesbill)
Poisonous plants *Atropa belladonna* (deadly nightshade); *Conium maculatum* (hemlock); *Daphne mezereum; Heracleum mantegazzianum* (giant hogweed); *Laburnum anagyroides; Taxus baccata* (yew).

BOTANIC GARDENS

Antibes (6,000 taxa); Dijon (5,000 taxa); Lyon (15,000 taxa); Nancy (4,500 taxa); Paris (24,000 taxa)

OAK FOREST AND SCENTED SCRUBLAND

A few thousand years ago most of the interior of France was covered by forests of beech (*Fagus sylvatica*) or mixed beech and oak (*Quercus robur*). Tracts of this woodland still exist, especially in the north and in parts of the Massif Central, together with a woodland ground cover of dog's mercury (*Mercurialis perennis*), sanicle (*Sanicula europaea*) and the dainty wood anemone (*Anemone nemorosa*). In the south and in parts of the mild west, evergreen oak (*Quercus ilex*) is now the dominant species.

Plants of the Mediterranean

The Mediterranean coastal area, with its hot, dry summers and rainy winters, provides ideal conditions for shortlived, sand-tolerant annual plants such as *Malcolmia* and *Lotus*. The extensive dune systems along this coastline have been colonized by resin-scented pines – mostly *Pinus pinaster*, along with the stone or umbrella pine (*P. pinea*). The Aleppo pine (*P. halepensis*), with its silver-gray branches and pale green foliage, is occasionally found here too. Beneath the pines grows an aromatic ground cover of thyme (*Thymus*) and lavender (*Lavandula*).

Farther inland, in the hot central southern area of France called the Midi, are extensive treeless areas of scrub known as garigue. Here aromatic plants, low shrubs, bulbs and orchids are typical. In many places, in a densely shrubby extreme of the garigue, rock roses (*Cistus*) predominate, giving rise to the term maquis – the name for this vegetation type, from the Corsican word for rock rose, *macchia*. Here *Cistus albidus*, with its gray, felty leaves and pinkish flowers, and white-flowered *C. monspeliensis* form dense bushes 1–2 m (3–6 ft) in height. Tucked in at the bases of individual plants grow the vivid yellow and scarlet clusters of *Cytinus hypocistus*, a parasite that lives off the roots of rock roses.

Military orchid *Orchis militaris* grows in woodland glades and limestone grasslands. Like many of the European orchids it has suffered from overcollection and changing patterns of land use.

Map of floristic regions. Like much of Europe, France is included in two of the major regions of the Holarctic Kingdom. The warm Mediterranean climate of the extreme southeast is reflected in the characteristic plant life there.

Island refuges

More than 2,000 plant species are native to Corsica, many of which are found nowhere else in the world. These endemic species mostly grow on the high mountains in areas that have heavy winter snow. They include the herbs *Silene requienii* and *Potentilla crassinervia*. Corsican plants that also grow on other islands usually thrive at lower altitudes, such as *Morisia monanthos* and *Stachys corsica*, both found in Sardinia.

Many mainland species became isolated up in the mountains by the ice ages thousands of years ago – a wealth of rare and endemic species can be found in the Massif Central, the Pyrenees and the Alps. In the volcanic Auvergne area to the northwest of the Massif Central species such as *Pulsatilla alpina*, five species of

Saxifraga, Dryas octopetala, and many more alpine plants flourish.

In the Cévennes to the southeast, at altitudes of 800–1,000 m (2,500–3,300 ft), well above the Mediterranean garigue communities, a rocky limestone plateau called Les Causses supports a rich community of plants. Many of these also grow in central Europe, including the garland flower (*Daphne cneorum*), *Ajuga genevensis* and *Aster alpinus*. Box (*Buxus sempervirens*) and the less common mespilus (*Amelanchier ovalis*) are also characteristic plants of this landscape.

The endemic plants of the Pyrenees generally grow at high altitudes. Typical species include *Ranunculus gouanii, Saxifraga aquatica, S. umbrosa, Aquilegia aragonensis* and *Borderea pyrenaica*, an isolated genus related to the tropical yam.

Floristic regions

Holarctic Kingdom

Circumboreal Region Extensive region covering Earth's cool northern temperate zone. Gesneriads, primula- and carrot-relatives provide links with other parts of region.

Mediterranean Region Hot, dry summers and winter rain. Typical trees include holm oak and cultivated olives; dominated by aromatic scrub with rosemary, thyme and heaths.

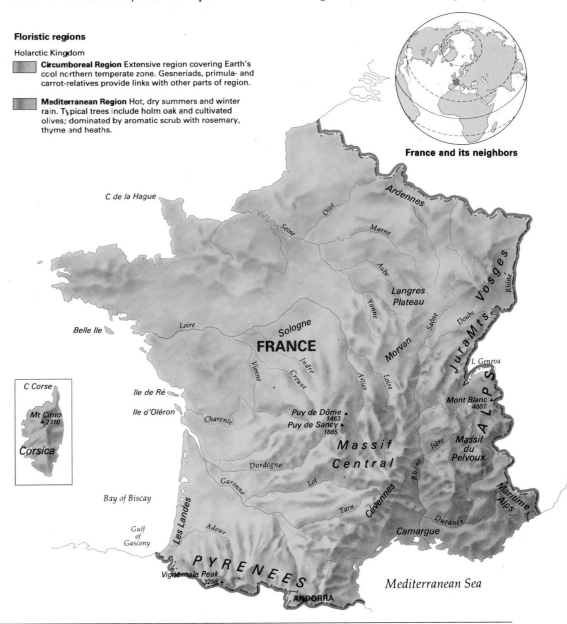

France and its neighbors

COPING WITH EXPOSURE

The herbaceous plants of the maquis produce a splendid show of color in the spring, attracting pollinators before the summer drought sets in. In the hot, dry conditions, the plants wither, but continue to conserve moisture and nutrients in swollen underground organs. The remains of old fruiting stalks of asphodels, mullein (*Verbascum*) and numerous thistles stand as sentinels to the long struggle for survival until the winter rains bring about new growth.

Woody plants, too, are well adapted to withstand extreme desiccation from a combination of the hot dry climate and frequent strong winds. Some of the rock roses and myrtle (*Myrtus communis*) produce leaves that are coated with waxes and oils to reduce transpiration. Other plants, such as juniper (*Juniperus oxycedrus*), broom (*Genista cinerea*) and the spiny buckthorn (*Rhamnus alaternus*) have very small, tough leaves, so little water is lost through evaporation. Tamarisk (*Tamarix*) is perfectly adapted to hot, dry and exposed places. It has minute linear leaves that give the plant a feathery appearance and keep water loss

Alpine toadflax (*above*) *Linaria alpina* is a perennial of mountain cliffs, scree and rock crevices. Its low, compact growth form offers little resistance to the wind, and traps heat during the day to moderate the cold temperatures at night.

Flowers above water (*below*) The greater bladderwort (*Utricularia vulgaris*) has the feathery leaves typical of many aquatic plants, which are structurally weak as they get their support from the water. Some are modified into bladders that trap tiny animals.

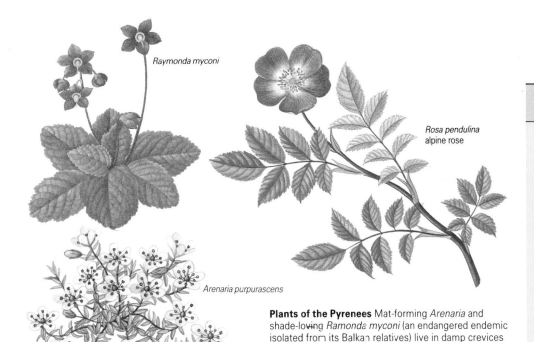

Raymonda myconi

Rosa pendulina
alpine rose

Arenaria purpurascens

Plants of the Pyrenees Mat-forming *Arenaria* and shade-loving *Ramonda myconi* (an endangered endemic isolated from its Balkan relatives) live in damp crevices of the alpine zone. The dwarf alpine rose grows in woods and open ground up to the treeline.

THE WHITE WATERLILY

The white waterlily (*Nymphaea alba*) grows wild in the ponds and waterways of France. Its rounded leaves grow on long, flexible stems and are supported on the water's surface. The upper surface of the leaves is glossy and water-repellent, while the underside seems to cling to the water by means of surface tension. This lower surface contains a blue pigment that concentrates the sun's energy in the leaves, where it is used to make food through the process of photosynthesis, rather than allowing it to pass through to the water below.

The waterlily is able to cope with an occasional drought – if the water dries up, it produces aerial leaves on short stems. When the water returns, floating leaves form once more. Seed dispersal is achieved by means of floating seeds that contain air pockets; the seeds drift away from the parent plant, gradually becoming waterlogged and sinking. In winter, when the leaves die back, the plant retreats into a fleshy rhizome.

The white waterlily is the parent of the ornamental waterlilies commonly grown in garden ponds. These were developed by Joseph Bory Latour-Marliac, who was determined to breed hybrids as colorful as the tropical species he had seen on his travels – but that would withstand the European climate. In 1878 he produced a pink waterlily, the first of more than seventy cultivars to bear his name. Few other plant breeders have been able to hybridize these notoriously difficult plants, and Marliac's methods have remained a mystery to this day.

to a minimum. The small pinkish flowers have a succulent disk-shaped nectary that attracts insect pollinators. The seeds have a tuft of hairs that catches the wind to aid dispersal, and may also combat desiccation. Tamarisk plants have very deep roots that tap water 10 m (33 ft) or more below the ground. This enables them to survive in brackish water or salty habitats along the Mediterranean coast.

Cliff dwellers

Cliffs of vertical rock are common in France; the plants that live on them need to be highly specialized to cope with water shortage, poor thin soil and exposure to a wide range of temperatures and searing winds. Cliffs can be found at all altitudes, on mountains and by the coast, along river gorges and in the isolated Puys in the Massif Central.

When the windblown seeds of trees become lodged in a cliff crevice, the plants that grow are often dwarfed and distorted. These stunted trees often develop flagged growth – branches grow only on the side of the trunk that faces away from the prevailing winds.

As the steepness of the cliff increases, the number of niches suitable for plants decreases. North-facing crevices are relatively shaded, cool and damp compared with those facing south. Limestone rock usually contains deep fissures – one reason why limestone cliffs are often rich in plant species. Other types of rock tend to weather into smooth shapes with fewer cracks and hollows.

Cliff-dwelling plants require an extensive root system both for anchorage and to seek moisture. As annual plants lack this feature, characteristic cliff plants are perennial, often longlived herbs with woody stems. Cushion plants dominate the more exposed locations. With the exception of the larger cracks and ledges, each crevice in the rock tends to support only one plant. As a result the population of most cliffs remains remarkably constant over the years. This may help to explain the many local endemics.

Plants that grow in limited, isolated and constant communities on cliffs are likely to evolve at different rates from those of closely related species living in environments where competition is more severe. Some of the cliff dwellers change relatively rapidly as a result of inbreeding; others scarcely evolve at all. A cliff face with the same climate as neighboring precipes may nevertheless carry a unique plant community.

Beneath the cliffs fallen rock debris forms scree slopes. Only specialized plants are able to colonize this unstable habitat; they include, for example, the burnt candytuft (*Aethionema saxatile*), which grows among the limestone foothills of the Mediterranean. This is a small, spreading perennial with leathery leaves, a woody base and tough roots – in the precarious scree it is vital for plants to form an extensive root system that can stabilize the loose rock and give them plants a secure foothold.

Colorful parasite *Cytinus hypocistus* grows on the roots of rock roses in the maquis of the Mediterranean. It is totally dependent on its host for food and water, as it lacks roots, leaves and chlorophylll, so cannot photosynthesize.

FRUITS, PERFUMES AND HERBAL REMEDIES

The once extensive native beech and oak forests of France have largely been destroyed by human activity over the centuries. However, surviving woodlands are still exploited for a variety of products. In the southwest, sweet chestnut (*Castanea sativa*) is the characteristic woodland tree. It is not in fact a native but was introduced several thousand years ago, along with the walnut (*Juglans regia*), from southwest Asia. The nuts are gathered each year in the fall by local people.

In Les Landes on the southwestern coast of France there are vast pine forests – the land was once a malaria-infested swamp, but was reclaimed on the instructions of Napoleon. The trees produce timber and resin, which is distilled to yield turpentine. Toward the Pyrenees the native cork oak (*Quercus suber*) is exploited for its cork (the bark).

Oaks are the main host for the mycorrhizal fungus that is the source of the black truffle, so esteemed by gastronomes. The trees are also used to make the casks in which cognac, most famous French brandy, is matured.

The wild ancestors of several familiar fruits and vegetables can still be found growing in the countryside. Wild raspberries (*Rubus idaeus*) and alpine strawberries (*Fragaria vesca*) are gathered in the woods to flavor liqueurs, for preserves and also for medicinal purposes. The wild pear (*Pyrus communis*) has been used to breed the numerous dessert varieties now available. Sugarbeet, beetroot and spinach beet have been developed from a single wild species, *Beta vulgaris*; and celery and celeriac are both derived from the wild celery (*Apium graveolens*).

Scents and essential oils

In the south of France the extensive areas of garigue, which have long been used for grazing, are also a bountiful source of scented plants. These are gathered for their essential oils, which are used in perfumery and to make herbal extracts. The great center of the perfume industry is in the southeast of the region at Grasse.

Many of the essential oils used in perfumes are taken from the flowers of local plants such as carnation (*Dianthus*), mignonette (*Reseda*), lily-of-the-valley (*Convallaria majalis*) and jonquil (*Narcissus poeticus*); some oils are taken from the leaves as well as the flowers – these species include *Viola*, lavender, mint (*Mentha*) and rosemary (*Rosmarinus officinalis*). In some plants, such as cranesbills (*Geranium*) and *Verbena*, the oils are derived from the leaves and stems; less frequently, the roots and seeds are the source, as in *Angelica*.

The quality and quantity of the oils can vary greatly, depending on the growing conditions, the time of day or night at which harvesting takes place, and how quickly the oils are extracted after picking. In general, the amount of oil is greatest immediately before the flowers open; it is vital to harvest them at exactly the right moment because, even under the best conditions, 2 kg (4.5 lb) of jasmine flowers, for example, will produce only 1 g of precious oil.

Folk medicine

Wild plants have been gathered for centuries for use in magic potions and herbal cures. Today, herbal remedies using native plants are much in demand, as are essential oils for use in aromatherapy. The burnt ashes of broom (*Cytisus scoparius*) are still used to treat rheumatism, gout, and liver and gall bladder complaints. Broom is also said to act as an antidote to the bite of a venomous snake. Melilot (*Melilotus officinalis*), a common plant of meadows and roadsides, was used in magic philtres in medieval times and later to treat colic, gout and abdominal pain. It has a mildly sedative effect, and is taken to combat insomnia and stress. Valerian (*Valeriana officinalis*) is also used as a herbal sedative.

Sweet chestnuts (*below*) The fruits, which are enclosed in a prickly capsule, are eaten roasted or pureed, and made into a luxurious sweetmeat, *marrons glacés*. The tree is easily identified by its spirally ridged trunk and glossy leaves.

Flowers in Provence (*above*) Spurge (*Euphorbia*) and lavender flourish in the dry, sunny climate of the south. The spurges are a varied group of plants, but all of them have tiny flowers surrounded by decorative bracts, and an irritant milky sap.

Lily-of-the-valley (*right*) is a shade-loving plant of dry woodland and hedgerows. Small bunches of this fragrant flower are exchanged on the first of May. It is also a symbol of good luck, though all its parts are extremely poisonous.

The attractive carline thistle (*Carlina acaulis*) grows on the slopes of the sunny Auvergne and in the poor grasslands of the south. The fleshy flowerhead can be cooked and eaten in the same way as a globe artichoke heart. In the past, its root was dug up in the fall and used to make a winter tonic. Fennel seeds (*Foeniculum*) are taken to relieve indigestion. The root of the yellow gentian (*Gentiana lutea*) also has digestive properties, and is used in a nerve tonic. *Arnica montana* is called sneezeweed in France because its flowers bring on bouts of sneezing. It is used to treat fever, vertigo and hearing problems. In the past the dried leaves were used as a tobacco substitute.

THE SECRET RECIPES OF LIQUEURS

The 17th century French philosopher Jean-Jacques Rousseau, after visiting the Grande Chartreuse monastery near Grenoble, wrote: "I have found here rare plants and virtues rarer still," presumably referring to the local liqueur. From early medieval times monks have been distilling the liquors obtained from fermented fruits and herbs to produce "eau de vie" (the water of life), an essentially medicinal beverage. By the mid-17th century distillation methods had greatly improved and numerous French monasteries were producing their own liqueurs.

Most French liqueurs have a brandy base infused with fruit or herbs, usually gathered from the countryside around the monastery. The famous DOM Benedictine, which has been produced at Fécamp on the Normandy coast since 1510, has a cognac base with honey and 27 herbs, including *Melissa*, *Hyssopus* and *Angelica*.

Chartreuse, which was in production in the 16th century, is made from a secret recipe that has been jealously protected and enhanced over the years; today some 130 herbs, including *Dianthus*, *Artemisia* and the buds of *Pinus*, are infused in cognac and then distilled to make this famous liqueur.

The spectacular flowers of the Alps

In all of France the most spectacular array of plants is to be found in the Alps. Most alpines are compact, cushion-forming plants that thrive in poor, well-drained soils and are tolerant of extremes of temperature. Their flowers are very diverse in form and color. Members of many plant families have become adapted to the specialized mountain environment. They include the pinks (Caryophyllaceae), cresses (Cruciferae), primroses (Primulaceae), mints (Labiatae), daisies (Compositae), saxifrages (Saxifragaceae), cranesbills (Geraniaceae) and gentians (Gentianaceae).

The alpine plant life begins above the deciduous forest zone at about 1,500 m (5,000 ft), in an area once extensively covered by coniferous woodland. The woods were cleared to provide land for settlement, hay meadows for winter pasture and, most destructively, for ski-runs.

Visitors to the Alps in early summer may be overwhelmed by the color and variety of the meadow flowers. Yellow blooms predominate, ranging from the dainty, dancing heads of buckler mustard (*Biscutella*), woad (*Isatis*), St John's wort (*Hypericum*), groundsels (*Senecio*) and other daisies, to the dense spikes of mignonette (*Reseda*) and yellow-rattle (*Rhinanthus*). The blues of bellflowers (*Campanula*) and viper's bugloss (*Echium*), the pinks of lousewort (*Pedicularis*) and the multicolored orchids all combine to create an unforgettable scene.

The dark coniferous forests provide a marked contrast to the sun-filled meadows. They grow on the steeper slopes, where they help to stabilize the soil. The two main species are the silver fir (*Abies alba*) and Norway spruce (*Picea abies*). Larch (*Larix decidua*), pines and yew (*Taxus*) also grow in these woods, which have an upper treeline of about 2,000 m (6,500 ft). Ground cover is sparse, but some attractive species include primulas, cyclamen and yellow foxgloves (*Digitalis grandiflora*), many of them growing on rocky outcrops within the forest.

In woodland clearings untidy yellow spikes of mullein, various members of the mint and borage families (Boraginaceae), yarrows (*Achillea*) and the distinctive yellow leopard's bane (*Doronicum*) flourish. Open spaces are also colonized by *Laburnum* and greenweeds (*Genista*). Other small deciduous trees found at these higher altitudes are rowans (*Sorbus*) and willows (*Salix*), but the dwarf juniper

(*Juniperus communis*) outstrips them all, surviving at over 3,500 m (11,500 ft).

Above the treeline and hay meadows, grassland extends to about 3,000 m (9,800 ft), and is used for summer grazing. Many of the flowering plants of lower altitudes grow here, though they tend to be smaller because of the constant grazing and increased exposure to the wind. Other plants appear on these higher pastures, such as gentians and the snowbell (*Soldanella*), whose violet flowers unfold above the late-melting snow.

Higher up the mountainsides the grassland gives way to rocky slopes, scree and cliffs. In sheltered crevices small rosette plants grow, the best known being the white-flowered saxifrages (*Saxifraga*) and the pinkish rock jasmine (*Androsace*). Alpine toadflaxes (*Linaria*) are especially adept colonizers of the scree slopes. At over 3,500 m (11,500 ft) the mountains are clothed only in icy glaciers, permanent snow and exposed rock. Not even the hardiest plants can survive this inhospitable landscape.

Dainty alpines (*right*) *Erinus* and *Saxifraga caesia*, with their low, creeping growth form, are both typical alpines of scree and rocky ground. The snowbell lives in damper places, where it starts into growth before the snow melts.

Soldanella alpina snowbell

Erinus alpinus

Saxifraga caesia

Deep blue gentian (*above*) The flowers of this clump-forming plant of moist, sunny places are almost stemless. *Gentiana* is a large genus, with blue flowers predominant among European species though red and white flowers occur elsewhere.

Vetches in a hay meadow (*right*) In the Alps hay meadows on a north-facing slope are found at about 1,500 m (5,000 ft), but on a south-facing slope, because of the more favorable climate, they can extend to altitudes of over 2,000 m (6,600 ft).

DUNE PLANTS AND FERTILE MEADOWS

SAND- AND SALT-TOLERANT SPECIES · ADAPTING TO DIFFERENT SOILS · EXPLOITING DIKES AND DITCHES

There is surprising diversity in the plant life of the Low Countries, where about 2,000 species have been recorded. Two main factors account for this: first, all the main soil types can be found here, ranging from the poorest sands along the coastal duneland, to fertile clay inland and, far to the south, the well-watered limestone valleys and sandstone ridges of the Ardennes highland; and secondly the temperate climate of the region allows most of the plants typical of western Europe to thrive (with the exception of species common only to mountainous areas or rocky shores). To the northwest, where the landscape is almost flat, the lush green meadows are able to take full advantage of the uniform sunlight and abundant rain. Toward the south, where the terrain begins to rise, woodlands predominate.

COUNTRIES IN THE REGION

Belgium, Luxembourg, Netherlands

DIVERSITY

	Number of species	Endemism
Belgium	1,700	very low
Luxembourg	1,200	very low
Netherlands	1,500	very low

PLANTS IN DANGER

Number of threatened species	12
Number of endangered species	3
Number of extinct species	1

Examples *Agrostemma githago* (corncockle); *Echinodorus repens*; *Eriophorum gracile*; *Halimione pedunculata*; *Hammarbya paludosa* (bog orchid); *Luronium natans*; *Petroselinum segetum*; *Pilularia globulifera* (pillwort); *Salvinia natans*; *Spiranthes spiralis* (spiral orchid)

USEFUL AND DANGEROUS NATIVE PLANTS

Crop plants *Avena sativa* (oats); *Pyrus communis* (pear)
Garden plants *Corydalis lutea*; *Helianthum nummularium* (rock rose); *Helleborus foetidus* (stinking hellbore); *Malva alcea*
Poisonous plants *Aconitum napellus* (monkshood); *Atropa belladonna* (deadly nightshade); *Conium maculatum* (hemlock); *Digitalis purpurea* (foxglove); *Solanum nigrum* (black nightshade)

BOTANIC GARDENS

Leiden University (8,000 taxa); National Botanic Gardens, Brussels (15,000 taxa); Utrecht University (12,500 taxa)

SAND- AND SALT-TOLERANT SPECIES

The plants of the Low Countries can be roughly divided into two areas: those of the coastal land to the west of the region, which benefit from the warming influence of the sea; and those of the landlocked east, with its colder winters. Centuries of land management have brought about dramatic changes to the landscape, leaving only pockets of woodland amid the heather and chalk pasture and the vast stretches of arable land.

Pockets of diversity

Along the coast the predominant southwesterly winds have built up a continuous border of parallel sand dunes. Few plants can tolerate the severe water deficiency and the high levels of salt and sand found in the soil here. Above the high tide mark a few well-adapted low-growing plants, such as saltwort (*Salsola*

Map of floristic provinces The Low Countries forms a small part of the extensive Circumboreal Region, the Earth's cool northern temperate zone. Plant groups such as the anemones and the buttercups provide links with the plants of other parts of the region.

The Low Countries

kali) and sea rocket (*Cakile maritima*), thrust their branching roots deep down to reach the mineral-rich subsoil. Sea holly (*Eryngium maritimum*) and sea bindweed (*Calystegia soldanella*) are typical plants of young dunes that contain large amounts of chalk in the form of crushed seashells.

Farther inland the wind has built up higher sand dunes that play an essential role in protecting the land behind them. Tall drought-resistant marram grass (*Ammophila arenaria*) helps to hold the dunes in place with its strong roots. Only a handful of other plants, such as creeping willow (*Salix repens*), sand sedge (*Carex arenaria*) and sea spurge (*Euphorbia paralias*), can survive the continual submersion under drifting sand or frequent uprooting by the wind.

The landward dunes contain progressively less chalk and salt. Here the sand is

West Frisian Islands
Schiermonnikoog
Ameland
Terschelling
Vlieland
Waddenzee
Texel
North Sea
IJsselmeer
NE Polder
E Flevoland
S Flevoland
Veluwe
Oude
NETHERLANDS
Lek
Rhine
Waal
IJssel
Meuse
Schouwen
Overflakkee
E Schelde
Walcheren
W Schelde
Kempenland
Schelde
BELGIUM
Hesbaye
Meuse
Botrange ▲ 694
Hautes Fagnes
Sambre
Ourthe
A r d e n n e s
LUXEMBOURG
Moselle

Floristic provinces

Holarctic Kingdom/Circumboreal Region

Atlantic-European Province Influenced by the maritime climate; typical plants include ivy and heathland species such as heather and gorse.

Central European Province Largely mountainous zone; eyebrights, louseworts and bellflowers reflect links with eastern parts of the province.

Pale pink trumpets and fleshy leaves (*below*) Sea bindweed, with its trailing stems and deep roots, is one of the early colonizers of a sand dune. It comes in after the pioneer grasses, such as marram, have begun to stabilize the loose sand.

Purple heathland (*above*) Hummocky heather or ling grows well in the moderate, moist climate of the Low Countries. It depends on an association with a fungus within its roots (mycorrhiza) to obtain nutrients from the poor sandy soil.

fixed by carpets of lichens and mosses such as *Tortula ruralis*. Shrubs such as sea buckthorn (*Hippophae rhamnoides*), creeping willow and common privet (*Ligustrum vulgare*), form dense thickets in undisturbed areas.

Behind the dunes

Behind the wall of protective sand dunes a great alluvial plain, the polders, stretches out across a large part of the Netherlands and western Belgium. Lying below sea level, the plain is dominated by marsh plants such as spurge (*Euphorbia palustris*) and marsh pea (*Lathyrus palustris*) in the northern Netherlands. Pollarded willows (*Salix*) and poplars (*Populus*), both of

which do well on wet soils, are also characteristic of this landscape.

Farther east the polders give way to poorer sandy soils that were once covered by heathland. Ling or heather (*Calluna vulgaris*) grows on this dry acid soil, together with plants similar to those of the fixed sand dunes nearer the coast. In wetter areas cross-leaved heath (*Erica tetralix*) can be found, with bog asphodel (*Narthecium ossifragum*), the insectivorous sundew (*Drosera*) and the beautiful blue marsh gentian (*Gentiana pneumonanthe*). In many areas the heathland has been planted with pine forest (*Pinus*), greatly reducing the diversity of natural species.

To the south, fertile soils support fields of wheat and sugarbeet. The few scattered beechwoods (*Fagus*) are carpeted with ramsons (*Allium ursinum*) in the east, bluebell (*Hyacinthella nonscriptus*) and black bryony (*Tamus communis*) in the west.

ADAPTING TO DIFFERENT SOILS

In southern Belgium, Luxembourg and the southern Netherlands, chalk pastures have been created by the felling of old woodlands. Subsequent grazing prevents the trees from reestablishing themselves, enabling chalk-loving plants rapidly to colonize the new pastures.

Several great European rivers (the Schelde, Meuse and Rhine) cut their way through the soft chalk subsoil. Their wide valleys are covered in species-rich grassland. The rivers flow mainly south to north, and this has enabled several central European species to migrate northward and westward; they thrive on the steep, sunbaked slopes. Typical examples are meadow clary (*Salvia pratensis*), field eryngo (*Eryngium campestre*) and white rockrose (*Helianthemum apenninum*), while along the sloping valleys of southern Belgium box (*Buxus sempervirens*) forms isolated dense thickets. Numerous orchids grow scattered on the slopes; even the beautiful and rare bee orchid (*Orchis apifera*) may occasionally be found.

The underlying rocks of Belgium alternate between those that easily split (schistous), and hard limestone (calcareous). This results in smooth slopes, often covered by woodland, and steep cliffs respectively. As the calcareous rocks break up, a thin layer of mineral-rich topsoil develops and is soon invaded by species such as bloody cranesbill (*Geranium sanguineum*), *Globularia punctata*, *Vincetoxicum hirundinaria* and stinking hellebore (*Helleborus foetidus*). Traveler's joy (*Clematis vitalba*) is one of the few climbing species of the region. In its quest for light it grows up to 30 m (90 ft) high in trees rooted in calcareous soils, and can cover a patch of woodland in a long curtain of tiny white flowers and feathery fruitlets.

In a small area in Belgium where the soil is rich in zinc, a specifically zinc-adapted plant life has developed. This includes *Viola calaminaria*, which forms distinctive yellow carpets of flowers in small, restricted pockets.

In such a landscape the plants can change abruptly over a very short distance; for example, common rock rose (*Helianthemum nummularium*) and fescue (*Festuca glauca*) may suddenly give way to ling, bracken (*Pteridium aquilinum*) and

foxglove (*Digitalis purpurea*) as the alkali/acid composition of the underlying soils changes across the terrain.

The Ardennes

The upland areas of Belgium and Luxembourg are characterized by a changing landscape, exposed to frequent strong winds. High above the vineyards of the limestone valleys of the Ardennes there is an elevated sandstone plateau that receives high rainfall and has a harsh climate. The soil here is poor in plant nutrients, and the area is largely covered

by woodland. High humidity encourages the growth of numerous lichens that flourish on the tree trunks. The original beechwoods of the plateau have mostly been replaced by conifers, which have increased the level of acidity of the soil.

The woodland is often carpeted with brambles (*Rubus*) or bilberry (*Vaccinium myrtillus*), but a few pockets of plants that represent survivors of a colder age can still be found. Species include the alpine spignel (*Meum athamanticum*), *Arnica montana*, northern bog whortleberry (*Vaccinium uliginosum*) and marsh andromeda

Bloody cranesbill (*above*) The attractive ferny leaves enable the plant to lose heat and are a response to the dry, sunny places this plant tends to inhabit.

The white rock rose (*left*) has narrow, inrolled leaves, also a response to dry, sunny sites. The foliage and the stems are covered in downy hairs that reflect sunlight.

Eranthis hyemalis
winter aconite

Aconitum napellus
monkshood

Trollius europaeus
globe flower

Sanguisorba minor
salad burnet

Well-adapted perennials (*right*) Monkshood and globeflower are tall herbs of streamside and woodland edge, lowly aconites inhabit the forest floor, and salad burnet creeps on dry grassland.

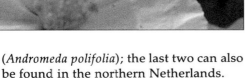

(*Andromeda polifolia*); the last two can also be found in the northern Netherlands.

The highest plateaus of the Ardennes, such as the Hautes Fagnes, are covered in fens and peat moors. The plants are similar to those of European heathland, but typical Atlantic plants such as bog myrtle (*Myrica gale*) and bell-heather (*Erica cinerea*) are replaced here by species such as chickweed wintergreen (*Trientalis europaea*) and cotton-grass (*Eriophorum vaginatum*), with its distinctive fluffy white inflorescences designed to catch the wind.

NATURE'S PLANT POISONERS

Certain plants contain poisonous substances as part of their defense system against predation by insects, grazing animals and humans. A number of the plant species of the Low Countries contain such potentially harmful chemicals. The plants tend to be widespread, either because they were deliberately cultivated in the past for their use in folk medicine, or because they have been able to flourish unhindered by grazing animals. Two families are especially common in fields, gardens and along roadsides: the carrot family (Umbelliferae) and the buttercup family (Ranunculaceae).

The plants of these families vary both in the degree of their toxicity to humans, and also in the way their poisons work when swallowed or touched. In the carrot family, hemlock (*Conium maculatum*) contains poisons that can be fatal within minutes of swallowing. In appearance it is very similar to the common herb parsley (*Petroselinum*), which has similar foliage but lacks the distinctive purple-spotted stems of hemlock. The sap of certain hogweeds, such as the enormous giant hogweed (*Heracleum mantegazzianum*), makes the skin ultrasensitive to sunlight, and can induce painful burns over a period of some hours.

Most members of the buttercup family contain irritants in their sap. They include monkshood (*Aconitum napellus*), wood anemone (*Anemone nemorosa*), marsh marigold (*Caltha palustris*) and celery-leaved buttercup (*Ranunculus sceleratus*), a small-flowered plant of damp fields and marshes that can burn the skin with its foliage.

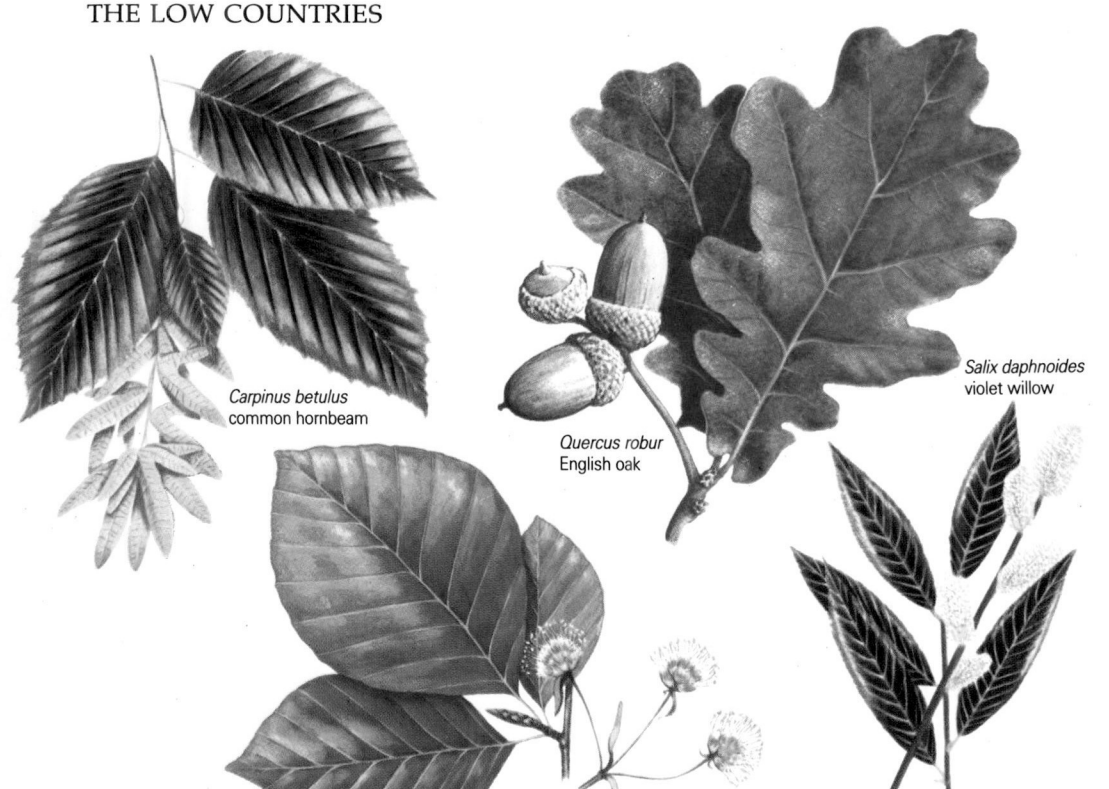

Carpinus betulus
common hornbeam

Quercus robur
English oak

Salix daphnoides
violet willow

Fagus sylvatica
European beech

Deciduous trees of the Low Countries (*left*) Large woodland trees once covered much of the region, and pollarded willows lined the waterways.

Remember the poppy (*right*) Once a common weed of wheatfields, *Papaver rhoeas* covered Flanders fields laid waste by the carnage of World War I.

EXPLOITING DIKES AND DITCHES

Human influence on the landscape is largely responsible for the plant life of the region. Certain wild plants have been used by the local inhabitants since time immemorial. Sweet woodruff (*Asperula odorata*), for example, a ground cover plant in woodland, is used to give a delicate taste to white wine. Willow is another versatile plant. Planted in rows along the banks of ditches, the trees form a characteristic feature of the landscape. Every few years they were pruned (pollarded) to encourage them to develop the flexible, whippy branches used in basketry. Glasswort (*Salicornia europaea*) and sea aster (*Aster tripolium*) are still collected from the wild, and cooked and eaten as vegetables.

Plants of the dikes
The continual process of land reclamation from the sea has not only shaped the coastal landscape, but has also affected the distribution of plant species in the region. The fertile lowlying polder plains exist because a combined system of protective dikes and drainage channels have removed excess water.

The riverbanks are also flanked by dikes, producing long stretches of flat land called forlands. In the drainage channels the water flows slowly and will only occasionally rise after heavy winter rains.

Apart from the plantiful light, there are

three main influences on the plant life of the dikes: moisture, nutrients and human interference. Along the banks there is a gradient in the distribution of plant species, determined primarily by their ability to survive submersion in water. The river forlands often become submerged in winter but dry out again in summer; the plants have had to adapt to these changing extreme conditions in order to survive.

The lower banks of the dike are usually dominated by tall, stout-stemmed species such as the reed *Phragmites australis* and lesser reedmace (*Typha latifolia*). Higher up yellow flag (*Iris pseudacorus*), meadowsweet (*Filipendula ulmaria*), *Carex cyperus*, wild parsnip (*Pastinaca sativa*) and great hairy willow-herb (*Epilobium hirsutum*) thrive in the slightly drier conditions.

comfrey (*Symphytum officinale*) and stinging nettle (*Urtica dioica*) are often found on the upper edges. On top of the dikes walnut trees (*Juglans regia*) were often planted in the past, but in Belgium these were mostly cut down for timber during World War I.

The availability of nutrients is another important factor that affects the distribution of the dike plants. Unpolluted areas usually support a wide range of species. In the channels themselves so many different plants flourish that the water that looks like a floating meadow. Species such as *Ranunculus aquatilis* and duckweed (*Lemna*), for example, can cover extensive areas. Where the water is shallow reedbeds often cover the entire surface, and numerous amphibians live on the floating carpets formed by the rhizomes of aquatic plants.

An overabundance of nutrients in the water, often the result of pollution, is commonplace in the Low Countries. The plants such as Canadian pondweed (*Elodea canadensis*) and various algae that thrive in such conditions act as indicators of polluted water.

Human maintenance of the dikes also influences the plants that grow there. If the plants are cut back mechanically, the diversity of species is maintained. However, if herbicides are used, this all too often leads to the survival of only a few dominant plants.

PROTECTING RARE PLANTS

Over centuries of land use, a balance has evolved in this region between the needs of the people and the plants. However, the relatively recent development of modern agricultural methods now poses a great threat to this delicate equilibrium. A 1969 survey of the Belgian plant life revealed that, since 1850, 75 of Belgium's mosses had become extinct and 50 more were endangered. About 60 species (5 percent) of the 1,300 ferns and seed plants had been lost, and another 70 seemed on the brink of extinction. The only species endemic to Belgium, and therefore found growing wild nowhere else in the world, *Bromus*

bromoideus, is now also extinct. Attempts to save the natural plant life of Belgium through the creation of nature reserves have mainly been private endeavors. In the Netherlands, in contrast, efforts have been made to protect the Dutch natural heritage by setting up national parks.

Many rare species are now protected by law throughout the Low Countries; these include all gentians, all species of *Primula* and all bellflowers (*Campanula*) in the Netherlands; orchids, monkshood, yellow whitlow grass (*Draba aizoides*), *Artemisia alba* and many other species throughout the region.

Specialized salt marsh plants

Salt marshes are commonly found where a river or stream runs into the sea. At a rivermouth or estuary the strength of the incoming seawaves is greatly reduced, enabling sand, clay and silt to accumulate over time. The continuous settling of organic detritus and silt is possible only in less exposed coastal areas, where the gathering particles are bonded together by algae, and a salt marsh develops.

The plants of salt marshes grow in distinct zones, from the wettest to the driest. At one extreme the small succulent glasswort, which resembles a miniature tree with thickened branches, is able to survive longterm inundation by seawater. It counters a low rate of assimilation by storing water in its stems. Other plants, such as thrift (*Armeria maritima*) and sea rush (*Juncus maritimus*) are restricted to the higher zones because they cannot tolerate prolonged waterlogging. The tidal rhythms greatly influence the type of salt marsh vegetation found. In areas that are dry only at low tide, just a handful of particularly well-adapted plants called halophytes can thrive.

Halophytes such as sea lavender (*Limonium vulgare*) and sea milkwort (*Glaux maritima*) can survive in salty habitats because they are able to absorb the saline water where other plants dry out. They store the salt in their cells, so their cell sap has a higher concentration of salt than the brackish water surrounding them; this enables them to absorb the salt water. Excess salt is disposed of through specialized salt glands on the surface of the plant.

Binding the silt

The grass *Spartina anglica* is a well-adapted species that was introduced into the salt marsh areas of the Netherlands because of its ability to bind silt with its roots, thus greatly aiding the recovery of land from the sea. However, its rapid spread has now made it a threat to other native species that cannot survive the competition of this greedy opportunist.

The gradual accumulation of silt eventually leads to the development of areas that remain almost permanently dry, being submerged only occasionally during heavy storms. Although the salinity of such areas is less extreme, a vegetation gradient relating to the height of the land above sea level can still be seen. Thrift, for example, is found only in the higher,

mainly dry, locations, along with the grass sea poa (*Puccinellia maritima*), which bears leaves that are wrapped tightly around the plant's stem to reduce water loss through evaporation. Sea lavender, sea aster (*Aster tripolium*) and sea arrowgrass (*Triglochin maritima*) all have leaves that are covered by a waxy secretion that reflects the light; sea wormwood (*Artemisia maritima*) reflects direct sunlight by means of a layer of small hairs.

The ecology of the salt marshes is delicate, and large sections can virtually be destroyed during heavy storms. In more protected areas, the invasion of the marshland by trees from farther inland has become a problem; in the past many salt marshes were kept free of colonizing trees by grazing animals.

Sea aster (*above*) Most plants of salt marshes are woody or herbaceous perennials. The sea aster is short lived and may complete its life in two or three years. It grows in the developing lower marsh.

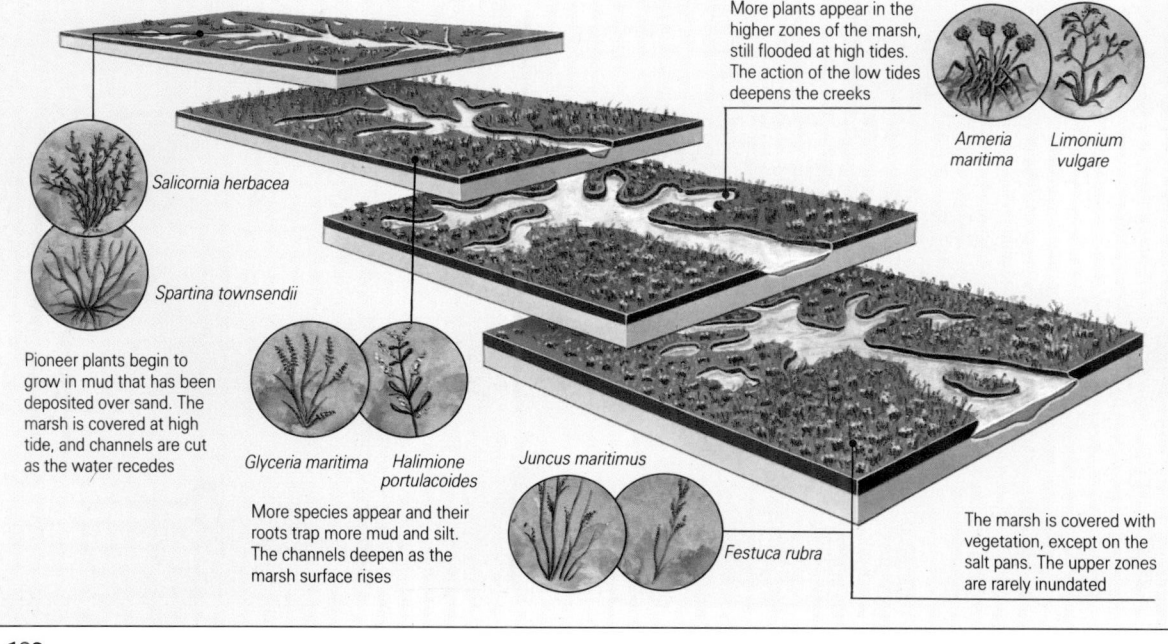

More plants appear in the higher zones of the marsh, still flooded at high tides. The action of the low tides deepens the creeks

Salicornia herbacea

Spartina townsendii

Pioneer plants begin to grow in mud that has been deposited over sand. The marsh is covered at high tide, and channels are cut as the water recedes

Glyceria maritima   *Halimione portulacoides*

More species appear and their roots trap more mud and silt. The channels deepen as the marsh surface rises

Juncus maritimus

Festuca rubra

The marsh is covered with vegetation, except on the salt pans. The upper zones are rarely inundated

Armeria maritima   *Limonium vulgare*

Development of a salt marsh (*left*) In sheltered bays and estuaries where the sea is shallow, fine particles of silt and mud are deposited on the seabed. Eventually they will form a mud flat exposed at low tide – an inhospitable surface, but one that certain specialist plants can colonize.

Sea lavender (*right*) When this longlived deciduous perennial sheds its leaves it also sheds excess salt that has accumulated in the foliage. Experiments have shown that it depends on sea water to break the dormancy of its seeds.

The Iberian Peninsula is a mountainous region – its average altitude is 800 m (2,600 ft). The plant life at low and mid-altitudes was once dominated by a mixed forest of pines and evergreen oaks, but as a result of several thousand years of human activity these forests have been largely replaced by scrubby shrubs. Together with the Balkans, this region is home to more kinds of plants than anywhere else in Europe, with at least 6,500 species. Their diversity is due in part to the wide range of environmental conditions found here, and in part to the topography. The numerous isolated mountain ranges and the scattered volcanic islands of Macaronesia – the Canaries, Madeira and Azores archipelagos – have resulted in the evolution of a large number of species, many of them found nowhere else in the world.

COUNTRIES IN THE REGION

Portugal, Spain

EXAMPLES OF DIVERSITY

	Number of species	Endemism
Portugal	2,600	4%
Spain	4,800	15%
Canary Islands	2,000	25%
Madeira	760	17%

PLANTS IN DANGER

	Threatened	Endangered	Extinct
Portugal	25	11	2
Spain	120	31	1
Canary Islands	200	126	–
Madeira	37	17	–

Examples *Aquilegia cazorlensis* (Cazorlon columbine); *Asphodelus bento-rainhae*, *Azorina vidalii* (Azores bellflower); *Eryngium viviparum* (Brittany eryngo); *Euphorbia handiensis*; *Loeflingia tavaresiana*; *Lotus kunkelii* (succulent birdsfoot trefoil); *Musschia wollastonii*; *Narcissus nevadensis* (Sierra Nevada daffodil); *Nautfraga balearica* (Balearic castaway)

USEFUL AND DANGEROUS NATIVE PLANTS

Crop plants *Allium ampeloprasum* var. *porrum* (leek); *Brassica oleracea* (cabbage); *Castanea sativa* (sweet chestnut); *Linum usitatissimum*

Garden plants *Antirrhinum* species; *Argyranthemum* species; *Campanula arvatica*; *Cistus* species (rock rose); *Narcissus rupicola*; *Ulex europaeus*

Poisonous plants *Arum maculatum* (lords and ladies); *Ilex aquifolium* (holly); *Ligustrum* species (privet); *Nerium oleander* (oleander); *Tamus communis* (black bryony)

BOTANIC GARDENS

Barcelona; Elvas (6,500 taxa); Valencia University

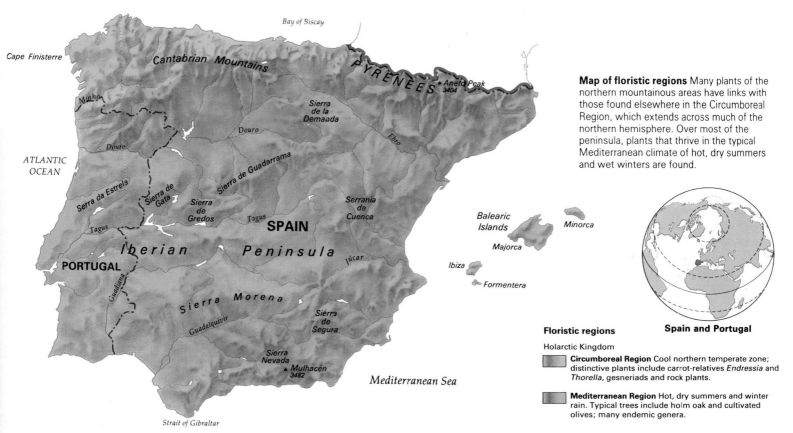

Map of floristic regions Many plants of the northern mountainous areas have links with those found elsewhere in the Circumboreal Region, which extends across much of the northern hemisphere. Over most of the peninsula, plants that thrive in the typical Mediterranean climate of hot, dry summers and wet winters are found.

Floristic regions   **Spain and Portugal**

Holarctic Kingdom

Circumboreal Region Cool northern temperate zone; distinctive plants include carrot-relatives *Endressia* and *Thorella*, gesneriads and rock plants.

Mediterranean Region Hot, dry summers and winter rain. Typical trees include holm oak and cultivated olives; many endemic genera.

A MEDITERRANEAN SHRUBLAND

Positioned between two seas, attached to the European mainland in the north and just grazing the African continent in the south, the plant life of the Iberian Peninsula has been open to influences from four sides. The result, not surprisingly, is a wide diversity of plants.

Links with the north

Many Iberian plants are related to those elsewhere in northern Europe. In northern Portugal and across north Spain these affinities are highlighted by forests of beech (*Fagus sylvatica*) and oak, especially the English oak (*Quercus robur*), *Q. petraea*, and the Pyrenean oak (*Q. pyrenaica*) which grows on acid soils. Members of the heath family (Ericaceae) such as St Dabeoc's heath (*Daboecia cantabrica*) and Mackay's heath (*Erica mackaiana*) are also widespread in the northwest, where the climate is moderated by the Atlantic.

In the eastern Pyrenees and the mountains immediately to the south, Scots pine (*Pinus sylvestris*), silver fir (*Abies alba*) and alpine species such as *Rhododendron ferrugineum* illustrate northern as well as

Matorral covers much of Iberia, in particular the Mediterranean and central regions. The thickets of mixed, often spiny, shrubs that cover the stony plateau are brightened by patches of perennial herbs, such as *Antirrhinum majus*.

alpine connections. Others grow alongside Mediterranean plants such as greenweed (*Genista legionensis*), stock (*Matthiola perennis*), the white-flowered *Ptilotrichum pyrenaicum*, and several members of the tiny *Petrocoptis*.

Evergreen cork oaks (*Quercus suber*) and the mastic tree (*Pistacia lentiscus*) once dominated the fertile lowlands of Portugal along the river Tagus. They have been superseded by spiny shrubs whose tiny leaves and compact shape protect them from water loss during the dry summers. These characteristic shrubs of the western Mediterranean include *Asparagus albus*, the partial parasite *Osyris quadripartita* and buckthorn (*Rhamnus oleoides*), as well as the olive (*Olea europaea* var. *sylvestris*) and the relict Portugal laurel (*Prunus lusitanica*).

Atlantic and African relatives

In addition to their prevalent Mediterranean vegetation, the extreme south of Portugal and the coastal area known as the Costa del Sol in southwest Spain have an interesting mixture of Atlantic and north African species. *Corema album*, an isolated member of the crowberry family (Empetraceae), and the fragrant lavender, *Lavandula viridis*, are found on Atlantic coastlines, while the carnivorous sundew *Drosophyllum lusitanicum* and the broomlike *Lygos sphaerocarpa* also grow in northwest Africa.

Some of the plants best suited to life in

hot, dry conditions are found in the extreme southeast of Spain. Many are related to those of northern Africa, especially Morocco, such as the cypress-like *Tetraclinis articulata*. The ancient forests of the Sierra de Cartagena, which flourished more than 3,500 years ago, were dominated by *Tetraclinis*, but now only a few individuals survive. Other African relatives include the spiny shrub *Maytenus senegalensis*, while *Zygophyllum fabago* and *Fagonia cretica* belong to a family (Zygophyllaceae) that is widespread in semidesert regions throughout the world.

The mountains that run from the Sierra Nevada northward along the eastern edge of Spain to the French frontier contain more endemic plants than anywhere else in the peninsula. This area encompasses the northern limit of the dwarf fan palm (*Chamaerops humilis*), the northernmost native palm in the world, which can reach a height of 2 m (6.6 ft) in areas where it is not subject to grazing by animals. This plant, together with the mastic tree, *Asparagus albus*, buckthorn and the grass *Hyparrhenia hirta*, is characteristic of warm coastal areas.

Many of the plants of the Canary Islands and the Azores are also related to those of Africa. Because of the isolation of these islands, certain plant species that are now extinct on the mainland have been able to thrive; some have even evolved to form new species.

PROTECTIVE MEASURES

The largely rugged topography of the Iberian Peninsula has created a wide range of environments, from coastal sands to bleak, snow-covered mountain peaks. This diversity has meant that many plants have had to develop special adaptations in order to survive.

Aromatic plants

Anyone walking through the Spanish shrublands is immediately aware of the aromas given off by the leaves of many of the plants. In some plants, such as sage (*Salvia officinalis*), thyme (*Thymus*), rosemary (*Rosmarinus officinalis*) and lavender (*Lavandula*), the scent is sweet and pleasant. In others, for example Spanish horehound (*Ballota hispanica*) and southernwood (*Artemisia abrotanum*) the odors may be pungent and less attractive. These smells come from aromatic essential oils, such as terpenes, which are released by increases in temperature or by trampling feet and hooves. Of all the species containing such oils, 49 percent occur in Mediterranean-type climates.

There are two main reasons why aromatic plants are so common in the region. The first is that the oils protect the plants from grazing animals, sap-sucking or leaf-chewing insects, and fungal attack. Sheep and goats, for example, will avoid bitter-tasting herbs, such as southernwood, but will graze the sweet aromatic upland thymes (from which they gain much flavor when cooked).

The second reason concerns the important role of fire in Mediterranean regions. Because the oils are volatile they cause the overground parts of the plants to burn quickly, so that the fire passes rapidly, leaving the roots undamaged. Most, if not all, plants that contain essential oils are able to regenerate quickly from their bases after they have been burned.

Coping with water loss

Many plants in the Mediterranean have developed adaptations that prevent water loss during the hot, dry summers. In sage, lavender, southernwood and horehound, for example, the leaves and stems are

Silver shrub *Artemisia arborescens* combats the heat and drought of the southern coasts with finely divided foliage covered in silvery hairs. This both reduces transpiration – the movement of water through the plant – and reflects radiation from the sun.

MEDITERRANEAN BIOCLIMATES

The Iberian Peninsula includes a wide range of microenvironments related to altitude, rainfall, temperature, aspect, exposure to wind and rock type. In mountainous zones the most dominant of these influences is altitude. Different plant communities grow at different altitudes in quite distinct zones that form "bands" up the mountain slopes. This is known as altitudinal zonation. The zonation differs according to whether a slope is north- or south-facing, because of the differing amounts of sunlight available to the plants growing there. For example, a certain vegetation zone on the northern slopes of a mountain will be approximately 300 m (1,000 ft) lower than the corresponding zone on the southern slope of the same mountain.

The effect of altitude on the distribution of plants in a region with such rugged topography is clear throughout the Mediterranean basin. However, because the communities of plants at the same altitude may be quite different from area to area, attempts have been made to define the type of environment more precisely. This has led to the concept of "bioclimates", which is applied particularly in Spain and France. Five categories have been established based on temperature, and six subcategories according to the amount of rainfall received.

wholly or partially covered in a mass of very fine hairs. The hairs enable the plant to retain moisture by slowing the rate of evaporation from the plant surface, their whitish color reflecting the sun's rays. A similar function is served by the long sticky hairs that cover the undersides of the leaves and branches of some rock roses, *Cistus monspeliensis* in particular.

In plants such as rosemary the adaptations have been further refined. The narrow leaves are protected by a leathery outer skin, or cuticle, and their edges curl under so that only the central part of the whitish underside from which water evaporates is exposed. This characteristic is shared by the heaths (*Erica*) and many other groups. The spiny or spinelike leaves of, for example, the thistles (*Carduus, Cirsium* and *Onopordum*) and some asparaguses such as *Asparagus horridus*, reduce water loss from the leaf surface; and the fleshy leaves of the tall, spectacular century plant (*Agave americana*)

Leaves like spines In northwestern Spain and Portugal the gorse *Ulex galli* is able to withstand the considerable force of Atlantic gales, and at the same time repels grazing herbivores by means of its small, tough leaves and dense growth form.

also conserve water. The century plant is not an Iberian native; it was introduced from the Americas in about 1560.

Adaptations to reducing water loss are not restricted to plants in the dry lowlands. At high altitudes, nightly or seasonal low temperatures also result in water stress. Houseleeks such as *Sempervivum nevadense* conserve water in their fleshy rosettes of leaves, which provide the maximum surface area for photosynthesis in the minimum space. The purple-flowered *Erinacea anthyllis* and the yellow-flowered *Vella spinosa* have spine-like leaves: the reduced surface keeps evaporation to a minimum. They are popularly known as "vegetable hedgehogs", a name derived both from their spiky leaves and their cushionlike form.

Nonspiny examples of high-altitude cushion plants include the yellow-flowered "primrose" *Vitaliana primuliflora*, several sandworts (*Arenaria tetraptera* and relatives) and a wormwood, *Artemisia assoana*, that is covered in a mass of short, white hairs that reflect sunlight and reduce water loss.

Cushionlike growth is a feature of many plants of the high sierras, as it is of plants that grow at high altitudes and latitudes throughout the world. In these cold, often windy "deserts" they are able to retain water in the dead tissues at the center of the cushions. Most cushion plants bear many small, individual flowers, which may account in part for their success: a cluster of flowers is easy for pollinating insects to see. In addition, since temperatures both at and very near the cushion's surface are higher than those in the surrounding air, any pollinators that reach them will linger rather than moving off to colder areas.

Convolvulus tricolor

Gladiolus illyricus

Centranthus ruber
red valerian

Narcissus bulbocodium
hoop-petticoat daffodil

Bee plants and bulbs (*left*) Valerian and *Convolvulus tricolor* are planted to attract beneficial insects such as bees and hoverflies. In Spain and Portugal there are many bulbous plants of interest, including *Narcissus, Gladiolus* and *Cyclamen*.

Rock roses for the garden (*above*) *Cistus* is a promiscuous genus, and natural hybrids occur in the wild. In addition the plant has been hybridized by nurserymen. The white or pink petals are often blotched with a darker shade.

Aromatic rosemary (*right*) Burned as an incense in medieval times, the leaves of this soothing herb are used as a herbal remedy to treat various stress-related and nervous conditions. The essential oil is used in toilet waters.

WOOD, FLAVORS AND FLOWERS

Many of the plants in Spain and Portugal have been used for many centuries, with the result that a number of plants have been overcollected or overused, and their range of distribution severely reduced.

Oaks, pines and junipers have long been a source of timber for construction. Holm oak (*Querus ilex*), for example, formerly covered extensive tracts of the central plateau of Spain. Its wood was once highly prized for shipbuilding, but it has largely been replaced by wheat and other cultivated crops. It has even been said that there is more of it lying off the west coasts of Scotland and Ireland, following the loss of the Spanish Armada in 1588, than remains in Spain today.

The cork oak (*Quercus suber*) is valued for the cork that is obtained by stripping the outer bark. It is widely grown com-

mercially (it is a major export product for Portugal), but its natural range has been restricted by human activity.

Several species of pine, especially the Aleppo pine (*Pinus halepensis*), are systematically tapped for their resin, which is used in the production of turpentine. The resin oozes from wounds made in the bark of the trees, and is collected in cups strapped to the trunks. Many of these pines are now grown in plantations.

Another important (though non-native) tree is the olive (*Olea*), grown principally for its oil. The decorative wood is carved into plates, trays, fruit bowls and other household utensils. Many olive groves are carpeted with the invasive bright yellow Bermuda buttercup (*Oxalis pes-caprae*).

Culinary plants

Iberian plants have made their greatest contributions in the kitchens and gardens of the temperate areas of the world.

Among the vegetable plants, asparagus, globe artichokes (*Cynara scolymus*), cardoon (*Cynara cardunculus*), leeks (*Allium ampeloprasum* var. *porrum*), black salsify or scorzonera (*Scorzonera hispanica*) and the asparagus pea (*Lotus tetragonolobus*) are all natives of the region. The fruits of the azarole hawthorn (*Crataegus azarolus*) are used to make jam and as a flavoring in liqueurs; those of the sweet chestnut (*Castanea sativa*) have a number of uses: they are ground into flour for use in soups, fritters, stuffings and stews, and are also eaten whole – boiled, roasted or preserved in syrup. The predominant fruit crops of the region – lemons, oranges, almonds and grapes – are not in fact native species, but originated elsewhere and were introduced here.

The five species of juniper (*Juniperus*) found in Spain range across Europe, illustrating the widespread distribution of these plants. The common juniper *Juniperus communis* grows throughout central and northern Europe. *J. phoenicea*, common to the Mediterranean region, sometimes grows as a small tree of up to 8 m (26 ft) tall, but is more usually a ground-hugging coastal shrub. *J. thurifera* is a pyramidal tree that reaches a height of as much as 20 m (65 ft) in the mountains of southwestern Europe. *J. oxycedrus*, a full-crowned tree, reaches 14 m (46 ft) tall, and grows around the Mediterranean as far west at Bulgaria. Finally, *J. sabina* is a lowgrowing shrub, less than 20 cm (8 in) tall, found in the mountains of south and central Europe and farther east.

Three of the species grow at mid-altitudes in the Iberian Peninsula, dominating considerable areas of open parkland. They are prized for their wood (particularly *J. thurifera*), which is used in construction because its high resin content makes it resistant to decay. The wood also yields turpentine when distilled, and the berries give gin its characteristic flavor.

A large proportion of the plants that make up the scrub vegetation of this region contain aromatic essential oils. It is not surprising, therefore, that a considerable number of them are grown all over the temperate parts of the world as culinary flavorings. In addition to thyme, of which there are no less than 26 species, bay laurel (*Laurus nobilis*), fenugreek (*Trigonella foenumgraecum*), cumin (*Cuminum cyminum*), fennel (*Foeniculum vulgare*), sage, marjoram (*Origanum*), rosemary, summer and winter savory (*Satureja hortensis* and *S. montana*), lemon balm (*Melissa officinalis*) and tarragon (*Artemisia dracunculus*) are among the most widely used. The unopened flower-buds of the straggly, spiny shrub *Capparis spinosa* are pickled to make the capers that add piquancy to sauces.

Many aromatic plants, including southernwood, chamomiles (*Anthemis* and *Matricaria*) and tansies (*Tanacetum*), are used to make a wide range of herbal teas (manzanillas). These are made by infusing the leaves, and sometimes the flowers, in boiling water, and are drunk for their actual or supposed medicinal properties. Herbal teas are thought to be particularly good for relieving tension, headaches and stomachaches.

Garden jewels

A large number of Iberian plants, including the alpine saxifrages and gentians, are highly regarded by specialist alpine gardeners, and many of the aromatic herbs are grown for their ornamental value. Perhaps the greatest contribution to horticulture comes from just three genera: snapdragon (*Antirrhinum*), rock rose (*Cistus*) and houseleek (*Sempervivum*). Sixteen species of snapdragons are native to the region, and two of these, *Antirrhinum majus* and *Antirrhinum hispanicum* (*A. glutinosum*), have been widely used in breeding the modern horticultural varieties. Similarly, the 12 native species of the shrubby rock rose are widely cultivated both as introductions and to produce hybrids with an extraordinary diversity of color and interesting foliage. The several species of houseleek, some of them Macaronesian, are also highly prized by gardeners for their colorful rosette leaves.

Migration to the Canary Islands

The Canary Islands are famous for the large numbers of plants that are endemic to the archipelago (more than 25 percent of the native species). Although these species are restricted to the islands, they show evidence of ancient affinities with plants elsewhere. In the more humid parts of Tenerife, for example, there are fragments of a type of forest called laurisilva, in which trees such as *Laurus azorica* and *Ocotea foetens* are linked, through fossil deposits, to the subtropical forests that covered Eurasia between 64 and 2 million years ago.

Many other groups of plants in the archipelago that seem to have ancient origins belong to a number of genera that are widespread in Europe, where they can often be aggressively weedy herbs. However, in the Canary Islands and elsewhere in Macaronesia they are woody plants. These plants have many primitive characteristics, so it is perhaps surprising that, far from surviving as isolated, remnant populations, many of them are currently evolving in different ways. The changes among these so-called "active epibionts" are particularly evident in such groups as the giant sowthistles (*Sonchus* subgenus *Dendrosonchus*), the buglosses (*Echium*) and the chrysanthemums that belong to the genus *Argyranthemum*. This last group contains no less than 22 species of woody shrubs that have diversified in isolation on different islands and even within the very different habitats that exist on individual islands. Many of these species are still diversifying as the local topography and environment change.

Woodland endemic (*above*) *Isoplexis canariensis* is becoming popular in cultivation. It is a small shrub that grows in laurel woodland. It has a close relative in *Digitalis obscura*, a less showy herbaceous plant of mainland Spain.

Bold flowering spikes (*right*) *Echium wildpretii* is a mountain plant with large rosettes of soft silvery foliage. Most species of *Echium* are endemic to the Canary Islands, but it does also have representatives throughout Europe.

Giant sowthistles evolve Starting with the treelike *Sonchus pinnatus*, several evolutionary trends in response to different habitats can be followed. In dry areas flowerheads become small, and the foliage more linear and finely divided. On mountains and cliffs small leaves form a rosette and the flowerhead is large, while in humid woods leaves are spreading and flowerheads large.

Dry montane forest and upland

S. platylepis

S. acaulis

Humid forest on lower slopes

S. jacquini

S. abbreviatus

Moist fertile valleys

S. pinnatus

Dry lowland

S. regisjubae

S. leptocephalus

Exposed low cliffs

S. radicatus

African origins

There has been much discussion about the origins of the plants on the Canary Islands over the years. It now seems clear that the islands were never part of the African continent, which means that the fruit, seeds or spores of these plants or their ancestors must have crossed at least 100 km (60 mi) of ocean to reach the islands. It has been estimated that 186 original colonizers would be enough to account for the present range of endemic plants. Of these, 63 seem to have been carried in the digestive tracts of birds, 35 were attached to birds' plumage by hooks on the seeds of fruit, 48 were carried on the wind, and 8 were borne on ocean currents. The method of travel of the remaining 32 migrants is not clear.

The distribution of the relatives of the modern Canarian endemic species shows overwhelmingly that they are related to plants on the African continent. It is very likely that the ancestors of the islands' plants migrated along a route that included south and eastern Africa. The genus *Micromeria*, for example, is found not only throughout the Mediterranean region but also in eastern Africa and the Indian subcontinent; on the Canary Islands it has evolved to form new species.

Among other groups of plants is *Aeonium*, a diverse genus in Macaronesia, with species distributed from Somalia to the south of the Arabian Peninsula. The very distinctive bellflower of the archipelago, *Canarina canariensis*, has its only relatives in eastern Africa, while Canarian genera such as *Phyllis* and the shrubby, cactuslike spurges (*Euphorbia* section *Diacanthium*) have their closest relatives in the Cape region of South Africa.

Canary bellflower, a tuberous-rooted clambering herb of cliffs and forests. Its east African relatives are also perennial, but they have more succulent stems and flowers that are dull yellow in color.

COLORFUL AROMATIC SCRUBLAND

EVERGREENS AND AROMATIC SCRUB · RESISTING THE DROUGHT · NATIVES AND ALIENS

From the woody herbs, bulbs and oak scrub of the lowlands, to the deciduous and evergreen forests of the hill slopes and mountains, the plants of Italy and Greece have evolved in response to two main factors: climate and human activity. The mostly hot and dry summers, even in the mountains, have led many plants to adapt to what are frequently drought conditions. Many of them favor the open habitats created over the millennia by human activity, which has destroyed much of the native forest. New species are evolving, while other plants have become isolated in pockets of the region's unusually dissected landscape, with its rugged mountains, jagged coastline and numerous islands. Where the terrain is flat it has long been intensively cultivated, and only fragments of the native plant life remain.

COUNTRIES IN THE REGION

Cyprus, Greece, Italy, Malta, San Marino, Vatican City

DIVERSITY

	Number of species	Endemism
Cyprus	2,000	6%
Greece	5,500	20%
Italy	4,800	3%
Malta	900	less than 1%

PLANTS IN DANGER

Number of threatened species	50
Number of endangered species	42
Number of extinct species	5

Examples *Adonis distorta* (Apennine adonis); *Alyssum akamasicum*; *Cephalanthera cucullata* (Cretan helleborine); *Chionodoxa lochiae*; *Delphinium caseyi*; *Globularia stygia* (Styx globularia); *Linaria hellenica* (Malea toadflax); *Paeonia rhodia* (Rhodes peony); *Primula appennina* (Italian mountain primrose); *Veronica oetaea* (Mount Iti speedwell)

USEFUL AND DANGEROUS NATIVE PLANTS

Crop plants *Brassica oleracea* (cabbage); *Crocus sativus* (saffron); *Cynara scolymus* (globe artichoke); *Lactuca sativa* (lettuce); *Lens culinaris* (lentil); *Olea europaea* (olive).
Garden plants *Campanula garganica*; *Crocus niveus*; *Genista cinerea*; *Helleborus atrorubens*; *Lilium candidum* (Madonna lily).
Poisonous plants *Euphorbia peplus* (petty spurge); *Hyoscyamus niger* (henbane); *Laburnum anagyroides*; *Solanum nigrum* (black nightshade)

BOTANIC GARDENS

Athens University; Bologna University; Naples University (15,000 taxa); Padua (founded 1545); Pisa (founded 1542)

EVERGREENS AND AROMATIC SCRUB

The once extensive native lowland forests of Italy and Greece now survive in only a few places in the region, such as the Kuri Forest in northern Greece and the wooded slopes of the island of Samothrace. They are dominated by species of evergreen or semievergreen oak (*Quercus*). In moister areas, elm (*Ulmus minor*) forms woodland and grows in hedgerows.

The original distribution of some native species has been obscured as they have been commercially planted in other areas. Examples include the Valonia oak (*Quercus macrolepis*) of southeastern Italy and Greece, and the cork oak (*Q. suber*) in parts of Italy and Sicily. Pine forests are common in the lowlands, especially along the coast, but like oaks they have frequently been planted. A characteristic pine of Italy is the umbrella-shaped stone pine (*Pinus pinea*), grown for its edible seeds – pine-nuts. The native Aleppo pine (*P. halepensis*) is also widely planted. On dry soils it is often accompanied by carob (*Ceratonia siliqua*).

Up in the mountains, at about 800 m (2,600 ft), more montane species grow. The Apennines of Italy and the Pindus Mountains of central and northern Greece are covered with deciduous forests of beech (*Fagus sylvatica*), chestnut (*Castanea sativa*), hornbeam (*Carpinus*), hop hornbeam (*Ostrya carpinifolia*) and oak.

At higher altitudes, where there is snow and frost in winter, the forests are dominated by conifers: pines, firs (*Abies*) and, on Cyprus, cedar (*Cedrus brevifolius*). In the mountains of Crete and Rhodes are stands of cypress (*Cupressus sempervirens*), which has spreading branches like the cedar.

Dense oak scrub

The uncultivated lowlands in Italy and Greece are covered with a dense scrub called maquis (or macchie), which ranges from woodland to scattered hummocks of grazed shrubs, often dominated by the Kermes oak (*Quercus coccifera*).

Macchie also contains many spiny plants, and others with leathery or hairy aromatic leaves. These include the aromatic bay laurel (*Laurus nobilis*), myrtle (*Myrtus communis*), the shrubby wild olive (*Olea europaea*) and the tree heather (*Erica arborea*).

In drier areas, particularly ones that are affected by erosion and grazing, lower growing scrub not more than 1m (3 ft) high persists, with additional species such as the mastic tree (*Pistacia lentiscus*) and several species of sunrose (*Cistus*). Further degradation of the vegetation by burning and grazing has led to the development of garigue (or phrygana), low, hummocky shrubs and a good deal of open ground where annual plants and bulbs produce a brilliant show of color in the spring.

Neighborly influences

The wealth of herbaceous plants in Italy and Greece includes both local species and those of neighboring regions. Many of the plants of the Greek mountains, for example, have affinities with those of Turkey. On the hot, arid southern and eastern coasts of Crete grow date palms (*Phoenix theophrasti*) and a shrubby violet with fragrant yellow flowers (*Viola scorpiuroides*), both of which have close relatives in North Africa.

The plants of this region are still evolving and producing new species. This is particularly true in disturbed habitats, where new environmental niches are being exploited by plants such as knapweeds (*Centaurea*), Aaron's rod (*Verbascum*), bellflowers (*Campanula*) and campions (*Silene*).

Umbrella-like canopies A forest of stone pines shelters a rich shrubby undergrowth on the coastal hillsides of the island of Elba. These distinctive conifers with their large glossy cones are characteristic of the sandy shores of the Mediterranean.

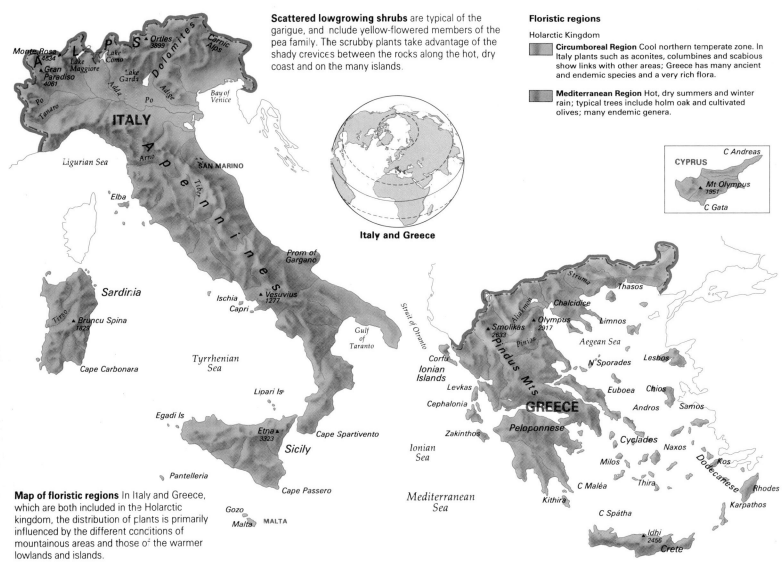

Scattered lowgrowing shrubs are typical of the garigue, and nclude yellow-flowered members of the pea family. The scrubby plants take advantage of the shady crevices between the rocks along the hot, dry coast and on the many islands.

Floristic regions

Holarctic Kingdom

Circumboreal Region Cool northern temperate zone. In Italy plants such as aconites, columbines and scabious show links with other areas; Greece has many ancient and endemic species and a very rich flora.

Mediterranean Region Hot, dry summers and winter rain; typical trees include holm oak and cultivated olives; many endemic genera.

Italy and Greece

CYPRUS
C Andreas
▲ Mt Olympus
1951
C Gata

Map of floristic regions In Italy and Greece, which are both included in the Holarctic kingdom, the distribution of plants is primarily influenced by the different conditions of mountainous areas and those of the warmer lowlands and islands.

RESISTING THE DROUGHT

The plants of Italy and Greece have adapted in several ways to survive the summer drought, which affects them even in the mountainous areas.

Some annuals try to avoid the summer altogether by remaining as seeds, which are filled with a store of food. Many herbaceous perennials, such as bulbs, flower during the spring but die down in summer, keeping a store of food safely below the surface of the soil.

In woody plants, especially those of the macchie, the most common means of minimizing water loss is by reducing the leaf surface area – any of the macchie shrubs and trees have compound leaves made up of leaflets that are often small or needlelike. The leaves may also have a thick, waxy outer layer or cuticle, or a dense covering of long, branched, star-shaped or sticky hairs. These reduce evaporation of precious water from the leaves in the hot, dry air. They also prevent excessive loss of water from the pores (stomata) through which the plants exchange gases during photosynthesis.

The three-lobed sage (*Salvia triloba*), widespread in southern Italy and Greece, is typical, with its leathery, three-lobed leaves and a dense covering of hairs on the lower side.

Some shrubby species, such as the joint pines (*Ephedra*) and various species of broom – *Cytisus*, *Genista* and *Spartium junceum* in particular – have either few or even no leaves, or shed their leaves during the summer months. These summer-deciduous plants often have green stems that carry out photosynthesis and respiration. The stems frequently have ridges or wings that increase their surface area, as in the winged broom (*Chamaespartium sagittale*). This is also true of a group of vetches (*Lathyrus*), in which the function of the leaves may also be taken over

A purple cushion (*above*) High in the Dolomites the perennial moss campion (*Silene acaulis*) puts on a massed display of bright flowers to attract pollinating insects, while its narrow leaves keep water loss to a minimum in such exposed conditions.

Sinister hooded *Arum pictum* (*right*) This shade-loving plant is a fly-trap. The phallic appendage gives off a strong odor to lure flies down into the chamber at its base. There, where the flower parts are concealed, they act as pollinators.

SICILY'S SPECIALTIES

Many of the plants that grow on the island of Sicily are found nowhere else in the world. The island lies in the middle of the Mediterranean Sea, close to both Europe and Africa, and for millions of years it has been a cross-roads between the two continents where its plant life originated. Some of Sicily's endemic species are thought to be very ancient, while others, such as those that grow on Etna, have evolved relatively recently.

In the Madonie Mountains in northern Sicily, which rise to almost 2,000 m (6,560 ft), there are a number of distinctive plants. These include a few surviving examples of Sicilian fir (*Abies nebrodensis*), all that remain of what was once an extensive forest. Higher up on the mountain rocks grow other endemic plants, including species of broom (*Genista cupanii*), everlasting (*Helichry-sum nebrodense*), whose flowers are popular with flower arrangers, thorow-wax (*Bupleurum elatum*), and a sub-species of the dusty miller, widely grown in gardens (*Senecio cinerea* subsp. *nebrodensis*).

On the Egadi Islands off the western tip of Sicily other unique collections of species grow. Another thorow-wax (*Bupleurum dianthifolium*), a small shrub, clings to a few cliffs on the island of Marettimo, together with a robust squill (*Scilla hughii*), with its deep violet flowers. Near the sea grows the Egadi cabbage (*Brassica macrocarpa*), a woody relative of the garden cabbage.

White wool The mass of hairs that covers *Ballota acetabuosa* reduces evaporation through the leaves during the hot summer. A plant of the stony lands of Greece, it has funnel-like sepals that surround small purple flowers. With age the leaves turn gray.

wholly or partly by leaflike appendages called stipules.

Many shrubby species and some herbs, such as members of the mint family, contain volatile essential oils and aromatic compounds that vaporize easily, giving them a distinctive smell on hot days. The chemicals may serve to reduce water loss, but they are probably also important in deterring predators.

Spines, too, are a means of defense against grazing animals, though they may also be a response to dry conditions, preventing the loss of water. They are found on many woody and herbaceous plants in the region, particularly in the macchie. Spines develop either from branches, as in spiny broom (*Calicotome villosa*), or from leaves, as in Kermes oak. Thistles have both spiny leaves and spiny bracts (leaflike structures that surround the flowers). They belong to either the daisy family, Compositae, or the carrot family, Umbelliferae (such as the genus *Eryngium*). Thistles are not totally resistant to grazing, and are eaten by goats and other livestock (the Greeks call them "donkey plants"), but they survive well enough to grow and set seed.

Another adaptation that discourages grazing is unpalatability. Nettles, with their unpleasant stinging hairs, fall into this category. In the lowlands several species of annual nettle grow around villages and on rich soils; in the mountains they are replaced by the perennial nettle (*Urtica dioica*).

Timed flowering
To avoid the dry summer months many plants flower in spring. The flowering year starts with crocuses, anemones and other bulbous species, and annuals such as campions and various members of the cabbage and daisy families. A great variety of herbaceous and woody species follow, the majority flowering in April in the lowlands. By the end of May many lowland plants die down.

The summer drought is less pronounced up in the mountains, especially in the northern parts of Italy and Greece; plant growth and flowering continue throughout the summer months. However, many mountain plants, such as species of *Alyssum* and *Saxifraga* still have drought adaptations because of the high levels of sunlight and the frequent drying winds that blow in the mountains. The plants usually have a compact growth habit and small leathery or hairy leaves.

Fall heralds the start of the floral year in the lowlands and on the lower mountain slopes. The first bulbs flower in October – autumn crocuses (*Colchicum*) and, in the Aegean area especially, species of the arum family with specialized spikes of flowers enclosed by a cowl-like bract. Many annual weeds flower in the fields. This is also the season when many of the trees of the Mediterranean zone are in flower, for example the carob, with its sickly-scented spikes. In the mountain woods sheets of cyclamen (*Cyclamen hederifolium*) cover the ground.

NATIVES AND ALIENS

The appearance of the landscape is the result of a long and complex interaction between people and the natural environment. Repeated phases of cutting and burning, and grazing by sheep and goats, have destroyed the forest cover and have led to the dominance of drought-resistant, often spiny, shrub species that are able to exist where the protective forest canopy has been removed.

The opening up of the forests did, however, create new possibilities for a far greater range of both native and alien species to become established. Some native species, such as cereals, began to be cultivated as crops.

Food plants and ornamentals

Many plants native to the region have been cultivated for centuries. The olive and the vine, together with various pulses, notably lentils (*Lens culinaris*), and vegetables such as cabbage (*Brassica oleracea*) and lettuce (*Lactuca sativa*), were popular with the people of ancient Greece and Rome just as they are today.

The rich, greenish olive oil has long been a staple item in the diet of the region. There are forests of olive in northern Africa, and spiny wild olives grow in the macchie scrub. However, recent studies of pollen preserved in wetland peat suggest that the olive spread to the Mediterranean from the Middle East, where cultivation began, during the last 10,000 years. This tree, now so much part of the landscape in both Italy and Greece, may have been deliberately introduced.

Another tree that is widespread, and undoubtedly native, is the Oriental plane (*Platanus orientalis*). It commonly grows on the banks of stony torrents, which are usually dry in the summer, around the Aegean Sea and in Sicily. From these native habitats it has been planted to give shade in village squares. Oleander (*Nerium oleander*) is also a plant of stony streams in the drier parts of the Mediterranean. Perhaps native in southern Greece, is now a familiar sight everywhere – beside streams, in gullies and along highways and boulevards.

Sulfur-yellow flowers and silvery leaves of *Senecio cineraria*, a shrub widely cultivated for its foliage that is at its most handsome before flowering. This color combination is common to many plants that grow by the Mediterranean Sea.

Many of the ornamental plants now naturalized in Italy and Greece came from the tropics – agaves, bougainvillea, Canary palm, false acacia, eucalyptus, mimosa, morning glory and trumpet vine, together with an array of herbaceous plants. Some of them, such as false acacia (*Robinia pseudoacacia*) and prickly pear (*Opuntia ficus-indica*), have invaded native plant communities and now pose a threat to them.

With the ornamentals came weeds, opportunists that exploit the niches created by human disturbance of the land. The modern landscapes of Italy and Greece suit them very well, especially where irrigation has eased the effects of the summer drought, and tropical weeds, particularly some grasses, are able to flourish. The ricefields of the Po valley in Italy and some of the deltas of Greece now have many distinctly tropical weeds.

Plants as a resource

The Mediterranean environment provides plentiful resources for its human inhabitants. Large numbers of oaks were once used in the construction of ships; the wood is still used for charcoal and the bark for tanning leather. The large acorn-cups of the Valonia oak are a source of black dye, and the thick bark of the cork oak is used to make cork. Herbs such as rosemary (*Rosmarinus officinalis*), sage (*Salvia*), thyme (*Thymus*), and lavender (*Lavandula*) are used in medicine and in cooking. The aromatic mints (*Mentha*) and other members of the family Labiatae are used to make teas and medicinal infusions, while many species provide nectar for honey in the flowering season; in the fall, especially in Italy, woodlands are a source of various edible fungi. Even roadside weeds – dandelions (*Taraxacum*), chicory (*Cichorium*) and other members of the daisy family – are eaten as salads in the spring.

Cercis siliquastrum
Judas tree

Arbutus undeo
strawberry tree

Castanea sativa
sweet chestnut

Populus alba
white poplar

Laburnum anagyroides

Flowers, fruits and foliage Some of the decorative trees of Italy and Greece that are now grown widely in parks and gardens (though the seeds of laburnum are poisonous). The fruits of *Arbutus unedo* are used locally in preserves and to flavor a liqueur.

ANISE – THE FLAVOR OF THE MEDITERRANEAN

In many Mediterranean countries alcoholic drinks are flavored with aniseed, the fruit of anise (*Pimpinella anisum*). This aromatic member of the carrot family (Umbelliferae) is an erect, slender and rather hairy annual, 10–50 cm (4–20 in) tall, with finely divided leaves and umbrella-shaped clusters of tiny white or pinkish flowers. Its ribbed brown fruits, up to 5 mm (0.2 in) long, are borne in pairs. It is probably native to Asia, but anise has been grown in Europe at least since the first century AD, particularly in the Mediterranean region. It is sown in spring and harvested in summer.

The active constituent of anise is an essential oil containing anethole, which is obtained by distillation of the seeds. It was first isolated more than 150 years ago. The anise-based drink called ouzo is popular in Greece; in the Greek islands it is often replaced by a stronger drink known as raki, which is also the principal alcoholic drink of Turkey. Similar drinks are found in Lebanon, Egypt, France and Spain.

Flowering bulbs of Greece

Many of the plants that survive the drought and intense heat of the Greek summer store food accumulated during the favorable growth period from fall to spring. These plants are bulbs, which are characterized by their compact, fleshy underground storage organs.

The specialized underground organs commonly called bulbs may not in fact be bulbs, but rather corms, rhizomes or tubers. They often lie well below the surface of the soil, where they are protected from heat and drought. Their outer layers are often hard and leathery, giving them additional protection.

The stored food is generally starch, but in some species – onions (*Allium*), for example – it is a soluble sugar such as glucose. This food may have a smell or taste that attracts foraging animals; to deter predators a number of species, such as the fall-flowering giant squill (*Drimia maritima*), contain noxious chemical compounds such as cardiac glycocides.

Bulbs usually divide, produce offsets or have spreading underground stems. These provide the plants with another means of reproduction in addition to seeds. This makes bulbs very popular as cultivated plants, but it also means they can be problematic weeds. The cornfields of Crete and the Peloponnese were once plagued by tulips (*Tulipa boetica*) and gladioli (*Gladiolus italicus*), though the increasing use of deep plowing and herbicides has considerably reduced their distribution. The Bermuda buttercup (*Oxalis pes-caprae*) is an introduced bulb that, during the last hundred years or so, has increased to such an extent that it now carpets huge areas of olive and citrus groves with its brilliant yellow flowers in March and April.

Most bulbs flower in the spring, though many, especially in Crete and on the other islands, flower in the fall and winter. The spring bulbs are at their best from February (when crocuses appear in great numbers even in the suburbs of Athens) to April, when a range of species is in flower. Even in late spring, pheasant's eye (*Narcissus poeticus*) can be found in damp mountain pastures, and in the hot summer sea daffodils (*Pancratium maritimum*) flaunt their richly scented white flowers on seashores.

Most of the Greek bulbs have colorful flowers, and many have a compact growth habit. These features, and the fact that they flower when relatively few other species are in bloom, have made the bulbs irresistible to gardeners throughout the world. Their popularity has led to the growth of a considerable trade in bulbs collected from the wild, especially large cyclamen plants, which has seriously depleted wild stocks. In many countries importing cyclamen is now prohibited, but the trade continues. Protective legislation and local education programs are needed to secure the plants' future.

Yellow flowers in the sun *Sternbergia lutea* is a true bulb found on the stony hillsides of Greece and the island of Crete, where like many other bulbs it flowers in the fall. Although it resembles a crocus it is in fact a member of the daffodil family.

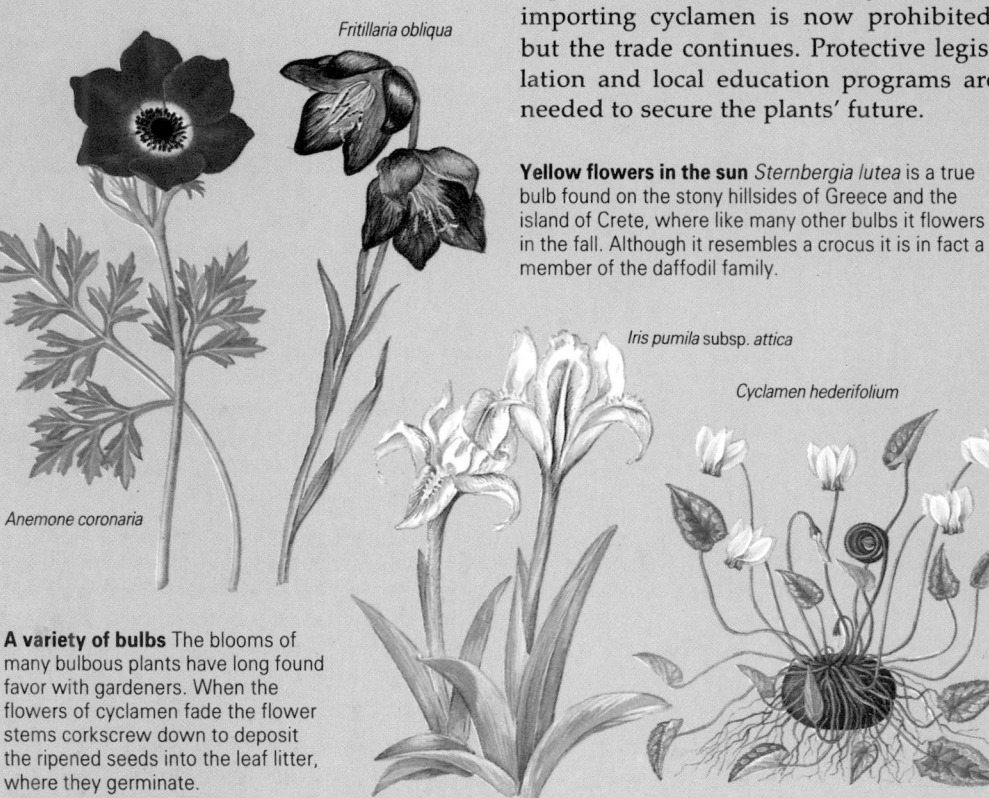

Fritillaria obliqua

Iris pumila subsp. *attica*

Cyclamen hederifolium

Anemone coronaria

A variety of bulbs The blooms of many bulbous plants have long found favor with gardeners. When the flowers of cyclamen fade the flower stems corkscrew down to deposit the ripened seeds into the leaf litter, where they germinate.

Deadly but beautiful

The oleander (*Nerium oleander*) is found growing throughout the Mediterranean region on stony ground near the coast, but most commonly in dried-up river courses and ravines. Here its roots can penetrate to take advantage of the water lying deep below the ground.

With its flamboyant, sweet-smelling flowers it is, not surprisingly, widely grown in sunny gardens and as a cut flower. However, one leaf can kill. All parts of the plant are poisonous, including the flowers, but the leaves are the most deadly. In these cardiac glycosides are concentrated; a small quantity is capable of killing an animal or a human being by cardiac arrest.

Oleander belongs to a small genus with one other very similar and equally poisonous member, *Nerium odorum*, which grows in Japan. This has pinkish white, perfumed flowers and is also cultivated. The oleanders are related to the periwinkles (*Vinca*), two species of which also grow in the Mediterranean region. These, however, are plants of shady banks and woodland floor. They contain indole alkaloids, which are used in various proprietary medicines but can have harmful effects if taken in quantity. Far more important are the alkaloids obtained from the closely related Madagascan pink periwinkle (*Catharanthus rosea*), which are used in the treatment of cancer.

A colorful shrub The pink flowers of oleander are in evidence for most of the year. A minimum of water is lost through the tough leathery leaves of this strong-growing shrub. Long hanging pods contain tufted seeds that, when discharged, are blown away.

PLANTS OF THE ALPS AND THE FORESTS

MIXED FORESTS AND GLORIOUS ALPINES · WOODLAND HERBS AND MOUNTAIN TREES · THE CHANGING FORESTS

The climatic conditions of most of Central Europe favor the growth of deciduous trees. The region was once covered in dense broadleaf forests; the plant life is characterized by trees and shrubs that formed part of this woodland canopy, and species that grew on the forest floor. To the north and at higher altitudes, coniferous trees predominate. The shortened growing season of these areas limits the distribution of deciduous woodland and its associated plant life. In the drier eastern areas lies the steppe – these vast grasslands support few trees, though patches of oak woodland grow in some places. The Alps provide a habitat for a large number of colorful mountain-dwelling species; they also form a barrier separating the plant life of Central Europe from the sub-Mediterranean vegetation found farther south.

COUNTRIES IN THE REGION

Austria, Germany, Liechtenstein, Switzerland

EXAMPLES OF DIVERSITY

	Number of species	Endemism
Austria	2,900	1%
Germany	2,600	less than 1%
Switzerland	2,700	less than 1%

PLANTS IN DANGER

	Threatened	Endangered	Extinct
Austria	18	1	1
Germany	16	5	2
Switzerland	11	1	1

Examples *Artemisia laciniata; Betula humilis; Crassula aquatica; Dracocephalum austriacum; Gentiana pneumonanthe; Jurinea cyanoides; Myosotis rehsteineri* (Lake Constance forget-me-not); *Oenanthe conioides* (Elb water dropwort); *Pulsatilla patens; Spiranthes aestivalis*

USEFUL AND DANGEROUS NATIVE PLANTS

Crop plants *Asparagus officinalis* (asparagus); *Brassica rapa* (turnip); *Ribes grossularia* (gooseberry); *Vaccinium oxycoccus* (cranberry)
Garden plants *Clematis alpina; Convallaria majalis* (lily-of-the-valley); *Gentiana acaulis; Geum reptans; Helleborus niger; Leontopodium alpinum* (edelweiss)
Poisonous plants *Aconitum napellus* (monkshood); *Arum maculatum* (lords and ladies); *Conium maculatum* (hemlock); *Convallaria majalis* (lily-of-the-valley); *Euonymus europaeus* (spindle tree); *Tamus communis* (black bryony)

BOTANIC GARDENS

Aachen (6,000 taxa); Berlin (18,000 taxa); Frohnleiten (10,000 taxa); Geneva (15,000 taxa)

MIXED FORESTS AND GLORIOUS ALPINES

Central Europe does not possess as many species of deciduous trees as the ideal climatic conditions – warm, frost-free summers, cool winters and steady supply of rainfall – would suggest. This relative poverty of woody species is the result of repeated glaciations: many trees were wiped out in the ice ages, and the mountain chains of the Alps, Pyrenees and Carpathians formed a barrier that prevented these species from migrating southward during glacial periods.

The main deciduous woodland trees are beech (*Fagus sylvatica*), oaks (*Quercus robur* and *Q. petraea*) and hornbeam (*Carpinus betulus*). Scots pine (*Pinus sylvestris*) grows on the drier, sandy soils, especially in the east. At higher altitudes, conifers such as spruce (*Picea abies*) and silver fir (*Abies alba*) – often mixed with deciduous trees, particularly beech – form the forest communities.

Growing in the shade
Typical Central European woodland shrubs are scrambling roses (*Rosa*), old man's beard or traveler's joy (*Clematis vitalba*), hawthorn (*Crataegus laevigata*)

Central Europe

Mountain gentian (*left*) *Gentiana burseri* may also be found with unspotted petals. Such local variations are common among mountain plants, and endemic subspecies may occur. These differences may be due to isolation, habitat or seedling variation.

and spindle (*Euonymus europaeus*). The many woodland herbs found in the region include wild arum, also known as lords and ladies (*Arum maculatum*), white helleborine (*Cephalanthera damasonium*), bittercress (*Dentaria bulbifera*), early dog violet (*Viola reichenbachiana*), bugle (*Ajuga reptans*), wood anemone (*Anemone nemorosa*), lesser celandine (*Ranunculus ficaria*) and greater stitchwort (*Stellaria holostea*). Clearings in the woodlands and the margins of the forests provided the original habitat for a variety of species that are now widespread, such as elder (*Sambucus nigra*), deadly nightshade (*Atropa belladonna*), teasel (*Dipsacus fullonum*), wood groundsel (*Senecio sylvaticus*), white campion (*Silene dioica*) and comfrey (*Symphytum officinale*).

Eastern Austria contains large areas of lowlands and hills that often support mixed woodlands of oak and hornbeam. Along the river Danube and its tributaries are remnants of fertile alluvial woods of willow (*Salix*), poplar (*Populus*), ash (*Fraxinus*) and elm (*Ulmus*), with a tangle of climbers and herbs growing beneath the trees.

Western Germany has a wide range of plant habitats. In the extreme north lie the saltmarshes and coastal dunes of the North Sea and the Baltic. Inland there are heathlands dominated by heaths (*Erica*), heather (*Calluna*) and juniper (*Juniperus*), and raised bogs and fens. Farther to the south, alpine and subalpine plants can be found on the mountain slopes along the northern edges of the Alps.

Above the treeline in the Alps are stretches of grassland, with patchy scree and rocky habitats at higher altitudes. The plant life here varies from area to area according to the type of soil. The alpine pasque flower (*Pulsatilla alpina*), for example, grows on limy (calcareous) soil, whereas its close relative, yellow alpine pasque flower (*P. apiifolia*), favors acid soil conditions.

Alpine plants such as *Saxifraga aizoides* and alpine toadflax (*Linaria alpina*) can occasionally be found at lower altitudes. The seeds of these species are carried down avalanche tracks or fast-flowing mountain streams, and germinate in river gravel and alluvium.

Map of floristic provinces The Circumboreal Region extends throughout the Earth's cool northern temperate zone. In Central Europe plant groups such as gentians and primulas provide links with other parts of this floristic region, which is the largest in the world.

Floristic provinces

Holarctic Kingdom/Circumboreal Region

Atlantic-European Province Influenced by the oceanic climate of the Atlantic, and characterized by ivy, holly, heather and gorse.

Cental European Province Largely mountainous zone distinguished by numerous species, such as monkshoods, various hogweeds, valerians, etc.

Steppe and saltmarsh

The plant life of eastern Germany reflects the drier, more continental climate of this area. Stretching south and east in Europe and on into the Soviet Union are vast grasslands known as steppe. Various species of feathergrass (*Stipa*) are prominent here – the tall wavy feathergrasses *S. capillata* and *S. pennata* sometimes take over abandoned vineyards. In the rain shadow of the Harz mountains and in other dry places, the steppe is dotted with patches of oak woodland, which provide

a habitat for plants such as the white cinquefoil (*Potentilla alba*).

The damp, saline soils of the Neusiedlersee area of southeastern Austria, on the Hungarian border, support large swards of sea aster (*Aster tripolium*), *Juncus gerardii* and sea plantain (*Plantago maritima*). Marsh samphire or glasswort (*Salicornia europaea*) and annual seablite (*Suaeda maritima*) are also found here – these salt-tolerant plants are common coastal species elsewhere in Europe and on the salt steppe of the Soviet Union.

WOODLAND HERBS AND MOUNTAIN TREES

A whole range of shade-tolerant shrubs and herbs have adapted to the dominant woodland environment of Central Europe. The forests provide relatively stable conditions, protecting the plants growing in them from extreme variations in temperature and in the humidity of the soil. As the plant life is shielded from the wind, more moisture is available here than in open habitats, even though the plants have to compete for water with the extensive root systems of the trees.

Before the leaves of the forest canopy develop fully, various herb species come into flower in the field layer. These plants flower at different times according to their ability to cope with low levels of light. In a typical mixed woodland of oak and hornbeam, for example, many of the plants flower and set seed before the canopy begins to block out the light. Species such as spring snowflake (*Leucojum vernum*), yellow star of Bethlehem (*Gagea lutea*), wood anemone, lesser celandine and hepatica (*Hepatica nobilis*) flower in April. Another flush of woodland plants capable of growing and flowering in dappled shade comes into bloom in May – wild arum, ramsons (*Allium ursinum*), goldilocks (*Ranunculus auricomus*), herb paris (*Paris quadrifolia*),

wood stichwort (*Stellaria nemorum*) and the early dog violet.

Several of these early flowering herbs, such as the spring snowflake and yellow star of Bethlehem, are bulbous or tuberous, allowing them to retreat underground for part of their annual cycle. They pass the worst of the winter below the surface, and as the early stages of flower and leaf development take place within the bulb, the plants are ready to take full advantage of the first spring sunlight.

Alpine adaptations

Alpine plants have adapted to shorter growing periods – reduced by about six days for every extra 100 m (330 ft) of altitude – and a colder climate. The plants also have to tolerate extremes of temperature, as solar radiation is more intense in the thinner air of high altitudes, and ground frosts are more frequent at night. To cope with these conditions, plants such as *Saxifraga aizoides* have small, tough leaves and form compact mats; other species, such as the alpine pasque flower, are hairy. Snow is important to alpine plants, as it provides a protective insulating blanket from hard,

Orange lily of the mountains *Lilium bulbiferum* has a useful method of vegetative propagation: it produces small bulbils in the leaf axils. When ripe these bulbils fall to the ground and develop into new plants identical to the parent, though they may not flower for two or three years.

damaging frosts. The foliage of alpine grasses and herbs remains alive and green under the snow cover, and these plants are ready to grow as soon as the spring thaw occurs. The flower buds of *Crocus albiflorus* and the snowbells (*Soldanella*) form under the snow – these plants are able to flower immediately the snow melts in order to make the most of the short growing season.

Two unusual dwarf trees are found in parts of the Alps and in some other European mountain ranges – dwarf mountain pine (*Pinus mugo*) and green alder (*Alnus viridis*). Although both species are lowgrowing bushes, adapted to the rigorous conditions above the treeline, they occupy different habitats. The dwarf mountain pine grows on limestone areas between 1,400 and 2,000 m (4,600–6,600 ft), often in the company of the hairy alpenrose (*Rhododendron hirsutum*). The green alder favors steeper, north-facing slopes with more acidic soils. This species is a stunted relative of two woodland trees, the common alder (*Alnus glutinosa*) and gray alder (*A. incana*), which are widespread in lowland forests in the region.

EDELWEISS – STAR OF THE ALPS

Edelweiss (*Leontopodium alpinum*) has come to symbolize alpine flowers. This plant is one of Central Europe's most threatened species – numbers have been much reduced by collectors – and is now protected in all alpine countries.

Edelweiss is found mainly on lime-bearing (calcareous) soils in the Alps (and in the Balkans, Carpathians and Pyrenees). Although it originally proliferated on dry alpine meadows, edelweiss is most likely to be found clinging to remote, rocky sites.

The tiny gray-brown flowers of edelweiss are inconspicuous, clustering together in little groups (capitula). What appears to be the flower is in fact a group of about twelve capitula forming a secondary head that is surrounded by a number of tongue-like modified leaves (bracts). The bracts are covered with white hairs, giving the whole flowerhead a soft, silvery appearance. This white covering is most prominent in plants growing at high altitudes, and disappears when they are replanted in lowland gardens. Collectors have tended to remove the larger-flowered specimens, leaving smaller and less spectacular examples behind.

Fruits for dispersal (*above*) The red hips of wild roses are popular with birds. They eat the fleshy outer layer and the seeds are voided. The plumed seeds of the aptly named old man's beard, a vigorous climber, are dispersed by the wind.

After the ice (*below*) The last ice age finished about 10,000 years ago; then over thousands of years woodland developed to cover much of the region.

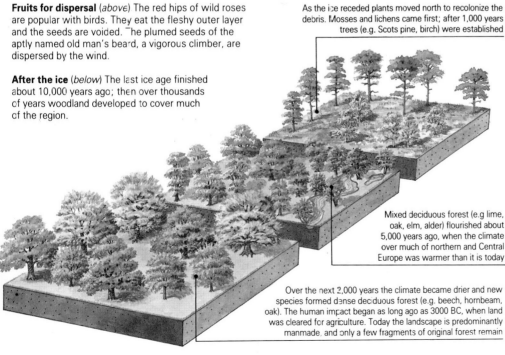

As the ice receded plants moved north to recolonize the debris. Mosses and lichens came first; after 1,000 years trees (e.g. Scots pine, birch) were established

Mixed deciduous forest (e.g lime, oak, elm, alder) flourished about 5,000 years ago, when the climate over much of northern and Central Europe was warmer than it is today

Over the next 2,000 years the climate became drier and new species formed dense deciduous forest (e.g. beech, hornbeam, oak). The human impact began as long ago as 3000 BC, when land was cleared for agriculture. Today the landscape is predominantly manmade, and only a few fragments of original forest remain

The Arolla pine (*Pinus cembra*) – one of the most interesting of Central Europe's native conifers – is found in the Alps and in the Carpathian Mountains to the east. This is an example of a five-needled pine: the short shoots bear tufts of five needles, unlike those of the widespread Scots pine (*P. sylvestris*) which has pairs of needles. It is known from fossil evidence that five-needled pines were widespread in much of the northern hemisphere before the ice ages. The Arolla pine is closely related to a common Siberian pine, *P. sibirica*; the European population appears to be a relict from a postglacial period, when northern continental species spread into Central Europe. In Switzerland the remaining communities of Arolla pine are rigorously protected, and in some places – notably around the base of the vast Aletsch glacier in the Bernese Oberland – conservation measures have enabled this species to regenerate.

Snowdrops in damp woodland (*above*) The hanging flowers of snowdrop (*Galanthus nivalis*) serve a practical purpose; they prevent the pollen getting wet. Snowdrops and the closely related snowflakes belong to the daffodil family (Amaryllidaceae).

Christmas rose (*below*) *Helleborus niger* is a widely cultivated plant of limestone woodlands in the foothills of the Alps. All hellebores are poisonous, as they contain glycosides, notably hellebrin, which are capable of killing an animal or a person.

THE CHANGING FORESTS

From ancient times the forest has yielded essential materials. Some of these, notably timber used for construction work, have remained of great importance right up to the present day, though other industries' that were once of great significance no longer survive. Charcoal-burning, for example, had a dramatic effect on the beech forests in the 17th and 18th centuries. Ash, elm and hornbeam were extensively cut as "leaf-hay" to provide winter fodder for animals.

As the trees suitable for timber, winter fodder and many other uses were selectively eliminated from the woodland, agricultural communities began to protect and cultivate copses and wooded areas near their settlements. A pattern of forest management known as "coppice with standards" developed, and many of these woods survived until the 20th century, when the system began to break down.

The understory of coppice – as trees managed by the practice of cutting regularly on a 15–25 year cycle are known – yielded fuel and poles. Not all native trees can stand regular coppicing – hazel, ash and hornbeam are among the most tolerant, and many woods today show traces of a coppice structure featuring one or other of these three tree species. The "standards" were mainly oak, which was grown in a spaced pattern and selectively felled for timber.

The need for conifers

In the 20th century forestry has increasingly tended to favor softwood (coniferous trees), rather than hardwood from the native broadleaf trees such as oak and beech. There are two reasons for this change. Saleable timber can be obtained from softwood trees on a relatively quick cycle of 20 to 40 years; and modern building methods make greater use of softwoods. The effect of this change is to favor coniferous over deciduous woodland, and to obscure the natural preference of the native softwoods – pine and spruce – for the poorer acid soils.

In some places purely artificial coniferous forests have been created, often using exotic imported species such as the North American sitka spruce (*Picea sitchensis*). These plantations are clear-felled (all the trees are felled and removed) after about forty years, though this method has been banned in Switzerland because of its destructive ecological effects – not least the increased risk of avalanches on the deforested slopes. Much coniferous forest on upland or mountain sites is, however, still composed of native trees, and the timber is selectively felled so that the forest can regenerate naturally.

Garden plants

Many of the flowering plants of this region have become popular garden plants. Alpine species, in particular, have been widely collected and propagated – gentians, saxifrages and alpine pasque flowers are examples of favorites. The edelweiss has not only been collected as a garden curiosity but has also played a role in traditional medicine – in the Tyrol and Carinthia, for example, it was used as a treatment for diarrhoea and to chase away ghosts and evil spirits.

The alpine areas are not only under pressure from the timber industry: leisure activities are also taking their toll. The increasingly heavy use of the ski slopes is damaging the habitat for many alpine plants, and areas of forest have been cleared to construct ski lifts. The meadows produced by these clearances provide a home for a variety of colorful tall herbs, however, including the purple

Adenostyles alliariae, the bright blue alpine sowthistle (*Cicerbita alpina*), white *Ranunculus aconitifolius* and the greenish-white false helleborine (*Veratrum album*), which contains toxic alkaloids that are used in the pharmaceutical industry. All these plants have been widely cultivated in temperate regions.

Other popular garden plants from Central Europe are the stately, tussock-forming ostrich fern (*Matteucia struthiopteris*) and the tall meadow rue (*Thalictrum aquilegiifolium*), both of which are characteristic plants of the wet alder woods covering river floodplains. The willow gentian (*Gentiana asclepiadea*) also favors damp woodland. Unlike any other European species, this plant has large, deep blue flowers held in the angles between the narrow leaves and the stem.

Tilia x europaea
common lime

Tilia platyphyllos

Fraxinus excelsior
European ash

Salix viminalis
common osier

Pinus nigra
Austrian pine

Trees of Central Europe The common lime or linden is a natural hybrid, with *Tilia platyphyllos* in its parentage. Willow grows on moist soil, while the Austrian or black pine, with dark gray bark, is found on poor, thin soils to the east of the region.

HOPS – FLAVOR FOR BEER

The wild hop (*Humulus lupulus*) is found throughout Central Europe, often growing as a climber in the shrub layer of wooded river valleys. Hops have long been valued as a medicinal herb, and are now cultivated on a large scale – especially in the lowlands of southern Germany – for the huge brewing industry. The plants are grown in fields of wooden frames that support the vines. The rich, golden-green, headily scented fruits are harvested in the fall.

Hop plants are either male or female, but only the latter are cultivated since it is the papery, catkin-like fruits (strobili) that are gathered for brewing. These fruits contain a bitter substance called lupulin that helps to give the beer its distinctive flavor. Lupulin also prevents bacterial growth, and thus acts as a natural preservative; this is probably how it first came to be included in the recipe. Hops are used to a lesser extent as a vegetable – the young shoots can be steamed – and also as a constituent of "hop pillows", which are said to promote peaceful sleep.

The Black Forest tragedy

The Black Forest is situated on the eastern bank of the Rhine, stretching for about 170 km (105 mi) between Basel in Switzerland and Karlsruhe in Germany. In the southern part, the main forest types are beech, silver fir and spruce, in approximately equal proportions, while spruce dominates in the north. Spruce was originally much less widespread, tending to grow at high altitudes, but has now been widely planted in areas where beech forest would be the natural community.

Growth and death of the forest

Two opposing problems have affected the Black Forest in recent decades: extensive afforestation and forest "dieback". A quarter of the 400,000 ha (990,000 acres) of forest have been planted over the last hundred years to combat soil erosion. The ecological consequences of this policy, however, have been grave. Increasing the amount of forest has affected the local climate, making it damper, and reducing tourism in an area previously noted for dry, healthy air. The whole character of the Black Forest has changed from a patchwork of open meadows, light valleys and cool forests, to a dark, monotonous blanket of fast-growing spruce.

At the same time the forests are also dying. The precise reasons for this are still not fully known, despite extensive research. Similar problems that occurred earlier were associated with fungal and insect infection, or with industrial dust and smoke. The problem nowadays is that the damage often occurs far from the sources of pollution, and trees growing in high-altitude exposed areas are particularly vulnerable. Recent estimates suggest that as much as 70 percent of the Black Forest is damaged.

The agents of attack

Some chemicals attack the foliage of the trees directly; other pollutants dissolve in atmospheric moisture and form what is popularly known as acid rain. The major air pollutants are sulfur dioxide (SO_2) and oxides of nitrogen NO_x. Sulfur dioxide, which is produced by burning fossil fuels such as coal and oil, is the most widespread. It dissolves in water to create sulfuric acid, resulting in a large increase in the acidity of the precipitation, and ultimately of the soil. Nitrogen oxides are derived mainly from automobile exhaust emissions and industrial processes.

Forests act as huge filters, and are able to absorb airborne substances. The effect is enhanced under damp, foggy conditions because the water droplets that condense on the leaves are rich in acidic compounds. Increased soil acidity reduces the uptake of essential elements, especially calcium and magnesium – both vital to the physiological processes of respiration and photosynthesis.

Heavy metals such as lead affect the symbiotic fungi (mycorrhizae) that grow on the roots of most trees and help them absorb water and minerals. The presence of these elements has been shown to be linked to dieback of forest trees.

Reducing the amount of airborne pollutants is essential if the forests are to thrive. The effects of acid rain can be reduced by adding carbonate to the soil, but this can only be a local, temporary remedy – a cleaner environment is the only longterm solution.

The changing face of the Black Forest (*right*) Somber stands of exotic sitka spruce have replaced large areas of native mixed woodland. Spruce is vulnerable to acid rain; the first symptoms of damage are yellowing leaves at the top of the crown.

The dying trees (*left*) Acid rain attacks the natural defences of trees, including eating away the protective coating of their leaves. The trees are then less able to cope with drought, lack of nutrients and attack from pests and diseases.

ALPINE AND MARSHLAND PLANTS

GRASSY PLAINS AND MOUNTAIN REFUGES · MOUNTAIN AND WETLAND PLANTS · MEDICINAL AND GARDEN PLANTS

The northern plains of Eastern Europe were once thickly covered with mixed woodland. With the exception of one remarkable remnant of natural forest in the northeast, most of this natural plant life has been cleared, managed or pushed back to the marginal land and the mountainsides to make way for agriculture. The eastern grasslands have also suffered, but the less sophisticated farming methods of this area have enabled certain groups of field plants to survive in abundance, in contrast to the west where they have died out or are endangered. Alpine plants flourish in the mountainous parts of the region, particularly in the Balkan Peninsula. Toward the east, dry grasslands gradually mix with salt-tolerant species as the soil becomes more saline, and they are even dotted with inland salt marshes.

COUNTRIES IN THE REGION

Albania, Bulgaria, Czechoslovakia, Hungary, Poland, Romania, Yugoslavia

DIVERSITY

	Number of species	Endemism
Albania	3,300	24
Bulgaria	3,500	53
Czechoslovakia	2,600	10
Hungary	2,400	11
Poland	2,200	3
Romania	3,400	41
Yugoslavia	4,750	137

PLANTS IN DANGER

Number of threatened species	39
Number of endangered species	10
Number of extinct species	2

Examples *Astragalus ornacantha; Cochlearia polonica; Daphne arbuscula; Degenia velebitica; Dianthus uromoffii; Forsythia europaea; Lilium rhodopaeum; Onosma tornensis; Pulsatilla hungarica; Rhinanthus halophilus*

USEFUL AND DANGEROUS NATIVE PLANTS

Crop plants *Corylus avellana* (hazel); *Juglans regia* (walnut); *Rheum rhaponticum* (rhubarb); *Vicia faba* (broad bean)
Garden plants *Cornus mas; Daphne blagayana; Syringa vulgaris; Tilia cordata; Verbascum* species
Poisonous plants *Atropa belladonna* (deadly nightshade); *Euonymus europaeus* (spindle tree); *Hyoscamus niger* (henbane); *Laburnum anagyroides; Solanum nigrum* (black nightshade)

BOTANIC GARDENS

Bucharest University (11,000 taxa); Komenshy University, Bratislava (14,000 taxa); Prague University (6,000 taxa); Vacrotot (12,000 taxa); Warsaw University (6,000 taxa); Zagreb (10,000 taxa)

GRASSY PLAINS AND MOUNTAIN REFUGES

The plant life of eastern Europe contains many species found throughout northern Europe. Much of the deciduous woodland that once covered the north of the region has been cleared, but patches still exist. These are dominated by oak (*Quercus*), beech (*Fagus sylvatica*) and hornbeam (*Carpinus*). Limes (*Tilia*), sycamore and other maples (*Acer*), ash (*Fraxinus*) and elms (*Ulmus*) also grow in these woodlands. Scots pine (*Pinus sylvestris*) is the characteristic tree of dry sandy soils. In upland areas and on mountainsides spruce (*Picea*) and fir (*Abies*) dominate the coniferous forest. These are intermixed with deciduous trees, notably oak and beech, on the lower slopes.

In the northeast a last fragment remains of the wildwood that once spread over most of lowland Europe. This ancient hunting forest of Bialowieza supports 26 species of trees, including oak, hornbeam and ash, with alder (*Alnus*) growing in the wetter areas. The dense, dark woods are broken up in a number of places by sphagnum bog and marshlands that support many aquatic plants.

In general, agriculture is not highly mechanized, nor are chemicals widely used in the region. On the North European Plain cultivated fields are full of colorful annual weeds such as cornflower (*Centaurea cyanus*) and poppies (*Papaver*). They also contain plants like shepherd's needle (*Scandix pecten-veneris*), which was well known throughout Europe before the agricultural revolution of this century changed the landscape. The second effect of the relatively unchanged, traditional methods of agriculture is that wetlands and rough pastures have not been drained, so plants such as ragged robin (*Lychnis flos-cuculi*) and greater meadow-rue (*Thalictrum flavum*) are still abundant here.

Mountain plants

The Tatra Mountains on the Polish–Czechoslovak border form the northern tip of the great arc of the Carpathian Mountains, which sweep round to the east and south. Here the plant life is rich in alpine species. The eastern Carpathians in Romania contain some remarkable endemics, such as the snow campion (*Lychnis nivalis*), with its large, pale

purple flowers; the plant is confined to a single mountain peak.

The Balkan Peninsula, the southern part of the region, is one of the richest areas of plant life in the whole of Europe. The reason is simply that the many high mountain ranges have long provided refuges for plants. Species dating from the ice ages include the moss campion (*Silene acaulis*). In the Rhodope Mountains in Bulgaria is a relict from the Tertiary period, an attractive member of the African violet family (Gesneriaceae), *Haberlea rhodopensis*. This neat rosette plant has central flowers on leafless stalks. The mountains also provide a haven for several endemic genera, such as the rock plant *Edraianthus*.

Between the high mountaintops and the Mediterranean and Adriatic coasts are areas of forest, with Turkey oak (*Quercus cerris*) and limes being the dominant trees. Farther south oriental beech (*Fagus orientalis*) grows with limes, including *Tilia tomentosa*, and spruce. Along the

High in the Tatra Mountains grows the rosebay willow herb (*Epilobium angustifolium*). This widespread and adaptable plant is able to grow at high altitudes, on scree and in forest glades, it also flourishes by roadsides and on waste ground.

southern coasts the scrubby plants include caper (*Capparis spinosa*) and species of mullein (*Verbascum*). A few species grow from the Atlantic edge of Europe to the Mediterranean, such as the black bryony (*Tamus communis*).

The eastern grasslands

In Hungary steppe grassland, known locally as puszta, stretches across the plain, the Great Alföld. Like most grassland in Europe, such areas are at most only seminatural; they have been used for centuries as grazing land. In spring they display a spectacular array of flowering perennials, including the yellow adonis (*Adonis vernalis*), a typical steppe flower of sunny grasslands in Eastern Europe, but rare in the west. Feather-grasses (*Stipa*) with their long, feathery bristles (awns)

are also characteristic of this area.

Farther east the climate becomes drier, and the ratio of evaporation to rainfall increases. This causes salt to accumulate in the surface layers of the soil, and in places has led to the formation of inland salt marshes. The dry-steppe plants of Hungary and northeastern Yugoslavia therefore mix with plants adapted to salty conditions (halophytes), which also grow on coastal salt marshes. They include the succulent annual sea blite (*Suaeda maritima*) and other species of the goosefoot family (Chenopodiaceae), sea aster (*Aster tripolium*) and species of sea lavender (*Limonium*).

Floristic regions

Holarctic Kingdom

Circumboreal Region Cool northern temperate zone; endemic species include pinks, anemones, poppies, hawthorns and milk-vetches.

Mediterranean Region Hot, dry summers and winter rain; typical trees include holm oak and cultivated olives. Some 150 endemic genera.

Map of floristic regions The plant life of Eastern Europe has many links with the rest of continental Europe, and almost all of it falls within the extensive Circumboreal Region of the northern hemisphere's Holarctic Kingdom. The mountainous areas – particularly those of the Balkan Peninsula – are rich in plant species.

Eastern Europe

[Map labels:]
Baltic Sea
Gulf of Gdansk
Oder
Notec
Vistula
Bug
POLAND
Sudetic Mts
Silesian Plateau
Vistula
Bohemian Forest
Vltava
CZECHOSLOVAKIA
Carpathian Mts
Tatra Mountains
Gerlach Peak 2665
Vah
Slovakian Ore Mts
Danube
HUNGARY
Tisza
Great Alföld
Somes
L. Balaton
Bihor Mts
ROMANIA
Mures
Transylvanian Alps
Mt Negoiu 2548
Kras
Istria
Krk
Cres
Sava
Drava
Olt
Pag
YUGOSLAVIA
Morava
Danube
Black Sea
Dinaric Alps
Dalmatia
Drina
Brac
Hvar
Korcula
Durmitor 2522
Mljet
Balkan Mountains
Adriatic Sea
L Shkoder
Musala 2925
Maritsa
BULGARIA
Rhodope Mts
L Ohridsko
L Prespa
Pirin
ALBANIA

151

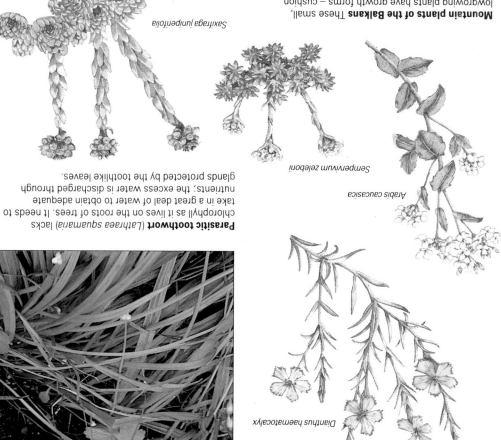

Saxifraga juniperifolia

Mountain plants of the Balkans These small, lowgrowing plants have growth forms – cushion, succulent rosette, downy mat and a tuft of linear leaves – that help them survive in exposed habitats.

Sempervivum zeleborii

Arabis caucasica

Dianthus haematocalyx

Parasitic toothwort (*Lathraea squamaria*) lacks chlorophyll as it lives on the roots of trees. It needs to take in a great deal of water to obtain adequate nutrients; the excess water is discharged through glands protected by the toothlike leaves.

MOUNTAIN AND WETLAND PLANTS

Most of the natural plant life of the flat lowlands in eastern Europe has fallen to the plow and hoe; the plants that have been least disrupted are found on the mountains (though even they have been seriously overgrazed). Many of these plants exhibit adaptations characteristic of mountain plants all over the world, enabling the plants to withstand the freezing temperatures of winter and the cold, drying winds.

On the upper reaches of the dry mountain slopes grows one of several endemic genera, *Edraianthus*. This neat, compact relative of the bellflower (*Campanula*) specializes in growing in rock crevices, where it is shielded from competition from scrub and woodland plants and from the attacks of grazing animals. Its long, stout roots can penetrate deeply along cracks in the rock to obtain moisture and provide a secure anchorage.

The hard, siliceous rocks near the summits of the mountains in northwestern Czechoslovakia support only a few species, including a number of endemic plants such as the shrubby whitebeam-relative *Sorbus sudetica*. The surrounding plants include heather (*Calluna*) and cowberry (*Vaccinium vitis-idaea*), which are widespread in upland areas of Europe where the soil is acidic, and *Sorbus sudetica*, which is an exceptional addition. Its pointed leaves resemble those of some variants of the widespread whitebeam (*Sorbus aria*); the upper surfaces are hairless and glossy, and the undersurface covered with dense grayish-hairs. This is possibly a modification the

plant has developed to reduce the amount of water lost through evaporation.

Many of the high mountain plants have large, attractive flowers, but numerous others, like the rare mountain sedges (*Carex*), have inconspicuous ones. These are adapted not to the visits of insects but to shed their abundant pollen into the air. Insects are scarce in the cold, exposed Arctic lands, and the proportion of wind-pollinated to insect-pollinated flowers generally increases toward the north – in the tundra at least a third of the plant species are pollinated by wind.

Plants without pollinators

Many mountain plants, living in conditions similar to those of the Arctic, have adapted to avoid the uncertainty of pollination and seed ripening in the short growing season. Some of them spread vegetatively; others produce offsets or bulbils. One important group of plants appears to have normal seed production, and may even have showy flowers attractive to insects, but the plants in fact set seed wholly or partly without any sexual fusion (fertilization) taking place. This phenomenon is known as apomixis; it is a complex process not yet fully understood. Common apomictic plants on Europe's mountains include the dandelions (*Taraxacum*), hawkweeds (*Hieracium*) and lady's mantle (*Alchemilla*).

Another apomictic plant is *Sorbus sudetica*, whose small white flowers are succeeded by red berries, as in the whitebeam; these attract birds, which eat the berries and then disperse the seeds. The plant does not require pollinating agents, as it reproduces asexually. This characteristic is common among species of *Sorbus*; it means that the plants are not dependent on insect pollinators; and also ensures that identical offspring are produced. It is likely that *S. sudetica* derived from a chance cross-fertilization between the tree whitebeam and a very different-looking dwarf shrub, *S. chamaemespilus*, which grows mainly in the central European mountains.

Surviving in the wetlands

There are several wetland areas in eastern Europe, including those of the lowlands, the Danube floodplain and its delta. Fens and marshes can prove very difficult places for plants to grow, principally because the ground is frequently water-logged or at least wet. Many plants adapt

by growing on islands of dead vegetation or on the relatively dry margins of these areas. One example is the great fen rag-wort (*Senecio paludosus*), which is still quite widespread in central and eastern Europe where the effects of agriculture have not yet been as extensive as in northwestern Europe. This distinctive plant, whose heads of yellow flowers often blossom above the reeds in high summer, is one of a small set of diminishing wetland rarities.

Another wetland species that is under threat in Europe is the carnivorous aquatic *Aldrovanda vesiculosa*. Little is known about this plant, partly because it is totally submerged when not in flower, and partly because it only blooms if the summer is particularly hot. Like many water plants of cool climates, it normally survives from season to season by producing winter buds (turions). These are little growing points with a foodstore. They detach themselves from the parent plant to spend the winter on the peaty bottom of the shallow lake or pond, and then start to grow into a new plant in spring. The plant traps insects in a similar way to that of its relative, the Venus' fly trap (*Dionaea*), by closing its leaves over them. The digested insect may provide nutrients lacking in the water or the acidic peaty soil. One of the few places in which this plant survives is the Masurian Lakes district in northern Poland.

Dog's-tooth violet (*Erythronium dens-canis*) Like other bulbs, this woodland plant can reproduce from seed and also multiply vegetatively from side buds within the storage organ. By the end of a growing season daughter bulbs will form, and eventually a large clump of plants will develop.

A TREE OF THE PAST – THE HORSE CHESTNUT

Some 20 percent of the plant species of the Balkans do not grow wild anywhere else. The most famous of these are the trees, particularly the horse chestnut (*Aesculus hippocastanum*), which now decorates streets and parkland in many temperate regions. It reaches 36 m (120 ft) and is particularly striking in spring, with its large compound leaves and tall candles of white flowers.

For a long time after the horse chestnut was first widely planted in Europe its origin remained a mystery. It is now known that as a native tree it is restricted to a relatively small area in the Balkan Peninsula (parts of south Yugoslavia, Albania, Bulgaria and northern Greece); it grows in sheltered, often steep, wooded valleys, with another relict tree, the walnut (*Juglans regia*).

The horse chestnut is an outstanding remnant of the rich plant life from the Tertiary period, 65–2 million years ago. Fossil evidence reveals that it was widespread across Europe during this time. Its only related species are all found outside Europe, in southeast Asia and North America.

MEDICINAL AND GARDEN PLANTS

Throughout medieval Europe, many widespread plants were known for their medicinal, culinary and horticultural value. In the industrial urbanized West much of the knowledge of medicinal plants, in particular, has gradually been lost, or survives only in folklore. However, in parts of Eastern Europe, where village traditions are still relatively unbroken and knowledge of the herbs of the countryside is widely disseminated, herbal remedies still survive.

In Bulgaria the gathering of wild plants for medicinal purposes is officially recognized. However, it is a practice that has recently become a cause for concern, as stocks of these plants in the wild are being depleted – particularly in areas where communities still practice large-scale collecting. Botanists have noted that some species are increasingly threatened by overcollecting. Wild gentians (*Gentiana*), to take one example, have been overexploited in several central and eastern European countries. The unusual and longlived species *Gentiana lutea* is particularly vulnerable. It is a tall plant, reaching 1.5 m (5 ft) in height, with yellow flowers growing at intervals up the stem. It contains bitter principles that aid digestion, and the fleshy underground stem is used to produce gentian brandy. Protection must be guaranteed if the plant is to survive.

Traditional plant remedies

Two plants that are particularly well known to herbalists are in fact extremely common perennial weeds that grow in lawns and along roadside verges throughout the region: they are the plantains *Plantago major* and *P. lanceolata*. The medicinal uses, particularly of the larger leaves of the broadleaved great plantain (*P. major*), go back to classical times; they were, and still are, used to ease coughs and digestive disorders, to soothe intestinal inflammation, to cure diarrhoea and as a diuretic.

Two other widespread herbs have specific names that give an indication of their long association with myth and medicine. One is the lady's mantle (*Alchemilla vulgaris*), whose kidney-shaped leaves were thought to hold "magic" drops of rain or dew – much used in medieval alchemical potions. The other is the lungwort (*Pulmonaria officinalis*), a

Spotted lungwort (*above*) A plant of woods and shaded places, which spreads by means of underground stems. The young lower leaves can be eaten in salads or added to cooked dishes. It is also used in herbal medicine.

Roses in Bulgaria (*below*) Between the towns of Klissura and Kazanluk lies the Valley of the Roses where oldfashioned roses, many of European origin, are grown like vines. Their dewy flowers are picked at dawn to produce attar of roses.

Clematis viticella

Syringa vulgaris
common lilac

Lathyrus grandiflorus
everlasting pea

Rosa rubrifolia

Decorative shrubs and climbers
Eastern Europe, particularly the
Balkan Peninsula, has many lovely
flowering plants that have made
their way into gardens. The lilac and
the clematis have been used in
breeding programs to develop
showier hybrids. The everlasting pea
and *Rosa rubrifolia* both grow
vigorously in cultivation.

PLUM BRANDY OF THE BALKANS

Slivovitz or plum brandy is a traditional and potent liqueur of the Balkans. This fiery alcoholic drink takes its name from the Slav word for plum. It is made from the fruits of *Prunus domestica*, a species that probably originated as a hybrid between the European blackthorn or sloe (*P. spinosa*) and the cherry-plum (*P. cerasifera*) of western Asia, whose range of distribution overlaps here in south-eastern Europe.

The fruits of these two species are markedly different. Those of the spiny blackthorn are small, blue-black and with an attractive bloom. Their flesh is bitter, as it contains tannin, responsible for the astringent tang of some plums. In contrast, the cherry-plum has bright-skinned fruits of red or yellow, juicy in texture but bland in taste.

Many of the best plums have been discovered growing wild, and because they freely produce shoots from the roots or lower stem (suckers), it has been an easy matter to cultivate them. This is probably how the old plum variety known as 'Quetsche' developed. It gives its name to the best-known plum brandy of central Europe. Similar old varieties are used in Bulgaria and Yugoslavia, where most of the slivovitz is nowadays made.

spring-flowering plant whose curiously mottled leaves were said to resemble lung tissue and were therefore deemed to be of value in curing diseases of the lung. It is still used in an infusion to treat asthma and catarrh.

The cranesbill, *Geranium macrorrhizum*, is another Bulgarian wild flower long believed to have medicinal properties; the thickened rhizome is still used medicinally. This species of cranesbill is sometimes cultivated in gardens in Western Europe.

Popular garden plants

Two Balkan shrubs of genera familiar in cultivation throughout Europe and North America are the lilac (*Syringa vulgaris*) and *Forsythia*. Lilac, now represented by many hybrids in gardens, was introduced by the Turks through Vienna in the 16th century. The wild shrub grows on dry hillsides in Bulgaria; usually lower growing than the garden varieties, it has the characteristic lilac scent. The spring-flowering shrub *Forsythia europaea* is

found in the wild only in northern Albania and neighboring parts of Yugo-slavia. This European species is smaller and less showy than its Asian relatives, from which the plants widely cultivated in the West are derived.

The thorny path of conservation

In a region where farming is so important, the native plant life suffers considerably. Nearly all the mountains of the Balkans, for example, are or have been heavily overgrazed, and the rare plants that remain cling to relatively inaccessible rock ledges. Attempts are being made to restore the natural plant life, but in so doing some thriving plant communities inevitably suffer.

In a certain area of the Tatra Mountains, for example, all human activities such as logging, haymaking and putting animals out to graze have been prohibited. This has resulted in the rapid disappearance of alpine meadow and pasture plants such as the globe flower (*Trollius*), in the shade of regenerating forests of spruce and

other woody colonizers. This type of natural succession creates a problem for conservationists trying to retain a balance in nature.

Tourism has led to the overpicking of a number of rare species, including lady's slipper orchid (*Cypripedium calceolus*). The Polish national park authorities have effectively reduced the threat to the famous mountain edelweiss (*Leontopodium alpinum*), partly through publicity about this legally protected species, and partly by cultivating and selling bunches of edelweiss to tourists.

The subtle ecological changes brought about by farming practices such as manuring are less easy to control. In the Tatra Mountains the rare white or purple-flowered *Dendranthema zawadskii*, a relative of the fall-flowering chrysanthemums popular in gardens, which grows in rock crevices, is suffering from the changes initiated by manuring; on the other hand, nitrogen-loving opportunists such as the common nettle (*Urtica dioica*) are thriving – often at the expense of the native plants.

The plants of Devinska Kobyla

The Devinska Kobyla Nature Reserve lies in the south of Czechoslovakia near the Slovak capital, Bratislava. Its unique plant life contains many species at the northern and western limits of their distribution. They include 275 species of algae, 132 lichens, 99 mosses, 26 liverworts and a staggering 1,157 species and subspecies of flowering plants. The plants in this reserve have long been studied, with the first records dating from 1583.

The reserve covers about 100 ha (250 acres) and is extremely hilly – it ranges in altitude from about 140 to 500 m (460–1,650 ft). The warm, fairly dry climate supports plants typical of such conditions, but the vegetation is only seminatural, having been disturbed for almost 7,000 years by human settlement and grazing animals. The varied topography and underlying rock types offer a range of plant habitats, the rich diversity of plants being enhanced by species that have been carried from the east and north along the Danube and Vah rivers respectively.

Mixed oak woods and grassland

In a few areas of the reserve the soil is mostly sand; one such place supports several threatened species that grow only on sandy soils, such as *Kochia laniflora*, *Gypsophila paniculata* (used by flower arrangers throughout the temperate world) and *Peucedanum arenarium*.

About two-thirds of the reserve is wooded. A mixture of oak (*Quercus*) and hornbeam (*Carpinus*) is most common, with plants such as the sedge *Carex pilosa* and comfrey (*Symphytum tuberosum*) growing in the understory. The mixed oak woods contain nine of the ten species of oak found in Slovakia, with the downy oak (*Quercus pubescens*) predominating. Typical plants are the shrubs Cornelian cherry (*Cornus mas*), St Lucie's cherry (*Prunus mahaleb*) and ground cherry (*P. fruticosa*); burning bush (*Dictamnus albus*) is also found. There are beech woods on the cooler northern slopes, where they grow at their lowest altitude in the western Carpathians.

Away from the woodland areas, the thin limestone soil supports sub-Mediterranean grassland plants. The dominant grasses are the fescues (*Festuca*) and upright brome (*Bromus erectus*), often with dwarf sedge (*Carex humilis*). Orchids grow in abundance, including toothed orchid (*Orchis tridentata*), early spider

Flowers on a leafless tree (*above*) The Cornelian cherry, a typical shrub of the reserve's woodland, flowers early in the year. The bright red fruits have an acid flavor but are nevertheless edible, and are used to make drinks and preserves.

Feathergrass meadow (*right*) The long, feathery inflorescences blow in the wind, which acts as the pollinating agent for this elegant grass. Such meadows of the thin limestone soils become bright with flowers in the spring.

orchid (*Ophrys sphegodes*), late spider orchid (*Ophrys fuciflora*) and a southern subspecies of the lizard orchid (*Himantoglossum hircinum* subsp. *adriaticum*). There are also two species of pasque flower (*Pulsatilla vulgaris* and *P. pratensis*).

Conserving endangered species

In all, 260 of Slovakia's 900 endangered plant species grow in this area, which was declared a reserve in 1964. In 1986 it was combined with the Sandberg (a paleontological site used for the study of fossils), and in 1988 a small nature trail was opened, with seven information points along its route.

To protect the reserve's botanical diversity, and to prevent the grassland and wooded steppe being succeeded by closed woodland, active management is required. Alien trees such as black pine (*Pinus nigra*) and robinia (*Robinia pseudacacia*) are being removed, and invading grasses such as *Calamagrostis epigejos* are discouraged.

Early spider orchid Despite its name, this orchid is pollinated by bees, which are sexually attracted to the flower. In their attempt to mate with it they collect the pollinia, which are deposited on the stigmatic surface of the next flower they are duped into visiting.

The common reed

This tall swamp perennial is most commonly found in shallow water growing with other aquatics such as cat's tail (*Typha latifolia*) and flowering rush (*Butomus umbellatus*). It is often the dominant species, and spreads by rhizomes to form dense stands. However, it also grows on the edges of mountain lakes, and in wet shallow scrub where it forms part of the herb layer with marsh marigold (*Caltha palustris*), *Iris pseudacorus* and bogbean (*Menyanthes trifoliata*).

Large stands on the water's edge accumulate silt and dead plant debris. In this way land is slowly reclaimed, to be colonized by terrestrial plants, while the reeds spread farther into the water.

The common reed, *Phragmites australis*, displays many characteristics of the grass family (Gramineae). It has a stem circular in cross-section, long narrow leaves with their lower part wrapped round the stem, and a terminal flowerhead made up of spikelets. In the case of the common reed this flowerhead is large and fluffy because of the silky hairs. Like all grasses it is pollinated by the wind, with the pollen-bearing anthers and the feathery stigmas hanging outside a pair of protective bracts. The upright growth form allows light to penetrate down to the lower leaves, which are held at an acute angle to the stem.

Reeds for industry In Romania the large reedbeds of the Danube delta are commercially exploited for use in making paper and chemicals. In other parts of Eastern Europe the common reed has lost many of its original habitats, but has invaded new ones.

A PLANT LIFE OF EXTREME CONTRASTS

TUNDRA, TREES AND GRASSLANDS · ADAPTING TO WATER STRESS · EDIBLE, HEALING AND DECORATIVE PLANTS

The plant life of the Soviet Union and Mongolia has developed in every niche in these vast territories, with their enormous variations in climate and landscape. The Arctic poppy clings to life in the frozen wastelands of the north; twisted and stunted shrubs struggle to exist in the bleak, boggy tundra; and many strange succulents are unique to the burning deserts of the southern Soviet Union and Mongolia. Across the land a belt of conifers and broadleaf trees grow together to form the taiga, the largest forest belt in the world. In the southeast subtropical trees support luxuriant vines and kiwi fruits, while in the south trees give way to great seas of grass, often bright with flowers in spring. The mountain ranges are home to many beautiful, much sought-after flowering shrubs, herbs and alpines.

COUNTRIES IN THE REGION

Mongolia, Union of Soviet Socialist Republics

DIVERSITY (Soviet Union)

Number of families	200
Number of genera	1,865
Number of species	21,000
Endemism	5.5%

PLANTS IN DANGER

Number of threatened species	361
Number of endangered species	167
Number of extinct species	22

Examples *Astragalus tanaiticus; Elytrigia stipifolia; Eremurus korovinii; Fritillaria eduardii; Iris paradoxa; Lilium caucasicum; Potentilla volgarica; Rhododendron fauriei; Scrophularia cretacea; Tulipa kaufmanniana*

USEFUL AND DANGEROUS NATIVE PLANTS

Crop plants *Cichorium intybus* (chicory); *Ficus carica* (fig); *Ribes nigrum* (blackcurrant); *Rubus ulmifolius* (blackberry); *Secale cerealis* (rye)
Garden plants *Bergenia cordifolia; Campanula carpatica; Elaeagnus angustifolia; Hemerocallis flava* (day lily); *Paeonia lactiflora; Tanacetum coccineum*
Poisonous plants *Daphne mezereum; Euphorbia peplus; Ligustrum species* (privet); *Taxus baccata* (yew)

BOTANIC GARDENS

Kiev State University (10,000 taxa); Komarov Botanical Institute (8,500 taxa); Moscow (16,000 taxa); Siberian Central Novosibirsk (9,000 taxa) ·

TUNDRA, TREES AND GRASSLANDS

The vast span of the Soviet Union and Mongolia, from both west to east and north to south, means that the region can support an enormous range of vegetation zones, and encompasses the various extremes of climate that accompany great latitudinal extent.

In the Arctic islands and the northern part of the Taymyr Peninsula the freezing temperatures enable algae, mosses and lichens to survive, along with only about 40 to 50 species of flowering plants. These include dwarf willow (*Salix polaris*) and Arctic poppy (*Papaver polare*), which grow in isolated patches.

Farther south the conditions in the treeless tundras are marginally less harsh. Mosses and lichens predominate, and sphagnum bogs occur in areas where drainage is poor. Dwarf shrubby birches (*Betula*), alders (*Alnus*) and willows (*Salix*) grow densely in places, along with perennial grasses. Scattered stunted trees, such as the ground-hugging Dahurian larch

A bright flower braves the harsh, hostile environment of a rocky gorge in Soviet Central Asia, in the south of the Soviet Union. This part of the country is largely dominated by desert and semidesert; in such an enormous country there are many favorable environments for plants as well as difficult ones.

(*Larix gmelinii*) and the crooked birch (*Betula tortosa*), form forest tundra.

Forest belts and mountain plants

In a wide belt south of the tundra is the taiga forest, which covers more than 50 percent of the Soviet Union. To the west the taiga is dominated by spruces (*Picea abies* and *P. obovata*), Scots pine (*Pinus sylvestris*), Siberian cedar (*Pinus sibirica*) and Siberian larch (*Larix sibirica*). The northern twinflower (*Linnaea borealis*) and the saprophytic ghost orchid (*Epipogium aphyllum*), which lives on decomposing matter, inhabit the forest floor, together with many mosses and lichens. In contrast, the eastern Siberian forest consists mostly of deciduous trees, in particular Siberian and Dahurian larch. The trees of the far east are mainly evergreen and similar to those of the Eurosiberian taiga, but with different species of spruce and fir.

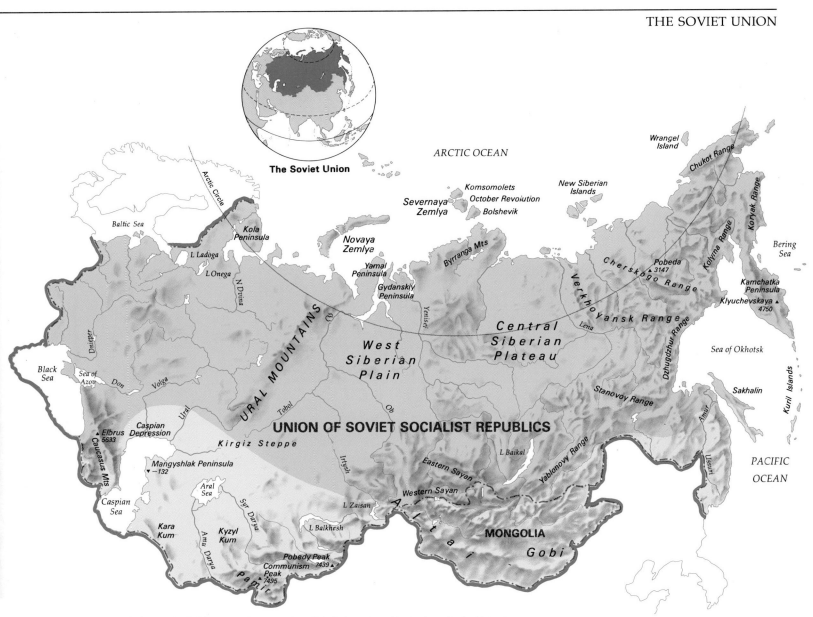

The Soviet Union

ARCTIC OCEAN

Wrangel Island
Chukot Range
Komsomolets
October Revolution
Severnaya Zemlya
Bolshevik
New Siberian Islands
Koryak Range
Baltic Sea
Kola Peninsula
Kolyma Range
Bering Sea
L. Ladoga
L. Onega
Novaya Zemlya
Cherskogo Range
Pobeda 3147
Kamchatka Peninsula
Klyuchevskaya 4750
N Dvina
Yamal Peninsula
Gydanskiy Peninsula
Yenisei
Verkhoyansk Range
Dzhugdzhur Range
Dnieper
URAL MOUNTAINS
West Siberian Plain
Central Siberian Plateau
Lena
Sea of Okhotsk
Black Sea
Sea of Azov
Don
Volga
Ob
Stanovoy Range
Sakhalin
Tobol
Amur
Caspian Depression
Elbrus 5633
UNION OF SOVIET SOCIALIST REPUBLICS
Kuril Islands
Caucasus Mts
Ural
Kirgiz Steppe
Irtysh
L Baikal
Ussuri
PACIFIC OCEAN
Mangyshlak Peninsula –132
Eastern Sayan
Yablonovy Range
Caspian Sea
Aral Sea
Western Sayan
L Zaisan
Syr Darya
L Balkhesh
Altai
Kara Kum
Kyzyl Kum
Amu Darya
MONGOLIA
Gobi
Pobedy Peak 7439
Communism Peak 7495
Pamir

Arctic Circle

Byrranga Mts

Birch forests and the rich field layer that grows beneath them characterize the Pacific forest-meadow of the Kamchatka Peninsula; in the south, the mixed forest contains such rarities as Korean pine (*Pinus koraiensis*) and the Japanese yew (*Taxus cuspidata*). The understory has a number of endemic shrubs such as the Amur lilac (*Syringa amurensis*) and a mock orange (*Philadelphus schrenkii*). Attractive lianas abound; they include wild kiwi fruits (*Actinidia*) and the fragrant-flowered *Schisandra chinensis*.

The various mountain ranges have a distinct plant life of their own. The lower slopes of the Caucasus are covered with oak-beech forest, Caucasian hornbeam (*Carpinus caucasica*), maple (*Acer*), elm (*Ulmus*) and lime (*Tilia*), with an under-story of *Rhododendron luteum*. The upper slopes are cloaked with conifers, mostly Caucasian fir (*Abies nordmanniana*), occasionally mixed with oriental spruce (*Picea orientalis*). Birch groves and thickets of *Rhododendron caucasicum* and *Juniperus hemisphaerica* grow at the treeline. The subalpine and alpine meadows beyond the treeline are home to some beautiful

Map of floristic regions Much of the Soviet Union is included in the Circumboreal region, the largest floristic region in the world. The desert areas in the south and the humid southeast each contain a quite different plant life.

endemic species of *Gentiana*, *Campanula* and *Primula*.

The high desert plateau of the eastern Pamir, in contrast, is dominated by scrubby *Artemisia pamirica* and *Ceratoides papposa*, with cushions of prickly thrift (*Acantholimon*) and thistly *Cousinia* also widely spread.

Steppe and desert

The steppes, which stretch from the river Danube in central Europe to the Ussuri in the far east of the Soviet Union, are dominated by drought-resistant grasses such as feather grass (*Stipa*). In the north, tall grasses and colorful flowers such as *Filipendula vulgaris*, *Salvia pratensis* and *Scorzonera purpurea* form the northern prairies and surround occasional islands of broadleaf trees. In the drier south, shorter grasses predominate, together with drought-resistant herbs such as sagebrushes (*Artemisia* and *Seriphidium*).

Floristic regions

Holarctic Kingdom

Circumboreal Region Cool northern temperate zone; here distinctive genera include cabbage-relatives, primula-relatives and the valerian *Pseudobetckea*.

Irano-Turanian Region Reflects in its plants the aridity of Central Asia; examples include many genera of saltbush, cabbage-relatives and bellflowers.

Eastern Asiatic Region Northeastern extremity of large region with humid climate, here containing heathlike and roselike genera, cucumber-relatives and skimmias.

In the dry deserts of the southern Soviet Union and Mongolia the plant life is sparse, though members of the goosefoot family (Chenopodiaceae) are widespread. Clay deserts support species of sagebrush or wormwood, whereas the sandy deserts of Soviet Central Asia are home to saxaul (*Arthrophytum*) and *Calligonum*. Here, plants such as the sedge *Carex physodes* and the grass *Stipagrostis pennata* bind the sand dunes, and giant fennels (*Ferula*) and a wild rhubarb (*Rheum turkestanicum*) colonize the lower slopes. These deserts contain some unique endemics, such as *Borszscowia aralocaspica* and *Alexandra lehmannii*, each so distinctive that it is the only species in its genus.

ADAPTING TO WATER STRESS

The two largest vegetation zones in the region are the taiga and the steppe. They have very different climates, but both are affected by a lack of water at certain times of year; in the taiga the water is frozen, while the steppe suffers from drought. The plants have adapted in different ways to survive this difficulty.

Living through low temperatures
The conifers are the dominant plants of the taiga because they have successfully adapted to very low winter temperatures and the consequent shortage of water. In contrast to that of many broadleaf species, the structure of their wood enables them to store water for use in times when groundwater is unavailable. This allows the trees quickly to replenish parts such as leaves that lose water through evaporation. The risk of water stress is greatest during early spring, when the sap begins to rise in response to the increased sunlight and warm air temperature, but the roots are still unable to take in water from the frozen soil.

The leaves of the evergreen conifers are also modified to reduce water loss by desiccation; they are small in size, have

Succulent glasswort of the salt steppe The branched stems and fleshy leaf "segments" are capable of absorbing moisture from the air, and even their inflorescence is fleshy to help cope with the permanent water stress.

sunken ventilating pores (stomata) and thick, protective, waxy cuticles.

Some trees, such as the Siberian larch, avoid damage from freezing by lowering the temperature of the water in their cells to at least $-70°C$ ($-95°F$) without it freezing. They can do this by increasing the amount of unsaturated fats, sugars, sugar alcohols and membrane proteins in their stored water, which lowers the freezing point. These and other changes in the leaf and bud structure take place just before and during the winter; this enables trees such as the Dahurian larch to survive as far north as 72° 50'.

The Dahurian larch additionally has a superficial root system that is specially adapted to grow in permanently frozen ground (permafrost), which melts in summer to a depth of only 50–100 cm (20–40 in). This root system is supplemented by adventitious roots that form above the base of the trunk. Individual larches can survive in the growing peat bogs for 300 to 400 years, whereas pines, which do not have adventitious roots, are suffocated by the moss growth after 50 to 70 years.

Survival in the steppe
The desert steppes of the eastern Pamir, the Altai, central Tien Shan and Mongolia are covered with needlegrass (drought-resistant species of *Stipa*) and needlegrass growing with subshrubs. North of the Gobi, Mongolian needlegrass and onion species (*Allium*) are common. They have all developed adaptations that enable them to reduce or tolerate the effects of drought, and so are able to survive and reproduce successfully.

On the Kirgiz Steppe the plants avoid water stress by completing their cycle of growth and reproduction rapidly, while water is available. These plants are ephemeroids – perennials that spend the dry season underground as storage organs such as bulbs or tubers – and ephemerals, which are annuals that spend the dry season as dormant seeds. Most of them flourish in spring, such as tulip

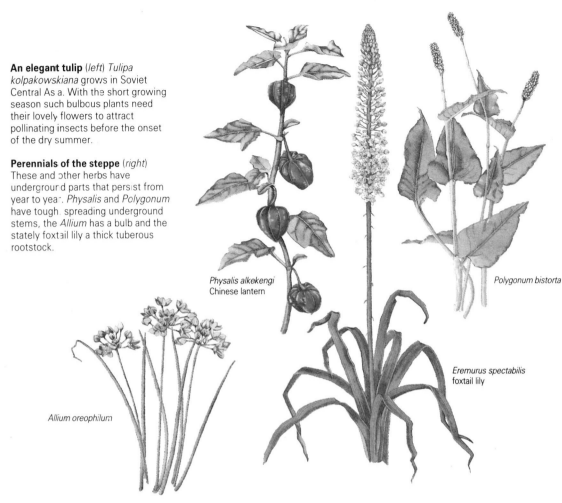

An elegant tulip (*left*) *Tulipa kolpakowskiana* grows in Soviet Central Asia. With the short growing season such bulbous plants need their lovely flowers to attract pollinating insects before the onset of the dry summer.

Perennials of the steppe (*right*) These and other herbs have underground parts that persist from year to year. *Physalis* and *Polygonum* have tough, spreading underground stems, the *Allium* has a bulb and the stately foxtail lily a thick tuberous rootstock.

Physalis alkekengi
Chinese lantern

Polygonum bistorta

Eremurus spectabilis
foxtail lily

Allium oreophilum

species (*Tulipa*), the grass *Poa bulbosa*, and *Valeriana tuberosa*; a few plants bloom in the fall, such as the autumn crocus (*Colchicum laetum*).

For those plants that live above ground during the dry season there are two main adaptations to overcome water stress. One involves taking in the maximum amount of water possible at times when it is available, the other the efficient use and conservation of the water that has been acquired. To take in water rapidly the plants need to have deep, penetrating root systems or abundant but shallow root systems, rapid root growth in the seedling and at the start of the growing season, and the ability to absorb what little water there is in fairly dry soils.

To conserve water some plants have a reduced leaf area, an increased ratio of green photosynthetic tissue to leaf area, or shed their leaves at the onset of drought. Plants that retain their leaves display other protective features such as sunken leaf pores (stomata), a thick, protective cuticle covering the leaf, a coating of wax on the leaf surface, hairs on both the leaf and the stem surface, inwardrolling leaves, and leaves that are positioned at an acute angle to the stem – all of which help to reduce water loss from the leaves through transpiration.

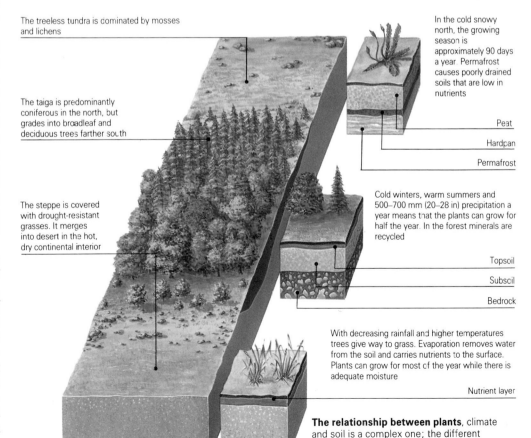

The treeless tundra is dominated by mosses and lichens

In the cold snowy north, the growing season is approximately 90 days a year. Permafrost causes poorly drained soils that are low in nutrients

The taiga is predominantly coniferous in the north, but grades into broadleaf and deciduous trees farther south

Peat

Hardpan

Permafrost

The steppe is covered with drought-resistant grasses. It merges into desert in the hot, dry continental interior

Cold winters, warm summers and 500–700 mm (20–28 in) precipitation a year means that the plants can grow for half the year. In the forest minerals are recycled

Topsoil

Subsoil

Bedrock

With decreasing rainfall and higher temperatures trees give way to grass. Evaporation removes water from the soil and carries nutrients to the surface. Plants can grow for most of the year while there is adequate moisture

Nutrient layer

The relationship between plants, climate and soil is a complex one; the different latitudinal zones of the Soviet Union provide a number of examples.

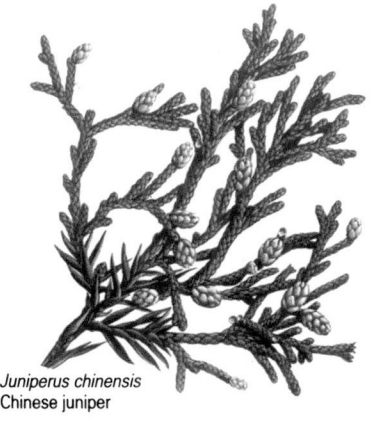

Larix decidua
European larch

Abies nordmanniana
Caucasian fir

Tilia mongolica
Mongolian lime

Juniperus chinensis
Chinese juniper

Zelkova carpinifolia

Larix gmelini
Dahurian larch

Trees of the taiga (*above*) The Caucasian fir, *Larix decidua* and *Zelkova* grow in western regions, the others are found to the east as their specific names suggest. Trees for timber and pulp are one of the Soviet Union's most important commodities.

Siberian iris (*below*) *Iris sibirica* is an elegant iris of streamsides and moist places; it is a variable species in the wild, and forms with white flowers instead of blue are sometimes to be found. It is cultivated in gardens and as a cut flower.

EDIBLE, HEALING AND DECORATIVE PLANTS

The very diverse plant communities of the region have always played a significant role in the life of the people. The steppes, meadows, deserts and tundra form a natural pasture and source of hay for livestock, and the forests are an extremely valuable source of timber. Wild plants also have innumerable uses, and are exploited both in medicine and as foods. In the Caucasus, for example, plants are the source of rodenticides and insecticides, vitamins, essential and industrial oils, soda and potash, gums and resins, rubber and gutta-percha (a latex obtained from trees), fibers and foaming agents; they have also been used in the manufacture of cheese, papermaking, wickerwork, tanning and metallurgy.

Culinary and medicinal plants
Many wild plants of the Soviet Union and Mongolia are of culinary importance. In the forests, wild fruits, nuts and edible fungi are picked and eaten throughout the region. The seeds of the Siberian cedar, known as cedar nuts, can be eaten both raw and cooked, and also yield an

PLANNED CONSERVATION

The long and intensive use of the natural vegetation for grazing, settlement, fuel, agriculture and industry has inevitably and dramatically changed or destroyed much of the plant life of the region. Most of the Kirgiz Steppe, for example, is now under the plow for cereal cultivation. For these reasons active conservation measures to protect the plants of the Soviet Union and Mongolia, the resources they represent and the environmental stability they engender, are an urgent priority.

Successful environmental conservation requires legal and political backing to restrict activities that degrade or pollute the environment. In a centrally planned economy such measures ought to be fairly straightforward to implement. In practice, ideological priorities

and ecologically uneducated central planning personnel have produced a number of major environmental disasters; the drying up of the Aral Sea and the destructive pollution of Lake Baikal are just two examples.

There have been some positive achievements as well, including the establishment in the Soviet Union of a network of protected areas such as national parks, sanctuaries and nature reserves, of which there are now more than 150. They incorporate many of the plant communities of the Soviet Union, and include 38 forest reserves, 11 Far East reserves – including the exceptionally species-rich forests of the Ussuri – 50 mountain reserves in the Caucasus and Central Asia, and 26 steppe and desert reserves.

edible oil. Important food plants of the southern mountains include the walnut (*Juglans regia*) and the pistachio (*Pistacia vera*), which is also a source of resin. In Mongolia, wild plants that are used include *Agriophyllum pungens*, which provides a cereal grain; wild rhubarb (*Rheum compactum*) is used as a vegetable, starch is extracted from the stems of the fleshy parasite *Cynomorium soongaricum*, and berries are collected from *Vaccinium*, *Rubus* and *Ribes* species.

The most important medicinal plants of the Soviet Union and Mongolia are species of the liquorice genus *Glycirrhiza*, especially *G. glabra* and *G. uralensis*. They are used in various forms as an expectorant (to relieve catarrh), and as an antiirritant and relaxant in the treatment of gastritis and stomach ulcers. In mixtures they are administered as a diuretic and laxative, and in powder form they are added to other medicines to improve their taste and smell.

A few more of the many hundreds of species with medicinal uses include *Ephedra equisetina*, a shrub from montane steppe that contains alkaloids such as ephedrine, which are used in the treatment of bronchial asthma, hayfever and low blood pressure. The daisy-flowered *Leuzea carthamoides*, a perennial herb endemic to the mountains of southern Siberia, is used as a stimulant in the treatment of nervous disorders and physical and mental exhaustion. *Ahnfeltia plicata*, a red seaweed harvested from the White Sea coasts, is a source of agar, which is used to treat chronic constipation, and also as a base for slowly dissolving tablets. *Sphaerophysa salsula*, a perennial herb of deserts and semideserts, has a long rhizome that yields alkaloids given to stimulate contractions of the uterus during childbirth.

Gifts to gardens

Many wild plants in the region are popular with gardeners. Some of the most decorative of all come from the Caucasus Mountains in the west. Here are found a great number of ornamental plants such as the Caucasian fir; *Zelkova carpinifolia*, a hardy deciduous tree of the elm family; *Daphne pontica*, a hardy evergreen shrub with fragrant yellow flowers; and *Elaeagnus angustifolius*, a decorative shrub with ornamental silvery white and yellow foliage. Ornamental grasses also grow in the mountains, such as *Erianthus purpurascens*, with plumes similar to those of pampas grass, and the feathergrass *Stipa pulcherrima*, whose flowerheads are dried for winter decoration.

The herbaceous perennials of the Caucasus Mountains include the gentian *Gentiana septemfida*, the shell-flower *Molucella laevis*, with its unusual green bracts, cornflowers (*Centaurea*) of many colors, large-flowered yellow *Paeonia* species and the giant hogweed *Heracleum mantegazzianum*, whose sap can cause a painful skin rash that is slow to heal. Other common perennials include red pyrethrum *Tanacetum coccineum*, which is now represented in gardens by numerous cultivars, and the bold daisy-flowered *Inula magnifica*, *Telekia speciosa* and *Doronicum macrophyllum*.

The bulbous plants of the Caucasus are very colorful; they include the lovely *Onocyclus* irises (of which *Iris reticulata* is perhaps the best known), tulip species such as the scarlet and blue-black *Tulipa eichleri*, and the autumn crocuses *Merendera trigyna* and *Colchicum speciosum*. From high altitudes come many rock garden and alpine plants: stonecrops such as *Sedum*, *Sempervivum* and *Rosularia*; and *Draba*, *Saxifraga*, *Androsace*, *Campanula* and the prickly thrift (*Acantholimon*).

Desert plants of the Gobi

The Gobi ("desert" in Mongolian) extends across the south of Mongolia. It is an extreme environment, combining arid, salty soils with temperatures that rise and fall dramatically from day to night and throughout the year. In the cold, dry, cloudless winters, conditions reach almost the lowest limits at which plants can survive. Much of the Gobi is stony desert, in places devoid of plants or with sparse vegetation cover. The number of plant species is small – there are only about 600 species of higher plants, of which few are dominant, and most have had to adapt to resist drought.

Drought-tolerant plants of the Gobi

The drought-tolerant plants (xerophytes) of the Gobi differ from those of hot deserts; in some respects, they combine the adaptations of mountain and cold zone plants with those of mountain and arid zones. Their productivity is comparatively lower than that of hot desert plants, and they are dormant for longer. The plants of the southwestern (Transaltai) Gobi are typical examples.

The southwestern Gobi is a sandy desert in which the plants are either sclerophytes or succulents. The sclerophytes are woody plants with small, thick leaves (or grasses with narrow, often inrolled leaves). These have a thick outer covering (cuticle), and a high proportion of cell wall in which water can be stored. Species include a wild rhubarb, *Rheum nanum*, the ornamental shrub *Caryopteris mongolica*, many species of milkweeds such as *Astragalus pavlovii*, and daisy-flowered plants such as *Ajania fruticulosa* and *Artemisia scoparia*. There are also grasses such as *Stipa gobica* and *Cleistogenes songorica*.

Unlike plants of average moist conditions, sclerophytes have compact photosynthetic tissue (mesophyll) with less space between the cells, but like them (except for some of the grasses), they photosynthesize in the usual way.

There are two types of succulents. Some have a low surface area, and a high proportion of water-storing tissue to reduce water loss through transpiration. They have evolved a specialized method of photosynthesizing (the photosynthetic tissues are differentiated into two types of cell), and they manufacture food by a process called the Crassulacean Acid Metabolism (CAM). The pores open at night (instead of during the day as those

In a dry and salty desert (*above*) *Anabasis brevifolium* belongs to the goosefoot family (Chenopodiaceae). Chenopods are predominantly herbaceous, and are adapted to live in saline soils. Many species, this plant included, exhibit xerophytic adaptations.

Tamarisk in the Gobi (*right*) *Tamarix ramosissima*, like its relatives in northern Africa and Europe, is salt tolerant and survives in dry and stony places. It is frequently cultivated in Californian gardens in the United States, where it forms an attractive small tree.

of other plants do). Through the pores the plant takes in and stores carbon dioxide, which is then released into the cells during the day when the plant is photosynthesizing. These plants include *Nitraria sphaerocarpa*, *Sympegma regelii* and *Reamuria soongorica*.

The second type of succulent is even more highly adapted to drought conditions. These are known as Kranz-syndrome succulents (their compact photosynthetic tissue is differentiated into three types of cell), and they have a special photosynthetic metabolism (C_4), which is more efficient in high temperatures and intense sunlight. Typical species include *Iljinia regelii*, *Haloxylon ammodendron*, *Anabasis brevifolia* and *Salsola arbuscula*. Some non-succulent plants, such as the grasses *Cleistogenes songorica* and *Enneapogon borealis*, also have C_4 photosynthesis.

Some plants survive the drought by having no leaves, such as the sclerophyte *Ephedra*, succulents with normal photosynthesis such as *Salicornia*, and Kranz-syndrome succulents such as *Haloxylon*, *Calligonum* and *Anabasis*.

These examples all demonstrate the extraordinary ingenuity of plants in finding ways to survive in the most inhospitable of environments.

THE FLOWERING DESERT

LUSH FOREST AND DESERT SCRUB · DROUGHT AND PREDATORS · FOOD, FUEL AND ENDANGERED BULBS

The plants of the Middle East have had to adapt to a land of extremes. They are fascinating and, at times, exceptionally beautiful. Their distribution is determined to a great extent by climate. Lush forests thrive in the northern coastal parts of the region, where the climate is mild and moist. In the hotter, drier south, where vast tracts of parched desert extend across the Arabian Peninsula, there may be no plants at all. Those that do live in these harsh conditions have developed highly specialized adaptations. Between these two extremes the plant life includes the seasonally flooded lowland forests of the Tigris–Euphrates basin, Mediterranean scrub, mangrove swamps, the shrubs and grasses of the central Asian steppes, heath and alpine plants on the mountains, and the spiny endemics of the island of Socotra.

COUNTRIES IN THE REGION

Afghanistan, Bahrain, Iran, Iraq, Israel, Jordan, Kuwait, Lebanon, Oman, Qatar, Saudi Arabia, Syria, Turkey, United Arab Emirates, Yemen

EXAMPLES OF DIVERSITY

	Number of species	Endemism
Afghanistan	3,000	25%
Arabian Peninsula (Oman, Saudi Arabia, UAE, Yemen)	4,500	less than 10%
Iran	7,000	20%
Iraq, Bahrain and Kuwait	2,937	6.5%
Israel, Jordan, Lebanon and Syria	3,500	11%
Turkey	8,000	25%

PLANTS IN DANGER

Little information available; at least 63 plants threatened and 7 endangered

Examples Alkanna macrophylla; Anthemis brachycarpa; Ceratonia oreothauma subsp. oreothauma; Dionysia mira; Erodium subintegrifolium; Ferulago longistylis; Iris calcarea; Iris lortetii; Rumex rothschildianus; Wissmannia carinensis

USEFUL AND DANGEROUS NATIVE PLANTS

Crop plants Cicer arietinum (chick pea); Ficus carica (fig); Lens culinaris (lentil); Olea europaea (olive); Phoenix dactylifera (date palm); Pisum sativum (pea); Triticum aestivum (wheat); Vitis vinifera (grape)
Garden plants Crocus kotschyanus; Cupressus sempervivum; Cyclamen persicum; Clematis flammula; Gladiolus byzantinus
Poisonous plants Heracleum mantegazzianum (giant hogweed); Narcissus species; Papaver somniferum (opium poppy)

BOTANIC GARDENS

Istanbul University (6,000 taxa); Tehran University; Tel Aviv University (2,000 taxa); Zaarfaniyah Arboretum, Baghdad

Stately crown imperials Fritillaria imperialis grows among the rocks and scrub of the maquis that covers much of the north of the region. Like many other bulbous plants, it flowers in spring after the winter rain, then dies down to spend the hot summer, dormant, underground.

LUSH FOREST AND DESERT SCRUB

Two factors dictate the distribution of plants in the Middle East – temperature and moisture. The mildest climates are found along the coastlines of the Mediterranean, which is characterized by mild wet winters and hot dry summers, and the Black and Caspian seas, which receive rainfall of up to 2,500 mm (100 in) a year.

In the Mediterranean region most plant growth takes place in the cooler months between October and March. However, trees and shrubs do survive the summer, the intensity of the summer heat being tempered by the proximity of the sea.

Originally, much of this area was covered by low forests dominated by pines such as *Pinus brutia*, oaks such as *Quercus calliprinos*, and a range of other small trees distinguished by leathery, evergreen foliage. Where people have cleared the forest, a secondary vegetation of dense, tall, evergreen scrubland (maquis) has developed. But where the land is subjected to intense burning and grazing, an open, low, thorn scrub (garigue) has formed. Both these habitats support a rich array of colorful spring- and fall-flowering plants, including annuals and

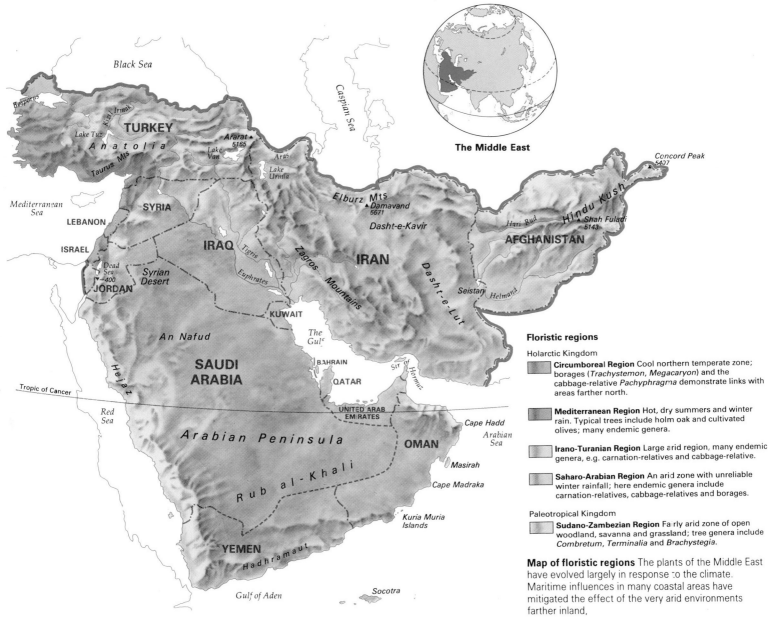

The Middle East

Floristic regions

Holarctic Kingdom

Circumboreal Region Cool northern temperate zone; borages (*Trachystemon*, *Megacaryon*) and the cabbage-relative *Pachyphragma* demonstrate links with areas farther north.

Mediterranean Region Hot, dry summers and winter rain. Typical trees include holm oak and cultivated olives; many endemic genera.

Irano-Turanian Region Large arid region, many endemic genera, e.g. carnation-relatives and cabbage-relative.

Saharo-Arabian Region An arid zone with unreliable winter rainfall; here endemic genera include carnation-relatives, cabbage-relatives and borages.

Paleotropical Kingdom

Sudano-Zambezian Region Fairly arid zone of open woodland, savanna and grassland; tree genera include *Combretum*, *Terminalia* and *Brachystegia*.

Map of floristic regions The plants of the Middle East have evolved largely in response to the climate. Maritime influences in many coastal areas have mitigated the effect of the very arid environments farther inland.

bulbous species such as *Fritillaria*, *Cyclamen* and ground orchids.

The plants that grow in the moist mountain ranges adjacent to the Black Sea and Caspian Sea coasts are typical of eastern Europe. The steep slopes of the mountains of northeastern Turkey are clothed with mixed forest dominated by oriental beech (*Fagus orientalis*), oaks (*Quercus*), hornbeam (*Carpinus*) and a range of firs (*Abies*) and spruces (*Picea*). Above them grow scrubby plants of the heath family Ericaceae, including species of *Rhododendron* and *Vaccinium*, which give way at the highest altitudes to subalpine pasture.

Plants of the steppes and desert

Away from the coasts and the moderating effects of the sea, the climate becomes progressively more extreme. In eastern Turkey, northern Iran and Afghanistan, temperatures soar in summer, when rainfall is virtually nonexistent, and then plunge again in winter. Only dwarf shrubs and grasses grow well here, protected by a blanket of snow in winter. This classic steppe vegetation extends into the heart of central Asia. Despite the extremes of the climate, the land supports a wealth of species, such as knapweeds (*Centaurea*) of which there are 200 species, milk vetches (*Astragalus*) with 900 species, and garlics and onions (*Allium*) with 150 species. Many of the species that grow here are endemic to the region.

Farther south lie the deserts and semideserts that cover much of the southern half of the Middle East. In Saudi Arabia the great sand seas of the "Empty Quarter" form a vast desert region. To survive here, plants must tolerate some of the highest temperatures in the world, extremely dry air, very low levels of rainfall and highly saline soil.

Coarse tufted grasses and dwarf drought-tolerant shrubs predominate, together with some annuals, but they grow only sparsely. No particular groups prevail and the overall diversity of species is very low: there are just 3,500 species in Saudi Arabia, whereas Turkey, which is three times smaller, and Israel, 116 times smaller, contain 8,000 and 2,317 species respectively.

Further diversity

Other types of plants bring greater variety to the region. For example, the Tigris–Euphrates lowlands still support seasonally flooded "ahrash" forests of tamarisk (*Tamarix*), poplar (*Populus*) and willow (*Salix*), whereas the *Acacia* bushland and mangrove swamps in western Yemen are more typical of Africa.

Isolated from Africa and Asia, the island of Socotra supports a unique plant life. Of the 680 species, 216 are endemic; more than a third of these are in danger of extinction, including the prickly *Lasiocorys speculifolia*, once so abundant that it covered much of the plains.

DEFENSES AGAINST DROUGHT AND PREDATORS

Specific adaptations are necessary for plants to survive in the Middle East, where there are some of the harshest environments on Earth. Most of the adaptations concern water retrieval, water retention or reducing water loss.

One obvious adaptation is the development of a deep root system that enables plants to tap water from the damp soil of lower levels. In the Arabian and Iraqi deserts, spreading mats of *Citrullus colocynthis*, a plant that is related to the water melon, remain green throughout the summer, and bear fruit the size of grapefruits. The leaves are poorly adapted to the harsh conditions, but the plant has developed extraordinarily long roots that enable it to survive.

The shrubby wormwood (*Artemisia herba-alba*), common throughout the steppes, adopts a different strategy. Its cell sap has very strong osmotic pressure that permits the plant to suck otherwise unavailable water from the driest of soils. Other plants have developed strategies to minimize water loss through transpiration (such plants are called xerophytes). In the driest areas many shrubs and trees have evergreen foliage that is leathery, with a very thick skin, and with very little sap. These plants, which are called scleromorphs, are characteristic of all Mediterranean-type plant life. They include the olive (*Olea*), myrtle (*Myrtus*) and oleander (*Nerium oleander*).

The most effective protection against an excessive loss of water is to have a greatly reduced surface area from which moisture is released. Many conifers have leaves that are reduced to needles, while other genera have evolved more bizarre adaptations. The spear grasses (*Stipa*), for example, an important forage plant of the steppes, have narrow leaves that lie flat during wet weather, but roll inward as the air becomes drier; this protects both the lower surface of the leaf, through which water loss occurs, and the green tissue on the upper surface. They also have seeds with a tail-like awn or bristle, which is coiled like a corkscrew. This twists and untwists as the humidity of the air falls and rises, helping the seed to free itself if it becomes caught up in vegetation, and to drive itself into the soil.

Certain plants have dispensed with

Although scattered across much of the northern hemisphere, roses are most diverse in central Asia and China, ranging from desert dwarfs just 50 cm (20 in) high to giants such as *Rosa gigantea*, which reaches 30 m (98 ft) tall. Roses are opportunists by nature. Distributed by birds, their seeds rapidly take hold wherever ground has been disturbed or vegetation opened up, aided by their rapid scrambling growth and protected by their fierce spines.

The 150 wild species bear single, five-petaled blooms. However, for over 2,500 years horticulturists have bred selected forms, and modern garden roses bear little resemblance to their wild ancestors. Among the old roses are the fragrant Damasks, reputed to have been brought to Europe from Damascus in Syria by medieval crusaders (though Roman texts suggest that they were well established in the Mediterranean much earlier than this). The Damasks, and other ancient garden roses such as the Gallicas and Chinas, in shades of red, pink and white, were the forerunners of the modern hybrids. *Rosa foetida*, and its varieties such as Austrian copper brier R. *foetida* var. *bicolor*) and Persian yellow (R. *foetida* var. *persica*), contributed tones of orange and clear bright yellow respectively.

A lone survivor (*left*) Normally lowgrowing, *Convolvulus lineatus* stands tall, much of its long taproot exposed by the wind, which has blown away the desert sand. The root penetrates deeply in its search for water.

Milkweed rosette (*above*) The ground-hugging growth form of milkweeds offers some protection against grazing animals. The dead midribs of the leaves become hard and sharp, and offer further protection by forming a cage around the short-stemmed flowers.

leaves altogether, relying instead on green stems to produce energy through the process of photosynthesis. Spanish broom (*Spartium junceum*), a native of the Mediterranean countries, is such a plant, a shrub with a mass of slender, almost leafless, whiplike green stems, and bright yellow, scented flowers in spring.

Annuals and bulbs

Some plants use the strategy of dormancy to survive periods when growth is impossible. The most notable of these are the annual and bulbous species that burst into colorful blooms in spring – a distinctive feature of the Middle East.

Annual plants complete their life cycle before the onset of summer. They survive the drought as seeds, and germinate rapidly when the fall rains moisten the soil. Generally poor competitors against coarse perennials, they thrive only in open grassy patches and disturbed soil where the intense drought inhibits the growth of perennials.

Plowed land provides an ideal habitat for opportunist annual weeds. In the absence of herbicides, cornfield species such as the cornflower (*Centaurea*), pheasant's eye (*Narcissus*) and love-in-a-mist (*Nigella damascena*) are common.

In perennial species, water storage organs such as the bulb, rhizome, corm and tuber enable the plants to lie dormant during the hot dry period. Many showy monocotyledons on the steppes and in Mediterranean habitats adopt this strategy. They include such genera as the garlics and onions, crocuses, fritillaries, bulbous irises, grape hyacinths (*Muscari*) and tulips (*Tulipa*).

Swollen underground organs are relatively rare in dicotyledons (the group of flowering plants that includes the woody plants); of these the region supports the cormous buttercup (*Ranunculus asiaticus*), cyclamen, members of the primrose family that display a wide diversity in Turkey, and even a range of grasses that includes the bulbous meadow-grasses *Poa bulbosa*

and *P. sinaica*. These two perennial grasses dominate vast tracts of both the desert and the steppes, growing rapidly from congested mounds of bulbs. Their growth point is at ground level, which means that they can regenerate rapidly after grazing. Livestock also help to spread the plant by disturbing the bulbs.

Warding off predators

In a land where plant growth is sparse and slow, plants cannot afford to suffer high levels of predation. The garlics and onions, for example, have developed strong flavors to discourage grazing.

Other plants exhibit physical defenses. The milkweeds are one of the most successful genera of this region. They owe their survival to the horrific spines that comprise the hardened, woody mid-ribs of their pinnate leaves. The desert species *Astragalus spinosus* is possibly even more ferocious: this plant is so well armed that it is often the only perennial that remains in the most overgrazed desert districts.

FOOD, FUEL AND ENDANGERED BULBS

For thousands of years people's basic demands for food, fuel and raw materials have fashioned their environments. In some places the face of the Earth has been irrevocably changed as a result.

Changing the landscape

In the Middle East, many of these changes took place long ago. Early civilizations were established in the Fertile Crescent – the plains and valleys drained by the Tigris and Euphrates rivers in what is now Iraq. About 10,000 years ago, when early farmers successfully domesticated plants, they began to clear and alter the natural vegetation. The effects were far-reaching: once damaged, the vegetation failed to reassert itself because of the harshness of the environment.

Today few "natural" landscapes remain. Some of the least changed areas in the region are the mountains of Turkey, Iran and Afghanistan, the salt flats of Iran, and the deserts of Arabia. It is not surprising that where the climate is least harsh and the plant life most rich, people have altered the environment the most.

The first major changes took place from about 6000 BC, in the foothills of the major mountain ranges, when communities cleared small areas for crops. Forest regeneration was weak because grazing animals hindered the growth of seedlings. In time, the forests were further reduced to provide fuel and wood for construction. The cedars of Lebanon were once found all along the eastern shores of the Mediterranean and in southern Turkey; they now occupy only a tiny proportion of their former range, having been prized since biblical times in many countries for their timber.

Few land management practices in the Middle East have enhanced the natural environment, and much of the human influence on individual plant species has also been harmful. On the other hand, certain species owe their current wide distribution to the activities of the farmers and gardeners of the past 6,000 years. Date palms (*Phoenix dactylifera*) are as much a feature of the oases of Iraq and Arabia as they are of their native home, south of the Sahara. Likewise, the fig (*Ficus carica*), now well naturalized in many parts of the Middle East, probably originated in the fertile areas of southern Arabia. Ornamental plants from this region have also been cultivated and distributed for so long that their origins can no longer be traced. The madonna lily

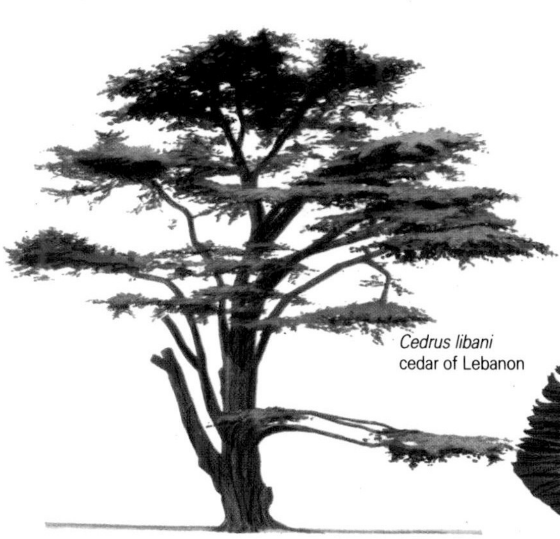

Notable trees of the Middle East Under pressure in its native land, the stately cedar of Lebanon is planted in many parts of the world. Fruits of Persian mulberry make a fine preserve. The bark of the manna ash yields a juice that has long been used in the eastern Mediterranean as a gentle laxative.

Sambucus racemosa red-berried elder

Cedrus libani cedar of Lebanon

Morus nigra Persian mulberry

Fraxinus ornus manna ash

A small fig tree This member of the mulberry family is capable of growing in the poorest of environments and may crop up to three times in any year. Fresh, dried or canned, figs are a sweet sustaining food and also a gentle laxative.

Clinging to the mountainside Thorny *Rosa ecae* grows high in the mountains of Afghanistan. Its flowers, which open in spring, are of the brightest golden yellow found among roses. Despite its obvious virtues it is not easy to establish in gardens.

(*Lilium candidum*) has been grown for at least 3,000 years, valued as much for its curative powers against boils and corns as for its beauty. It is now difficult to be sure of its original home, though this may have been southern Turkey.

Equally prized for its beautiful, if somewhat somber flowers, the mourning iris (*Iris susiana*) was once widely cultivated in the Middle East; it was introduced into Europe via Constantinople (now Istanbul in Turkey) before 1573. It is no longer found as a wild plant.

Bulbs in danger

Today, the collection of wild bulbous species for cultivation in the West forms an industry of economic importance in Turkey. The quantities removed are large – more than 540,000 kg (1,900,700 lb) were exported in 1981 alone, though most were relatively common species.

Nonetheless, conservationists are concerned by this continued plunder. Once common species have become rare, and certain varieties have also suffered. During the 1970s tens of thousands of corms of the very localized *Cyclamen mirabile* were inadvertently sold as the common *Cyclamen hederifolium* in chain stores in Britain, arousing a storm of protest among conservationists. The deliberate collection of the white-flowered *Sternbergia candida* from southwestern Turkey, first described by botanists in 1979, may have eliminated it in the wild.

Fortunately, many Middle Eastern countries recognize the need for conservation, both for sound economic reasons and for scientific study and pleasure. Forest parks have been created in many countries to protect the remaining stands of virgin woodland, while reafforestation is vastly increasing the forest cover, with major environmental gains. The trade in endangered bulb species is now strictly regulated under an international treaty, the Convention on International Trade in Endangered Species (CITES), while certain countries have passed strict laws to protect their rare plants. Much remains to be done, however – many countries have not yet documented their rare species, and work is needed to boost populations of critically endangered species.

MAKING THE DESERT BLOOM

Israel is a country noted for its rich diversity of landscape and plants. A relatively tiny area, it contains deserts, limestone mountains, fertile coastal plains and a rift valley containing salt seas, and its climates range from temperate through to tropical. The land also bears the scars of centuries of deforestation, overgrazing, and the ravages of the plow and of war.

Following the creation of the state of Israel in 1948, Israel's first prime Minister promised: "There will be cottages all over the country ... we will have ... villages filled with flowers and trees." To a large extent, this promise has been fulfilled. An aggressive irriga-

tion strategy, together with major reafforestation, has indeed made the desert bloom. There have been casualties, however. Much natural vegetation has been destroyed, and with it some of Israel's most precious plants. The lovely oncocylus irises, for example, have suffered greatly: *Iris lortetii*, with its huge pink flowers shot with a tracery of veins, is now restricted to just four tiny populations.

Fortunately, the current agricultural boom is matched by Israel's commitment to conserving its wild plants. Legislation and land management will attempt to conserve Israel's botanical heritage for future generations.

The grape and the grain

"And the ark rested in the seventh month, on the seventeenth day of the month, upon the mountains of Ararat ... And Noah began to be a husbandman, and he planted a vineyard." Whether or not we believe the biblical story that mount Ararat, on the borders of Turkey and the Soviet Union, represents the original home of the grapevine (*Vitis vinifera*), botanists are certain that the wild vine occurred naturally in an area that extends eastward from Ararat to Afghanistan. At first nomads marked the trees that supported vines with superior fruit. As forests were later cleared for agriculture, some vines were spared along field boundaries, and eventually fields were set aside specifically for vine growing.

From these early years of domestication, in about 4,000 BC, the grape was carried westward to Europe. It was grown in Greece during the last thousand years BC and had spread westward and northward to France and southern Germany by 55 AD. An estimated 10,000 varieties of grape have now been bred, and the world's vineyards occupy approximately 15 million ha (37 million acres).

Cereal crops

The Middle East is thought to be one of the principal areas in which agriculture first developed. The people of the early civilizations of the region – particularly the lowlands of what was then Mesopotamia, known as the Fertile Crescent – have left the temperate world with a legacy of invaluable crops.

Wheat (*Triticum aestivum*), for example, is one of the most important crops today. The earliest-known grains of domesticated wheats were found in prehistoric sites within the Fertile Crescent; this lies in the center of the natural range of the wild wheats from which they originate. Two types of wild wheat were of special importance: einkorn and emmer. With large seeds and an annual growth habit, they were ideal for domestication as a field crop, germinating and completing their life cycle during the short wet season. These species crossed naturally, producing new strains that were selected by early farmers. It is likely that the "durum" (pasta) wheats appeared as a mutation of cultivated emmer at this time, while today's ubiquitous bread wheats arose as a hybrid of the two.

Archeological finds dating from 6000 BC have identified such bread wheats

from the Mediterranean island of Crete; they later spread to the Nile basin in Africa, to western Europe and to India.

Early selection processes of strains with the best characteristics produced wheats with fruiting spikes that remained on the stem when ripe, making harvesting easier and more efficient. Subsequent breeding has introduced a range of characteristics such as early maturation, cold- and disease-resistance, and higher yields. In fact, wheat is now so highly bred that it would not survive in the wild: bread

wheat, for example, is no longer capable of natural seed dispersal.

Useful species

Many other useful plants originated from the Middle East. Grain crops (oats, barley and rye), pulses (peas and lentils), vegetable crops (most brassicas and onions); oil and dye plants (safflower); flax, used for making linen; fruit (apples, pears, figs, mulberries and olives) and fodder crops (lucerne) all developed from wild species of this region.

Almond blossom (*above*) The wild almond (*Prunus dulcis*) originated in the Middle East, though now it is cultivated in western Asia and around the Mediterranean Sea. It has two useful varieties: the sweet almond grown for its edible nuts, and the bitter almond, the source of almond oil.

Saffron thistle (*left*) more commonly known as safflower (*Carthamus tinctorius*), this tall annual is becoming important as a crop plant. Its seeds are rich in a light oil that is increasingly in demand in India and the West. The flowerheads are a traditional source of red dye.

Grapes on the vine (*right*) Grapevines form large plants. in the wild they will climb to 18 m (60 ft) or more, holding on to any supporting branch by means of tendrils. Cultivated vines would be equally vigorous, but are severely pruned to encourage fruit production.

A carpet of flowers

Crocus gargaricus and *Scilla biflora* are just two of the many bulbs of the Middle East. The perfect flowers of crocuses, hyacinths, fritillaries, lilies and, perhaps most of all, tulips that flourish in this region have long been admired. They were often featured in the decoration of tiles and in carpet patterns, and were planted in the gardens of the Ottoman Turks and Mughal princes. The Turks, in particular, were hybridizing wild species some four hundred years ago. They took a particular interest in tulips, and developed hundreds of cultivars – many splashed and streaked with different colors, others with long pointed or frilled petals – that astounded European travelers in the 17th century.

Bulbs have a major virtue for the plant collector in that they are easy to transport. When the leaves have died down, the dry bulb or corm will survive out of the ground for several months if it is treated with care. So it was easy to introduce these exciting flowers to Europe where, in the Netherlands in particular, they exchanged hands at enormously high prices for a time.

Today the bulbs of Turkey and the Middle East are still collected for sale in other parts of the world, often in great numbers. Some, such as *Sternbergia candida* and *Cyclamen mirabile*, are now very rare. However, bulbs are plants of sunny, open places, well adapted to living in hot, dry conditions, and with the clearing of many forests their habitats have been increased, though at the expense of many woodland plants.

Spring bulbfield The bright yellow *Crocus gargaricus* occurs over most of Asia Minor, while the blue-flowered squill, *Scilla biflora*, is found around the north of the Mediterranean region. Their meeting place is here in northwestern Turkey.

PLANTS OF THE SAHARA

THE DESERT AND ITS MARGINS · DESICCATING WINDS AND DROUGHT · USE AND OVERUSE

The plant life of northern Africa has had to adapt to extremely diverse conditions, ranging from the vast and arid Sahara to the ancient rainforest in southern Sudan. Remnants of broadleaf and coniferous forests grow in the Atlas Mountains, Mediterranean scrub in the north, hardy grasses and palm oases in the desert, and reed swamps in the south around the Nile. In Roman times the Mediterranean coastal plains supported well-irrigated crop plants, but these lands have become exhausted, and plants have had to adapt and specialize in order to withstand the barren conditions. The savanna to the south of the Sahara supports a variety of grasses and shrubs, which are subject to overgrazing by livestock. Land mismanagement and an increasingly dry climate are turning this area into desert too.

COUNTRIES IN THE REGION

Algeria, Chad, Djibouti, Egypt, Ethiopia, Libya, Mali, Mauritania, Morocco, Niger, Somalia, Sudan, Tunisia

EXAMPLES OF DIVERSITY

	Number of species	Endemism
Egypt	2,085	3%
Ethiopia	6,280	10%
Morocco	3,600	14%
Western Sahara	330	very low

PLANTS IN DANGER

Little information available for much of the region; there are many endangered and some extinct species

Examples *Allium crameri*; *Biscutella elbensis*; *Centaurea cyrenaica*; *Cordeauxia edulis* (ye-eb); *Cupressus dupreziana*; *Cyclamen rohlfsianum*; *Cyperus papyrus* subsp. *hadidii*; *Euphorbia cameronii*; *Gillettiodendron glandulosum*; *Olea laperrinei*

USEFUL AND DANGEROUS NATIVE PLANTS

Crop plants *Lawsonia inermis* (henna); *Pennisetum glaucum* (pearl millet); *Phoenix dactylifera* (date palm); *Sorghum bicolor* (sorghum)
Garden plants *Acanthus mollis*; *Chrysanthemum carinatum*; *Convolvulus mauritanicus*; *Cytisus battandieri* (Moroccan broom); *Ilex aquifolium* (holly)
Poisonous plants *Euphorbia* species; *Ilex aquifolium*; *Nerium oleander* (oleander)

BOTANIC GARDENS

Cherifien Scientific Institute, Rabat (1,200 taxa); Soba Arboretum, Khartoum (150 taxa); Tripoli (400 taxa); Zohria, Cairo (1,000 taxa)

THE DESERT AND ITS MARGINS

The plant life of northern Africa has been greatly affected by the impact of people and their livestock; climatic changes in the past have also had a major influence.

Colonizing the desert

About 100,000 years ago the climate was more humid in northern Africa, and Mediterranean plants such as maple (*Acer*) and pine (*Pinus*) penetrated the Sahara along the seasonal rivers as far south as the Ahaggar and Tibesti mountains in the center of the region.

With increasing temperatures and drier conditions, hardier, drought-resistant (xerophytic) plants such as olive (*Olea*) and mastic tree (*Pistacia*) established themselves. These Mediterranean species were then almost entirely replaced by species of *Acacia*, which came from the south of the Sahara. When the climate rapidly became even drier 50,000 years ago, the plants of the Saharan mountains became isolated and some unique species (endemics) evolved there, including the now endangered Duprez's cypress (*Cupressus dupreziana*) and a wild olive, *Olea europaea* subsp. *laperrinei*.

The western Mediterranean coast of Morocco was formerly covered with evergreen and deciduous forests. However, most of these were destroyed by human activities. Much of the northern coastal area is now covered by extensive grasslands of esparto grass, or alfa (*Stipa tenacissima*) in Morocco and Tunisia, albardine (*Lygeum spartum*) and diss grass (*Ampelodesmos mauritanicus*).

Moving farther south, into the transition zone between the Mediterranean and the Sahara, the plant life consists of spiny scrub forest, which in Morocco is dominated by stands of the bushy, generally evergreen, argan tree (*Argania spinosa*). The zone also supports *Acacia*

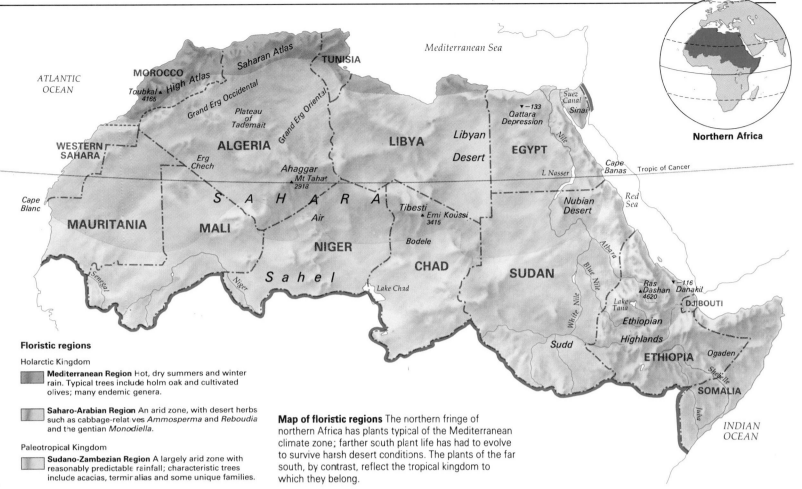

Mediterranean Sea

ATLANTIC
OCEAN

MOROCCO
Saharan Atlas
TUNISIA
High Atlas
Toubkal ▲
4165
Grand Erg Occidental
Grand Erg Oriental
Plateau
of
Tademait
Suez
Canal
Sinai
▼ −133
Qattara
Depression

WESTERN
SAHARA
ALGERIA
LIBYA
Libyan
Desert
EGYPT

Erg
Chech
Ahaggar
Mt Tahat
2918
Nile
L Nasser
Cape
Banas
Tropic of Cancer

Cape
Blanc
S A H A R A
Air
Tibesti
Emi Koussi ▲
3415
Nubian
Desert
Red
Sea

MAURITANIA
MALI
NIGER
Bodele
CHAD
SUDAN
Atbara
Ras
Dashan ▲
4620
▼ −116
Danakil

Senegal
Niger
Sahel
Lake Chad
Blue Nile
Lake
Tana
DJIBOUTI

White Nile
Sudd
Ethiopian
Highlands
Ogaden
SOMALIA

Sudd
ETHIOPIA
Shebelle

Juba
INDIAN
OCEAN

Northern Africa

Floristic regions

Holarctic Kingdom

Mediterranean Region Hot, dry summers and winter rain. Typical trees include holm oak and cultivated olives; many endemic genera.

Saharo-Arabian Region An arid zone, with desert herbs such as cabbage-relatives *Ammosperma* and *Reboudia* and the gentian *Monodiella*.

Paleotropical Kingdom

Sudano-Zambezian Region A largely arid zone with reasonably predictable rainfall; characteristic trees include acacias, terminalias and some unique families.

Map of floristic regions The northern fringe of northern Africa has plants typical of the Mediterranean climate zone; farther south plant life has had to evolve to survive harsh desert conditions. The plants of the far south, by contrast, reflect the tropical kingdom to which they belong.

The fragrant desert rose *Adenium obesum* belongs to a genus of African "bottle" trees, all of which have swollen trunks. This particular species grows in rocky places and dry areas to the east of the region. It has a subspecies found only on the island of Socotra in the Indian Ocean.

gummifera–Ziziphus lotus bushland, and succulent shrublands dominated by euphorbias and by saltmarsh communities of *Atriplex*, *Salsola*, *Suaeda* and other halophytes, which grow around the shallow, saline lakes.

The Sahara supports desert grasses such as *Cornulaca monacantha*, valued as camel fodder. Its southern limits are encroached by the grasses of the Sahel – the semiarid zone on the southern fringe of the Sahara – such as cram-cram (*Cenchrus biflorus*), and shrubs such as *Commiphora africana* and *Boscia senegalensis*.

Annual grasses predominate in the northern Sahel, with an open mixture of bushes and bushy trees such as *Acacia tortilis* growing increasingly vigorously and densely toward the south.

The sandy soils of the southern Sahel are characterized by small trees such as hashab (*Acacia senegal*), the principal source of gum arabic. *Acacia mellifera* scrub dominates the clay soil plains and depressions between the dunes.

Woodlands and swamp

Woodlands stretch from the Atlantic coast across to eastern Sudan, creating a fairly open canopy on the sandy soils. The characteristic small trees include hashab and *Lannea humilis*. The clay plains support *Acacia seyal*, with its conspicuous flaking red bark; it forms pure stands or grows together with the tree *Balanites aegyptiaca*. The grasses on these plains include the aromatic and unpalatable *Cymbopogon nervatus*. Farther south, tall bunch grasses cover the ground, together with trees such as *Anogeissus leiocarpus*, which yields an insoluble gum eaten locally, and species of *Terminalia*.

In southern Sudan lies the extensive Sudd reed swamp – the largest inland wetland in the world. The banks of permanent swamp are dominated by tall-stemmed papyrus sedge (*Cyperus papyrus*), which often forms pure stands; the other marsh plants include the grasslike herb reed-mace (*Typha latifolia*), reed (*Phragmites australis*), which can grow to 9 m (30 ft) tall, and the rare but very distinctive endemic grass *Suddia sagittifolia*. The reeds and grasses occasionally break away from the banks to form floating islands.

Much of Somalia is covered with *Acacia* and *Commiphora* deciduous bushland and thickets, with extensive areas of *Acacia bussei* scrub woodland growing in the north. The evergreen leguminous shrub ye-eb (*Cordeauxia edulis*) is endemic to central Somalia and the Ogaden desert plateau of Ethiopia.

DESICCATING WINDS AND DROUGHT

Plants of arid environments, such as that of northern Africa, have to contend with a short rainy season with sparse and erratic rains, and a long dry season, the effects of which are aggravated by strong, drying winds and high daytime temperatures. Such plants can survive the harshness of these conditions only by conserving water and using it to maximum effect, and to achieve this they have developed specialized adaptations.

Preventing water loss

Most plants manufacture food by photosynthesis – during the day they fix carbon dioxide and transform it into sugar. However, some plants, such as succulents, fix their carbon dioxide at night, so that they lose very little water to evaporation during the process; their breathing pores (stomata) remain closed during the day to reduce transpiration still further. The annual succulent herb *Mesembryanthemum forsskalei* fixes carbon dioxide at night during periods of drought or salinity, but reverts to normal daytime photosynthesis when the stress of these conditions is removed.

In dry environments plants must also be able to make rapid use of the little rain that falls. Some survive the long dry season as seeds, germinating rapidly when the rains start and completing their life cycle within a few weeks, or even days. The pillow cushion plant (*Fredolia aretioides*) has seeds that germinate as soon as the rain falls on them, sending down roots and starting photosynthesis in their seed leaves only 10 hours after they were first wetted. Their leathery leaves ensure that transpiration is reduced to a minimum, and the combination of widespread shallow roots and deep roots allows the plant to make full use of all the available soil moisture.

Some plants, such as the herb *Cassia obtusifolia*, adapt to the erratic rains by producing seeds that absorb varying amounts of moisture. Some germinate

Spiny succulent *Euphorbia echinus* has ridged water-storing stems, similar to those of some cacti of the Americas. This is an example of parallel evolution to cope with desert conditions by geographically distant and completely unrelated plants.

Redbarked thorntree The red-colored bark of *Acacia seyal* is due to the orientation of the cork cells in the bark. When young these absorb all light but red (so the bark looks red), but with age all light is absorbed, so the bark turns black.

during light rains; others wait until there is enough rain for moisture to penetrate the seedcoat. The plant is more likely to reproduce successfully, as not all the seed is spent at once. Another variation on this tactic is adopted by the woody plant *Neurada procumbens*. Each dry, spiny fruit contains several seeds; when a light rain falls only one seed germinates, sending a fine taproot deep into the sand. If no further rain falls the seedling dies. Subsequent showers will trigger more seeds to germinate, one at a time, giving greater chances of success.

The herb *Blepharis ciliaris* retains its seeds in a capsule throughout the dry season. When rain wets the capsule the seeds eject explosively. The seeds germinate within an hour of first being moistened. The moistened hairs on the slimy seedcoats become erect and angle

A dramatic response to rainfall Some ephemerals, such as *Convolvulus*, may complete their life cycle in three to six weeks. In the desert the effect of rainfall is restricted to the depressions where it is possible for moisture to accumulate.

the seeds so that the root can penetrate the ground immediately. *B. ciliaris* is valued as a fodder plant; however, heavy grazing during the dry season has led to its eradication in certain areas.

In some plants it is the stems that carry out photosynthesis instead of the reduced (vestigial) leaves; these include many members of the goosefoot family (Chenopodiaceae), as well as species of *Genista*, *Tamarix* and *Pituranthos*. Many of the grasses, such as *Panicum turgidum* and esparto grass, have inward-rolling leaves that protect the leaf pores (stomata) from the sunlight and thus reduce water loss.

The succulent *Euphorbia echinus* stores water in its stem; its low transpiration rate ensures that sufficient water is retained until the next rains. Other perennials, such as *Erodium guttatum*, store moisture in their tuberous roots.

Some species, including *Acacia tortilis*, are able to transpire throughout the dry season, though at a greatly reduced rate, by tapping underground water; these plants are called phreatophytes.

Trapping the sand

The coastal plain of northern Africa is characterized by vegetated mounds called nebkas, formed when windblown sand is trapped by plants, especially *Ziziphus lotus*. This plant is normally a small tree, but if it is subject to heavy browsing by animals it develops into a densely branched, hemispherical bush. As sand builds up around the bush, it sends out lateral branches that grow from the lower stump. These produce both adventitious roots and aerial shoots that, if undisturbed, will eventually cover the dune with growth, trapping more sand in the process.

Some nebkas in the Sahara are formed by the grasslike *Traganum nudatum*, and others in the Sahel by the shrub *Ziziphus spina-christi*. A number of shrubs, such as *Leptadenia pyrotechnica* and species of *Tamarix*, have dense lower branches that trap windblown sand, which may accumulate around the plants until they are completely buried. In order to survive, the buried branches are also able to develop adventitious roots.

PLANTS IN ARAB CULTURE

Plants have a major role in Arab culture. The seeds of the harmal bush (*Peganum harmala*) are the source of the pigment Turkey red, used for dying the caps known as tarbooshes. Its pounded roots and seeds are mixed with tobacco and smoked in a pipe to ease toothache, and the roasted fruits provide an incense. Another small tree, henna (*Lawsonia inermis*), is widely distributed throughout sub-Saharan Africa. The leaves and young shoots of this small tree yield the orange dye known as henna, which is commonly used for personal adornment, as a hair colorant and for dyeing fabrics. The ancient Egyptians produced a sweet-smelling oil by boiling the seeds in olive oil, and the shoots can be used for basketry.

The fruits of the tamarind tree (*Tamarindus indica*) are used for preserves, in cooking, and for making a popular, refreshing drink that is much appreciated during the fast of Ramadan (the ninth month of the Muslim year). Many of the herbs and spices of ancient Egypt are still eaten throughout northern Africa. They include white mustard (*Sinapis alba*), coriander (*Coriandrum sativum*), dill (*Anethium graveolens*) and safflower (*Carthamus tinctorius*); the black eyepaint known as kohl is made of soot from burned safflower.

USE AND OVERUSE

Among their many attributes, plants are a source of food, fuel, fiber and medicine, as well as providing fodder, shade and shelter for animals. Plants also protect the soil against erosion. The dilemma in this region is that human and animal dependence on the native plants has in many cases caused their destruction, and consequently that of the environment.

Multipurpose trees

The hashab tree exudes a substance known as gum arabic, which has a wide range of uses in the food, pharmaceutical, cosmetic, paint and dyeing industries. The gum is obtained during the dry season by tapping the tree trunk. The bark is cut by an ax or a tapping tool and is peeled back for about 40 cm (16 in) on both sides of the incision. Large drops of gum ooze from the exposed tissue within three to eight weeks of tapping. During the cropping cycle the amount of gum produced declines; this is a major constraint on this industry, and means that more and more plants are tapped to maintain levels of output. Efforts are now being made by all the producing countries to establish gum plantations and to include the tree in agroforestry projects.

The argan tree, which in many areas is grown intercropped with cereals during the rains, is now being replanted in much of its former range of distribution. The olive-sized fruits contain one to three seeds, the kernels of which provide the thick, greenish, rancid-smelling argan oil (similar to olive oil). This oil was once exported to Europe, but is now used locally for cooking and as a fuel for lights. The tree provides fodder for browsing animals, and its hard, durable wood is used as fuel and charcoal, poles and to make implements.

The ye-eb, another evergreen shrub browsed by camels and goats, is mainly valued for its delicious thin-shelled nuts, which could well have a potential market in Europe. Efforts are being made to cultivate the tree in Somalia.

The date palm (*Phoenix dactylifera*), which may have originated in Arabia, is

The date palm is frequently grown in desert oases with the aid of irrigation. Although these palms do not produce much fruit until they are between 10 and 15 years old, they may then continue to crop regularly for 60 years or more.

Papyrus in the Sudd Huge stands of this perennial grasslike sedge grow in the Sudd. Plans have been drawn up to drain this vast swamp; if they were ever carried out they could be the deathknell to this historic plant in Sudan.

its native predators it has spread rapidly to such an extent that it seriously interferes with navigation in the Sudd and threatens to block the irrigation canals of the Gezira – the zone between the Blue Nile and the White Nile.

The baobab tree (*Adansonia digitata*) is widely used throughout Africa. It is extremely versatile, as all the parts of the plant can be used, from the edible roots to the use of pollen for glue. But it regenerates very slowly, and pressure on the land for cultivation and grazing mean that its present populations are under serious threat.

Apart from the threats of invading and overused plants, the northern African plant life is generally being severely degraded. Arid and semiarid environments are particularly fragile; even when they are helped by irrigation and soil and water conservation techniques, they support a plant life limited in both quality and quantity. Overcultivation, overstocking and deforestation, together with the present drought in the Sahel, have led to desertification in sub-Saharan Africa. The only permanent solution, though it is impracticable, would be to maintain animal and human populations below the carrying capacity of the environment.

PAPYRUS AND PAPER

The papyrus reed (*Cyperus papyrus*), formerly the dominant species of the Nile valley and delta, is virtually extinct in Egypt as a result of agricultural development. For centuries it provided the basis of a paper industry and was also widely used for boat-making, cordage, matting, food and medicine. Papyrus paper was invented between about 3000 and 2500 BC. It is prepared by peeling and splitting the leafless green stems lengthways, and lying them side by side for the required width with other strips laid across them. After being soaked, the woven strips are bonded together by being beaten with a mallet and then dried in the sun. Papyrus is too brittle to be made into books, so it took the form of long scrolls, which were carefully rolled up on wooden or ivory rods.

Esparto grass provides the fiber for a good quality, flexible paper used in printing. It is pulped by being heated under pressure in a caustic soda solution. The sparte or albardine (*Lygeum spartum*) and diss grass (*Ampelopsis mauritanicus*), which also grow in the Mediterranean coastal areas, can be used for papermaking too.

so widely cultivated along the valley of the Nile in Egypt and northern Sudan that it has replaced the natural vegetation. The number of male trees that are cultivated is insufficient to pollinate the female, fruit-bearing trees by natural wind pollination, so they are pollinated by hand. Apart from the dates, the trees are valued for their trunks, which provide valuable timber, and their leaves, which are used for thatching and basketry.

The seeds of the harmal bush (*Peganum harmala*), a clump-forming perennial that grows up to 1 m (3 ft) high, contain harmine. This substance stimulates the central nervous system and is used in the treatment of Parkinson's disease and encephalitis. The oil from the seeds is also used to treat some infectious eye and skin diseases, rheumatism, and to make soap.

The legacy of cultivation

Not all plants in northern Africa are useful to its inhabitants. The water-hyacinth (*Eichhornia crassipes*), a native of South America introduced into Africa as an aquatic ornamental, escaped into the Nile during the 1950s. In the absence of

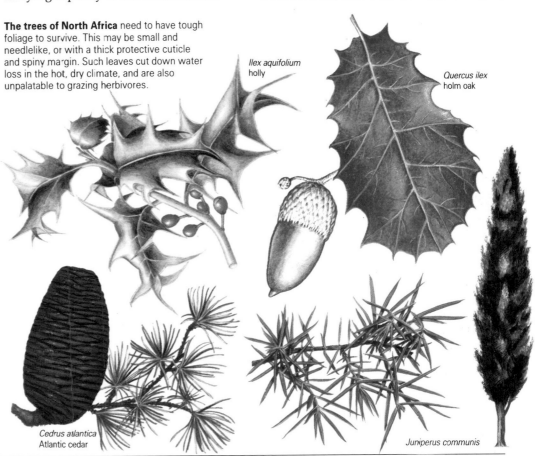

The trees of North Africa need to have tough foliage to survive. This may be small and needlelike, or with a thick protective cuticle and spiny margin. Such leaves cut down water loss in the hot, dry climate, and are also unpalatable to grazing herbivores.

Ilex aquifolium holly

Quercus ilex holm oak

Cedrus atlantica Atlantic cedar

Juniperus communis

The last forests of Somalia

Since the 1950s an increase in human population has led to a catastrophic reduction in the number of trees in Somalia. Many have been used for fuelwood and charcoal, and many more have been destroyed in the wholesale clearance of land to provide pasture for grazing animals. The once widespread forests have been reduced to relic forests alongside the Juba and Shabelle river systems of southern Somalia and the northern "tugs" or seasonal wadis. They are the only forests of their kind still in existence in northern Africa.

Relic trees

The upper reaches of the Juba and Shabelle rivers, in the Ethiopian Highlands, are lined by a narrow belt of trees and shrubs in which the fan palm (*Hyphaene compressa*) predominates. Other species in the upper stretch of the Juba are *Delonix elata*, *Tamarix aphylla*, *Maerua farinosa* and *Maytenus senegalensis*. The river's lower reaches are lined by woodland that includes acacias, *Balanites somalensis*, *Terminalia* species, *Mimusops* species and *Euphorbia bilocularis*. An endemic tree, especially of the smaller wadis, is *Mimusops angel*, which often forms open, pure stands. Its abundant edible fruit are sold locally in the markets.

The lower floodplains of the Shabelle were first cleared for cultivation in the 1930s, and the deforestation has continued. Many of the characteristic trees that remain are armed with spines. They include species of acacia, *Albizia anthelmintica*, *Balanites aegytiaca*, *Salvadora persica*, the succulent *Euphorbia ruspoli*, and the desert rose *Adenium obesum*, together with the scrambling *Acacia schweinfurthii* and *Cissus rotundifolia*.

All these trees, apart from *Acacia zanzibarica* and *Salvadora persica*, are cut down to make charcoal, an important commodity in dry regions that is often a major cause of deforestation. All of them are also browsed by goats and camels.

The characteristic tree of the drier northern tugs is the evergreen *Balanites orbicularis*, which grows interspersed with the scrambling shrub *Maerua oblongifolia* and the bushy *Zygophyllum hildebrandtii*, which forms a straggling, narrow border along the seasonal rivers.

In more favorable areas, where the groundwater lies not too far below the surface, the more frequent, but often thinly scattered, trees and shrubs that can tap the water (phreatophytes) include the deciduous *Acacia tortilis* subsp. *spirocarpa* and *Ziziphus mauritiana*, the evergreen *Balanites orbicularis* and *Salvadora persica*, and thickets of *Tamarix nilotica*, together with the creeping *Pentatropis cynanchoides*. The grasses include the tussocky

Panicum turgidum and *Saccharum ravennae* and the reeds *Phragmites australis* and *Typha domingensis*.

The salt-tolerant endemic *Conocarpus lancifolius*, which grows up to 30 m (100 ft) tall, is an important plant of the tugs, where it is often a dominant tree. It survives by tapping the water that lies far underground. The massive trunks, which can reach 2.5 m (8 ft) in diameter, were once exported to Arab countries for building the traditional boats (dhows), with specially shaped pieces being used for the keels. However, due to overexploitation, large-diameter trunks are no longer available. *Conocarpus* has now been introduced into a number of Asian and African countries as a street tree and for fuel and timber plantations.

A dark spreading canopy Nothing much is known about *Mimusops angel*, and its fruits are unknown to science, though they are produced in plenty and are eaten by the local people. These mysterious fruits contain flat, white, oval seeds.

Desert dates (*right*) Local people collect the edible fruits of *Balanites aegytiaca*, one of several species of this genus that are characteristic of the remaining forests. These fruits contain diosogenin, a substance that is used in birth control drugs. Although *Balanites* is not yet a commercial source, it is being considered as a candidate for commercial cultivation.

The last of the riverine forests (*below*) A remnant of the original vegetation that once clothed not only the banks of the rivers Shabelle and Juba and their tributaries but also much of the surrounding hinterland areas. Now most of these waterside forests have disappeared, having been cleared to make way for irrigated cultivation.

PLANTS OF FOREST AND GRASSLAND

FOREST GIANTS AND GRASSES · ASSOCIATIONS WITH ANIMALS · CULTIVATION AND DESTRUCTION

The plant life of western, central and eastern Africa almost all falls within two zones – the equatorial zone, which is wet almost throughout the year and is dominated by rainforest, and the drier, more seasonal tropical zone that has a monsoon-type climate in which woodland, thornbush and tall grasses are found. Some parts of eastern Africa, although they lie on the Equator, are dry, with two short and unreliable rainy seasons each year. Grasses therefore predominate rather than trees, which would also find it difficult to survive on the thin volcanic soils, though acacias are scattered across grassland areas. The plants found to the east of the Great Rift Valley and extending northward into Somalia have close links with those of the Yemen, and are also similar to the plants of Pakistan and northwest India.

COUNTRIES IN THE REGION

Benin, Burkina, Burundi, Cameroon, Cape Verde, Central African Republic, Congo, Equatorial Guinea, Gabon, Gambia, Ghana, Guinea, Guinea-Bissau, Ivory Coast, Kenya, Liberia, Nigeria, Rwanda, São Tomé and Principe, Senegal, Seychelles, Sierra Leone, Tanzania, Togo, Uganda, Zaire

EXAMPLES OF DIVERSITY

	Number of species	Endemism
Cameroon	8,000	2%
Gabon	8,000	20%
Nigeria	4,600	1%
Tanzania	10,000	10%
Zaire	11,000	30%

PLANTS IN DANGER

Little information available for this region. Many species lost through deforestation and many more threatened.

Examples *Aeschynomene batekensis; Drypetes singroboensis; Justicia hepperi; Memecylon fragrans; Pitcairnia feliciana; Saintpaulia ionantha* (African violet); *Sclerla shellae; Temnopteryx sericea; Uvariodendron gorgonis; Vernonia sechellensis*

USEFUL AND DANGEROUS NATIVE PLANTS

Crop plants *Cajanus cajan* (pigeon pea); *Chlorophora excelsa* (iroko); *Coffea canephora* (coffee); *Cucumis melo* (melon); *Dioscorea rotundata* (yam); *Elaeis gunineensis* (oil palm); *Gossypium species* (cotton); *Sesamum indicum* (sesame); *Sorghum bicolor* (sorghum)

Garden plants *Gloriosa species* (glory lily); *Hibiscus schizopetalus; Saintpaulia ionantha* (African violet); *Thunbergia gregorii*

Poisonous plants *Blighia sapida* (akee); *Cryptostegia grandiflora* (purple allomanda); *Gloriosa superba* (glory lily)

BOTANIC GARDENS

Entebbe (3,000 taxa); Nairobi Arboretum (450 taxa); University of Ghana (1,000 taxa); University of Ibadan

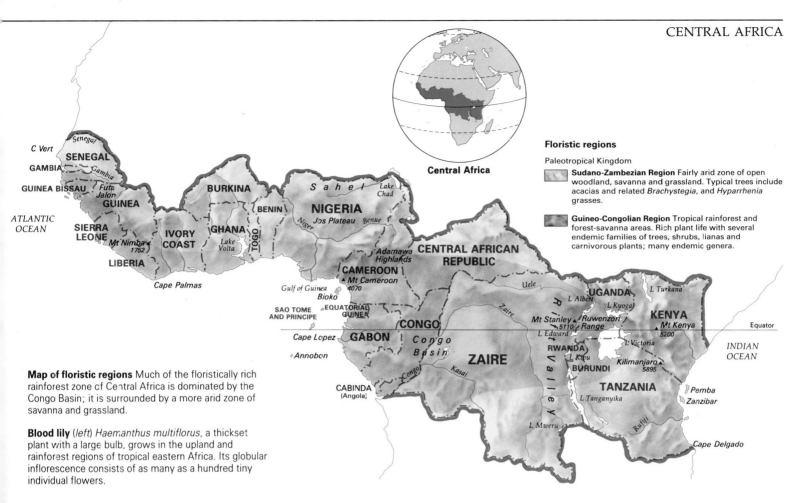

Floristic regions

Paleotropical Kingdom

Sudano-Zambezian Region Fairly arid zone of open woodland, savanna and grassland. Typical trees include acacias and related *Brachystegia*, and *Hyparrhenia* grasses.

Guineo-Congolian Region Tropical rainforest and forest-savanna areas. Rich plant life with several endemic families of trees, shrubs, lianas and carnivorous plants; many endemic genera.

Map of floristic regions Much of the floristically rich rainforest zone of Central Africa is dominated by the Congo Basin; it is surrounded by a more arid zone of savanna and grassland.

Blood lily (left) *Haemanthus multiflorus*, a thickset plant with a large bulb, grows in the upland and rainforest regions of tropical eastern Africa. Its globular inflorescence consists of as many as a hundred tiny individual flowers.

FOREST GIANTS AND GRASSES

The equatorial forests of Africa are separated into two parts. The larger portion occupies the basin of the Zaire river (the Congo Basin) and extends north into Nigeria. The Dahomey Gap, a break of about 600 km (375 mi) wide that corresponds with Benin, divides this forest from the smaller area that covers the south of Ghana and Ivory Coast, extending west to parts of Guinea and Sierra Leone.

Equatorial rainforest and its margins

The rainforests of the Congo Basin are the most extensive in Africa, and contain about 8,000 plant species. Almost 80 percent of these are unique (endemic) to this region. In these and many other African forests, the most abundant plants are members of the pea family (Leguminosae) and mahogany family (Meliaceae). Large stands of trees are sometimes formed by a single species of the pea family, which is unusual in tropical forests, where species are usually mixed. Among them are Uganda ironwood (*Cynometra alexandri*), which forms the climax forest type (the stable stage in ecological succession) in most of western Uganda and eastern Zaire, and *Gilberiodendron dewevrei*, which grows in the Zaire basin, in Cameroon and Gabon. The Dahomey Gap, which divides the forest, is

characterized by short- to medium-height grassland with scattered thicket clumps in which *Capparis tomentosa* is often found. Its isolated rocky hills are covered with dry forests.

On the Uganda–Zaire border, the mountains of the Ruwenzori Range rise out of the dense equatorial forest. Their upper slopes are thickly covered with large tree heaths, members of the heath family (Ericaceae). Above this belt grow giant groundsels (*Senecio* species) and giant lobelias (*Lobelia* species) that survive the daily extremes of climate found on an equatorial mountain. Similar vegetation grows on Mount Kenya and Mount Elgon.

To the north and south of the forest zone lie vast woodlands of small trees and thornbush standing in tall grass that is burned in most years. In the north there are more tree species, with *Terminalia* and *Combretum* both common in the wetter parts. In the south the woodland is known locally as *miombo*, and is made up of species of *Brachystegia* and *Isoberlinia*.

The plateau regions of eastern Africa are covered by open grasslands scattered with acacias, the most characteristic of which are the flat-topped *Acacia tortilis*, and the yellow-barked fever tree, *Acacia xanthophloea*, which grows along rivers. The lowlying coastal regions are mostly covered in bush vegetation such as *Commiphora* (of the myrrh family, Burseraceae) and acacia.

Impoverished islands

The Seychelles lie in the Indian Ocean about 1,400 km (875 mi) off the east coast of Africa. Remnants of wet forest clothe the moister upland areas, with dry forest on the mountain slopes and a few mangrove swamps in the lowlands. Only 480 species of plants grow on the islands; of the 233 native species at least 72 are endemic. These have been greatly threatened by human activity and by competition from introduced species, which now outnumber the native species. Among the endemic plants are the takamakea tree (*Calophyllum inophyllum*), *Medusagyne oppositifolia* (now almost extinct), and six species of palm, which include the coco de mer (*Lodoicea maldivica*), or double coconut, with the largest seed in the world.

The Cape Verde Islands off the coast of Senegal in the west have a total of 651 plant species, also with a low level of endemism (15 percent). More than half the species are tropical in origin and include such plants as cotton (*Gossypium capitis-viridis*) and morning glory (*Ipomoea sancti-nicolai*). Species of *Fagonia* and *Forsskaolea* probably originated in the more arid Saharan areas, while the presence of the dragon tree (*Dracaena draco*) suggests links with the Canary Islands off the west coast of Africa. Much of the natural vegetation has been replaced with cultivated forests of acacia, eucalyptus and pine.

ASSOCIATIONS WITH ANIMALS

Many African plants have adapted to accommodate plant-eating animals. The grasses that form the main food of the plains antelopes, for example, have leaves that, like those of most grasses, grow from the base; if the tip of the leaf is bitten off, the leaf can continue to grow.

The seeds and fruits of many plants are a plentiful source of nutrients for the animals that eat them. In turn these animals act as dispersal agents, particularly for hard seeds, which pass unharmed through the animal. Elephants are particularly effective seed-dispersers, because they eat a great deal but do not chew their food thoroughly. They can eat fruits as large as those of the Borassus palm, which are 15–20 cm (6–8 in) in diameter and contain large seeds about 10 cm (4 in) long. Many forest trees, such as Balanites and Panda oleosa, have similar fleshy fruits with huge seeds. They are eaten only by elephants, so if these animals become extinct the seeds will have no means of dispersal and eventually the plants will die out too.

In some species of acacia, particularly the umbrella acacia (Acacia tortilis), the relationship with animals is more complex, and provides an excellent example of the way plants and animals benefit each other. The pods and the seeds they contain are eaten by antelopes and elephants. The seeds are normally heavily infested by the larvae of a particular beetle (Bruchidius), which grow fast once the fruit has dropped off the tree and kill virtually every seed they attack. However, as the seeds pass through an animal's digestive system both eggs and larvae are killed, so the seeds that survive the journey are also cleared of the infestation. Some of the plant defences that people find formidable seem to leave animals unaffected. Nettles that can sting a person badly through two layers of clothing apparently cause no discomfort to gorillas, which eat them readily. Similarly, the long sharp spines of the whistling thorn (Acacia drepanolobium) are no deterrent against giraffes, which use their long, flexible tongues to pick the leaves from among the spines.

Bats and sunbirds

Many plants depend on animals to pollinate their flowers. In Africa several trees are known to be pollinated by bats. These include Maranthes polyandra, Parkia and the sausage tree (Kigelia aethiopica). Their white or cream flowers open at night, produce abundant nectar, and often have an unpleasant musty smell. Bats are clumsy creatures and are also relatively heavy, so they need a firm base on which to land. The flowers that depend on them for pollination usually hang clear of the leaves and are either large and strong, or grow in dense groups that provide a convenient landing platform.

There are no hummingbirds in Africa, and their local equivalent, the sunbirds, are much less agile and hover with difficulty. The flowers they normally pollinate, like flowers pollinated by bats, need to be strong enough to bear the weight of the birds as they perch and probe for nectar. Many of these flowers are red – a color easily seen by birds. They include the Uganda coral, or flame tree (Erythrina) and the African tulip tree (Spathodea).

Defensive ants

Some African plants form associations with ants, to the benefit of both. One of these is the whistling thorn, a small tree that grows up to about 5 m (16 ft) tall on seasonally waterlogged black clay soils. This plant is known for its pairs of spines, whose bases form hollow spherical chambers up to 3 cm (1.25 in) in diameter. When the wind blows them across it

SACRED AND USEFUL TREES

Throughout Africa there are trees long revered by the local people. In Uganda *mvule* trees (*Chlorophora excelsa*) are said to shelter the spirits of the departed from the midday sun. For this reason they are often left when forests are cleared, even though their timber is valuable. Trees may also be revered merely because they suggest a likeness: in Ghana *Distemonanthus benthamianus*, a forest tree with red bark, is known as the devil tree for the rather tenuous reason that Ghanaian devils are thought to have red hair.

Some trees are also worshipped just for their size; this may be the case with large baobabs (*Adansonia digitata*). The baobab, which can live for as long as 2,000 years, is not only extremely distinctive, having an enormously thick trunk covered with smooth gray bark and topped by a relatively small crown, but it also has many practical uses. The leaves are widely eaten as a green vegetable and are used as a thickener for soup; the acid pulp of the fruit is made into a refreshing drink as well as being used as a remedy for many diseases. The outer bark is softened, pounded and dried into barkcloth, which is used as a fabric, and the inner bark is twisted into rope.

African tulip tree (*left*) The large, urn-shaped blossoms are rich in nectar to attract birds and bats to act as pollinators. On the west coast this evergreen tree is known locally as *baton de sorcier*, for it is used by witch doctors in magic ritual.

Powderpuff flowers (*below*) Acacia trees are well adapted to life on the African plains, armed and able to survive periods of drought. Their fluffy flowers contain numerous stamens. Even tiny seedlings possess sharp thorns to deter grazing animals.

makes a low whistling sound. These chambers also provide a home for ants, which defend the plant (and their home) by rushing out and stinging or biting any animal that disturbs it.

Another small forest tree, *Barteria fistulosa*, is also defended by large ants, which live in its hollow stems. Scale insects suck the sap of the plant that provides the ants with food. The ants patrol around the plant, removing any rubbish that falls on it and biting off the tips of climbers and other plants within a radius of 2 m (6.5 ft) of the base of the tree. They too sting if the tree is disturbed; the effects of the sting are so painful and longlasting that the tree is usually left standing even when the forest is cleared for farming.

CULTIVATION AND DESTRUCTION

Africa is the original home of several grain crops that are today grown all over the world. However, most of the crops that now support Africa's populations originate from other continents.

Major crops

One of the most important food crops of African origin is the yam, a staple crop in the moister woodlands and grasslands of western Africa. (In the United States the sweet potato (*Ipomoea batatas*) is often referred to as yam, but it is in fact a different species.)

There are several species in cultivation; in Africa the most important are the white yam (*Dioscorea rotundata*) and the yellow or Guinea yam (*D. cayenensis*). There are also several wild yam species that are collected. Most species of yam produce one large tuber, though some have several smaller ones. Yams consist almost entirely of starch, so there is a possibility of protein deficiency among people who rely on them as a staple food. They have the advantage of storing better than most other tropical root crops.

Sorghum (*Sorghum bicolor*), with its wild relatives, grows in the seasonal tropics of Africa, and is now the world's fourth most important cereal, grown both in its native African home and in the United States, India and southern Europe. Like the millets, it is tolerant of poor soils and resistant to drought. Bulrush or pearl millet (*Pennisetum typhoideum*) is another grain crop that has been developed in the sub-Saharan zone of Africa, because it can tolerate an annual rainfall as low as 250 mm (10 in), it can be grown farther north than sorghum.

Another crop from the same climatic zone is African rice (*Oryza glaberrima*), which has both dry-land and flood-tolerant types. It is said by some to be hardier than the Asian cultivated rice, but it has spread little from its original home. Much less well known, but widely grown, are hungry rice or fonio (*Digitaria exilis*), which has very small grains with an

Finger millet (*above*) This drought-resistant grass is cultivated in the drier areas of Africa and India; it is productive even on poor soils. Locally the grain is pounded into meal to make porridge, and also fermented to make beer.

Fruits of the oil palm (*below*) Each palm develops several large fruit heads. These are studded with some 150 fruits, which are rich in vitamin A. The sap, a source of vitamin B, is tapped and then fermented to produce an alcoholic drink.

many species that yield valuable wood, such as the hard, heavy iroko from the *mvule* tree (*Chlorophora excelsa*), sapele (from *Entandrophragma cylindricaum*), afrormosia (from *Pericopsis elata*) and obeche (from *Triplochiton scleroxylon*).

Large areas of forest remain in Zaire, Gabon and Cameroon, but to the west of the Dahomey Gap the situation is very different. Excessive cutting for timber and clearance for oil palm and rubber plantations means that little undisturbed forest remains. The worst affected areas are Ivory Coast and Liberia. A rapidly expanding population and a desire for land in the better-watered areas threaten the forest further; once logged, the forest is often cleared for agriculture rather than being left to regenerate.

Research is being undertaken to develop ways of obtaining sustained yields of timber from these forests, rather than destroying them completely for the sake of a single crop. One glimmer of hope is that in many parts of western Africa small areas of forest are still regarded as sacred: they are used as burial grounds and a source of medicinal plants, and are believed to shelter the ancestral spirits. Such sites, often jealously and strictly guarded against outsiders, are refuges for many forest plants.

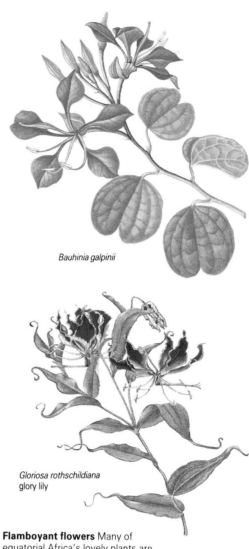

Bauhinia galpinii

Gloriosa rothschildiana
glory lily

Flamboyant flowers Many of equatorial Africa's lovely plants are now widely planted in towns and gardens throughout the tropics.

excellent flavor, and finger millet (*Eleusine coracana*), which is a staple grain crop in some parts of Africa.

The oil palm (*Elaeis guineensis*) is another important plant. In the wild this tree can reach a height of 20 m (65 ft), but cultivated plants are generally cut down long before they reach this height, as yields decline and harvesting becomes very difficult. The bright orange palm oil, which is the food reserve for the seed, is squeezed from the fruits. It was formerly used mainly for making soap and candles, but improvements in its quality now mean that it is widely used to make margarine and other edible fats. Palm kernel oil is of a lower grade.

Oil palms are grown in large plantations. The trees require high rainfall, so most plantations are established by clearing large areas of rainforest.

The state of the forests

The tropical rainforests of the Congo Basin are less severely threatened than those in other parts of the world, but they are felled for their timber. They contain

SUPERSTITION AND MEDICINAL PLANTS

An enormous number of African plants have one or more medicinal uses – perhaps as many as 80 percent of them. Many African countries, short of foreign exchange with which to buy Western medicines, are again turning to their own traditional remedies. The problem is to distinguish those plants used for superstitious reasons, often based only on similarities in appearance between the plant parts and human organs, from those that have genuine pharmacological effectiveness.

In parts of western Africa diviners would determine people's guilt by making them eat a calabar bean (the fruit of *Physostigma venenosum*); if they died they were guilty, and if they merely vomited they were purged of blame. The bean does contain a poison, physostigmine, so it was not surprising that some of those who ate it should die. It is now being used as an ingredient in eye drops, as it makes the pupils dilate. Another example is the sausage tree, which has huge sausage-

like fruits, used in Uganda to flavor beer. In western Africa they are given to cure dysentery and, contradictorily, as a purgative. The fruits are also used as a charm to make one rich. However, the fruit's resemblance to an elongated female breast or an enlarged male sex organ led to fears that it might affect these organs.

The curious sausage tree is sacred in central Africa. Its dangling fruits may weigh up to 7 kg (15 lb) and are used in medicine and magic.

Fire in the grasslands

The climate of the tropical grasslands makes them particularly susceptible to fires, and every year large areas of grassland and woodland are burned. Huge amounts of vegetation, mostly grass, grow during the wet season; in the dry season the hot winds desiccate the grass, which becomes a mass of highly inflammable hay. Lightning or volcanic eruptions sometimes start fires, but they are usually started by people.

Grass fires look horrifyingly destructive, but in fact the material that burns contains little nourishment; most of the protein moves to the roots before the stems and leaves dry up. These reserves are used to produce a quick flush of new growth when the rains arrive, and also enable the plant to regenerate after a fire. Many of the grasses form clumps or tussocks, and only the dead upper parts burn; the stem bases inside the tussock are protected, and survive to produce buds the following year.

Most of the trees that grow here have thick corky bark. Bark is dead tissue, and its air-filled cells form an excellent insulator for the delicate wood- and bark-producing tissues (cambium) beneath. The bark protects these tissues during a fire, as the fire moves quickly through dry vegetation and heats any one point for only a minute or two.

Survival of seeds and seedlings

Tree seedlings are particularly vulnerable to fire. Woodland fires that burn every year toward the end of the dry season, when the grass is very dry and the fire very hot, gradually kill the adult trees, and any seedlings that have established themselves have little chance of survival. Annual fires that take place early in the dry season, when the grass is still damp, burn less fiercely; they allow seedlings to become established and do not kill the mature trees.

Many of the woodland trees, such as the West African shea butter tree (*Butyrospermum*), have seedlings that grow in a most unusual way. After the seed has germinated the shoot grows downward into the soil. Only later does it turn upward and emerge into the light to photosynthesize. By then it has produced buds beneath the soil, so that if the tip is killed by fire one of the buds can grow to form a replacement shoot.

Many grasses have long, spirally twisted bristles called awns attached to their seeds. These awns are sensitive to moisture, and twist and untwist with the changing humidities of day and night. By twisting they move the seed about until its sharp point enters a crack in the soil. The awn continues to move, driving the seed into the soil where it is protected from the heat of fires.

A number of plants flower soon after fires have taken place, when the tall grass has been burned away and no longer conceals them. Their flowers, which are large and conspicuous, can easily be seen by pollinating insects. Some species even flower before their leaves appear. Many species have, therefore, evolved in ways that enable them to take advantage of the particular conditions that fires, with all their destructive power, can create in the grasslands.

Bush fire (*right*) When grassland is burned the germination of certain grass species is stimulated by the heat. The resulting seedlings grow strongly, as competition from other plants has been destroyed.

A phoenix from the ashes (*right*) A new shoot grows from a bud in a charred stump. Fires take place so regularly in some areas that the vegetation has gradually adapted to become almost exclusively fire tolerant.

Delicate mauve flowers (*below*) Wild ginger (*Siphonochilus aethiopicus*) is one of the plants that flowers after a fire. Fire not only clears away competing vegetation, triggering a burst of activity in the plants that survive, but is also a means of unlocking nutrients from dead plant material.

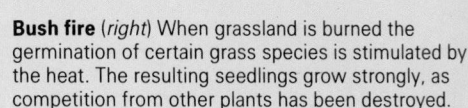

PLANT WEALTH OF THE SOUTH

THE KINGDOM OF THE CAPE · SURVIVING FROST, FIRE AND DROUGHT · BUSH FOOD AND HERBAL REMEDIES

Southern Africa has both an enormous number of plant species – 35,000 of them – and great diversity. They range from the sparse plant cover of the incredibly dry Kalahari Desert, to tropical rainforest in northern Angola and eastern Madagascar. Between these extremes are vast expanses of plateau grassland (highveld), a unique colorful bushland called fynbos, the semidesert succulents and shrubs of the Karroo tableland, the alpine plants of the Drakensberg mountains, and the coastal forest of the south. Southern Africa is ringed by islands, from the dry Ascension island in the mid-Atlantic to the Comoros and the densely forested islands of Réunion, Mauritius and Madagascar in the Indian Ocean. The last has about 10,000 species of higher plants, 80 percent of which are unknown elsewhere.

COUNTRIES IN THE REGION

Angola, Botswana, Comoros, Lesotho, Madagascar, Malawi, Mauritius, Mozambique, Namibia, South Africa, Swaziland, Zambia, Zimbabwe

EXAMPLES OF DIVERSITY

	Number of species	Endemism
Angola	5,000	20%
Madagascar	10,000	80%
Mauritius	900	30%
Mozambique	5,500	4%
South Africa, Lesotho and Swaziland	23,000	80%

PLANTS IN DANGER

	Threatened	Endangered	Extinct
Madagascar	33	3	–
Mauritius	100	65	19
South Africa, Lesotho and Swaziland	333	110	39

Examples *Allophylus chirindensis*; *Aloe polyphylla* (spiral aloe); *Dasylepis burttdavyi*; *Encephalartos chimanimaniensis*; *Hyophorbe amaricaulis*; *Jubaeopsis caffra*; *Kniphofia umbrina*; *Nesiota elliptica* (St Helena olive); *Protea odorata*; *Ramosmania heterophylla*

USEFUL AND DANGEROUS NATIVE PLANTS

Crop plants *Citrullus vulgaris* (watermelon); *Cucumis melo* (melon); *Sesamum indicum* (sesame)
Garden plants *Clivia miniata*; *Crassula arborescens*; *Delonix regia* (flamboyant tree); *Erica* species (Cape heaths); *Gazania linearis*; *Gerbera jamesonii* (Barberton daisy); *Gladiolus blandus*; *Kniphofia* species (red hot poker); *Mesembryanthemum* species; *Nerine bowdenii*; *Pelargonium* species
Poisonous plants *Adenia digitata*; *Bowiea volubilis*; *Datura stramonium* (thorn apple); *Euphorbia* species

BOTANIC GARDENS

National Botanic Gardens, Kirsten Bosch (6,200 taxa); National Botanic Gardens, Salisbury (2,000 taxa); Pretoria (5,000 taxa); Tananarive, Madagascar (2,000 taxa)

Wild flowers in Namaqualand, in southwestern Africa. *Aster* and orange *Gerbera* cover the ground after the winter rain. These are just two of the beautiful perennial flowers of the Cape that relieve the monotony of the lowgrowing scrub.

THE KINGDOM OF THE CAPE

The extreme south and southwestern area of Cape province in South Africa has an exceptionally large number of highly decorative plants, many of which do not grow anywhere else. About 7,000 species are found here, creating a colorful bushland, or fynbos. Some of the plant families are unique to the Cape, such as the Bruniaceae and Penaeaceae; others, such as the heath family (Ericaceae), which is widespread, have undergone evolution-

ary explosions here. About 650 species of this family are unique (endemic) to the area, 520 of which are in the genus *Erica* – these are the Cape heaths. Another well-known Cape genus is *Protea*, which has 85 species. The large, cup-shaped flowers are surrounded by spiky bracts. They are the national emblem of South Africa.

In the center of the long southern

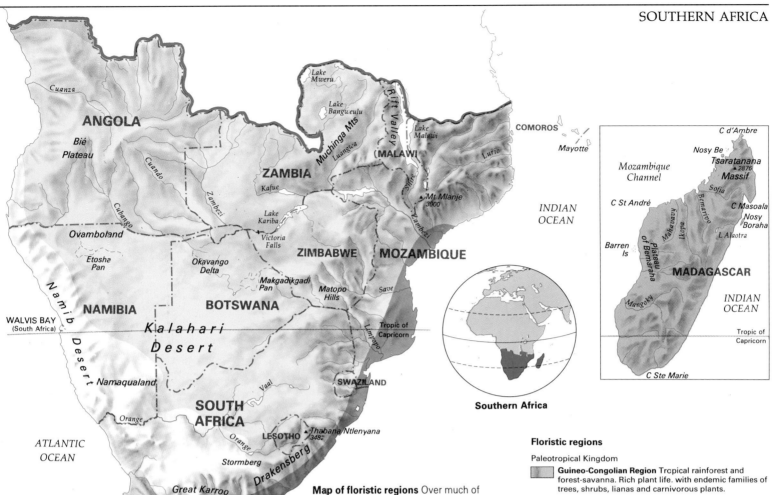

Map of floristic regions Over much of southern Africa plants have evolved in arid conditions; Madagascar and the extreme south and southeast are floristically distinct.

Floristic regions

Paleotropical Kingdom

Guineo-Congolian Region Tropical rainforest and forest-savanna. Rich plant life. with endemic families of trees, shrubs, lianas and carnivorous plants.

Sudano-Zambezian Region Open woodland, savanna and grassland. Few endemics; characteristic plants include acacia-like trees, *Colophospermum* (mopane tree) and *Julbernardia*.

Karoo-Namib Region Desert and semidesert; 1 endemic family, some 80 endemic genera, including mesembryanthemums, irises and lilies.

Uzambara-Zululand Region Less arid coastal zone characterized by high number of endemics – 40 percent of species found nowhere else.

Madagascan Region Relatively moist, stable climate and long isolation have resulted in 12 endemic families, 400 endemic genera and about 8,000 endemic species.

Cape Kingdom

Cape Region Hot, dry summers and winter rain. Highly distinctive plants; 73 percent of the 8,550 species are endemic, including heaths, proteas and rushlike herbs.

---

coastline is the last remaining strip of indigenous coastal forest, the Tsitsikamma Forest, which includes trees such as yellowwood, stinkwood and mboya, as well as cycads and tree ferns. It is now a protected reserve.

Succulent survivors

North of the Cape lie the Karroo and the Namib Desert. The plant life here ranges from open bushland to the sparse cover of the Namib. A few scattered trees are found, such as *Acacia karroo*, but succulents are more common. They include *Aloe* and *Crassula*, which are leaf succulents; *Euphorbia* and *Caralluma*, which are stem succulents; and "mesems" (succulents of the family Aizoaceae), which include *Mesembryanthemum*, with its daisylike flowers in brilliant pinks, oranges and yellows.

Dwarf, drought-tolerant shrubs such as *Euryops* and *Helichrysum* are common. So too are plants with underground storage organs (bulbs, corms and rhizomes) such as *Haemanthus*, *Oxalis* and *Babiana* (named after the Afrikaans term for the baboon, *bobbejaan*, which digs up the corms for food).

In the Namib Desert proper, where less than 100 mm (4 in) of rain falls a year, even the succulents are smaller, sparser

and, in the case of those known as "living stones" (*Lithops*), quite bizarre: in their camouflage they resemble pebbles. Here too is found the living fossil, *Welwitschia mirabilis*.

Woodland and thornveld

Woodland is the natural vegetation of much of Zambia, Malawi, Zimbabwe, and parts of Angola and Mozambique, an area sometimes known as the Great African Plateau. Grassland takes over where seasonal drought is most severe, or where the soils are either seasonally flooded or toxic to many other higher plants because they contain large amounts of naturally occurring heavy metals. Factors such as rainfall, altitude or the frequency of fires determine the type of woodland to be found; these range from thorny thicket to dense evergreen or deciduous woodland. *Brachystegia* is a characteristic tree, and there are many species in the region. Also typical are *Isoberlinia* and *Julbernardia*, particularly on the plateau, where they form what is locally called miombo woodland. On lowland plains the woodland is dominated by *Colophospermum mopane* and is even known as mopane woodland.

In the Kalahari Desert wooded grassland called thornveld is widespread, containing principally *Acacia* species, which

bear large thorns as a protection against grazing animals. These grow together with other trees such as *Combretum*, with its red or yellow spikes of flowers, and the evergreen *Terminalia sericea* with its downy shoots. The grasses are most commonly *Themeda* and *Stipagrostis*.

The high interior plateau of South Africa, from the eastern Cape to the Drakensberg mountain range, is covered by grassland vegetation known as highveld, which grades into grassy thornscrub. *Themeda* is the dominant grass, but a large number of colorful herbaceous plants grow with it. A good example of highveld vegetation is found in the Malobotja Nature Reserve in Swaziland. Approximately 2,000 species of plants flourish here on the land around Mount Ngwenya; they include the rare red hot poker (*Kniphofia umbrina*).

SURVIVING FROST, FIRE AND DROUGHT

Southern Africa is a region of many habitats and climates – dry and hot, wet and cool, coastal and montane. The large number of plant species it supports have all adapted to survive in their particular ecological niche.

Mount Mlanje, at the southeast tip of Malawi, is the highest point between the Drakensberg mountains on the southeast coast of South Africa and the Ruwenzori mountains in Uganda. It rises to almost 3,000 m (9,843 ft). Even though the mountain lies within the tropics, plants at such an altitude have to adapt in order to survive frost. Night frost occurs regularly, and many of the trees that grow in the high altitudes are sensitive to it. However, the freezing conditions never last long: the tropical sun can rid the landscape of all traces of hoarfrost within an hour or so of rising.

The high rainfall creates a very humid atmosphere, which is accentuated by dense cloud (called "chiperone") that can hang over the mountain for days. In this climate a varied plant life has developed that thrives in moist conditions. Characteristic plants are the tree ferns *Cyathea capensis* and *C. dregei*, the attractive purple-flowered shrub *Dissotis johnstonii*, and the glory of Mlanje, the Mlanje cedar (*Widdringtonia whytei*), which scents the air with cedar oils. This tree has a highly inflammable trunk, which makes it vulnerable to fire. However, it does not

regenerate under the close canopy of evergreen broadleaf trees around it, such as olive (*Olea*) and *Rapanea*, and to ensure its survival needs occasional fires to thin the surrounding vegetation.

Fire in the fynbos and grasslands

The plants of the fynbos are all adapted to periodic fires. To maintain the greatest richness of species, the fynbos needs to be burned about every 15 years; natural fires caused by lightning are now added to by ones started by humans.

The proteas are characteristic small trees of the fynbos. Some, such as *Protea cynaroides*, have a woody underground trunk (lignotuber), which survives fires; *Protea nitida* has stout woody heads that release seeds after burning; *Protea piscina* has underground branches that remain protected from fires when the tops of the plant burn off. The seeds of many proteas will germinate only after the heat of fire has cracked the hard seedcoat.

The large areas of grassland in southern Africa also need fire to regenerate. The extensive underground root systems of grasses enable them to regenerate after the blades above ground have been burned off, scorched by the summer sun or even frozen by the severe frosts found at higher altitudes.

Coping with drought

One of the difficulties plants in southern Africa have to deal with is lack of water. In the dry woodlands, trees such as marula (*Sclerobarrya birrea*) have seeds that contain two or three embryos to

Creeper of the Kalahari (*above*) Duwweltji (*Tribulus terrestris*) grows along rivers and in lowlying areas, where its long taproot can draw on water deep in the ground. Like other desert plants, it flowers after the rains.

Madagascan orchid (*left*) *Angraecum sesquipedale* leads an epiphytic life, perched on a branch where the air is more buoyant than on the woodland floor. The heavy rainfall ensures that its spongy roots can absorb enough water for it to survive without soil.

Succulent euphorbia (*right*) Fleshy, fingerlike stems capable of storing water and an absence of leaves are the perfect adaptation to life in the desert. The inflorescence is composed of bracts surrounding tiny flowers.

A BOTANICAL PUZZLE

There are only seven species of palms in the whole of mainland southern Africa – in contrast to Madagascar, which has some 100 species of native palms, most of them unique to the island. One explanation for the paucity of southern African palms is climatic change. Ten million years ago virtually the whole of southern Africa, including those areas that are now desert or scrub, was covered with forest. Palm pollen has been found in sediments laid down at this time, even in places where no palms now grow. But since then the region has become very dry, which possibly caused the extinction of many African palms, leaving a rich relict population of palms in Madagascar.

Among those on the mainland is the pondoland dwarf coconut (*Jubaeopsis caffra*), which grows only on the banks of the Mtentu and Msikaba rivers in the area between the Drakensberg mountains and the east coast of South Africa. Another palm restricted to a small area is the Kosi raffia palm (*Raphia australis*), which is found only around Lake Kosi, in Natal province in the far northeastern corner of South Africa, and over the border into southern Mozambique. The Borassus palm (*Borassus aethiopicum*) is more widespread; in the Transvaal in South Africa, on the edge of the upland plateau to the northwest of Swaziland, there is a large specimen. It is now dead, but is a national monument that is said to have contained the spirit of a former great chief, Magoeba, in the bulge of its trunk.

increase their chances of survival. The baobab tree (*Adansonia digitata*), found in seasonally dry grassland and thicket on the mainland and in Madagascar, has a massive swollen trunk of soft, spongy wood that provides a reserve of carbohydrates and water in dry weather.

In the thornveld many of the trees have small leaves, which reduce the potential loss of water through evaporation; their thorns protect the plants from being overgrazed by animals.

The heathers in the fynbos have small, tough leaves pressed close to the stem, which protects them from the desiccating southwesterly winds and from the heat of the summer sun. The proteas have leathery, often narrow leaves that also help them to reduce water loss.

Two strange-looking genera of the milkweed family (Asclepiadaceae), *Stapelia* and *Hoodia*, grow in the dry areas of southwest Africa. Both have leafless, water-storing, succulent stems. Their fleshy flowers, in livid shades of reddish brown, pink or purple, smell of rotten meat; this attracts flies, which act as pollinators. The pollen forms a sticky mass known as a pollinium (a characteristic shared only with orchids), which sticks to the fly. Once pollinated the seeds may not ripen for up to a year, to be dispersed speedily when conditions are favorable for germination.

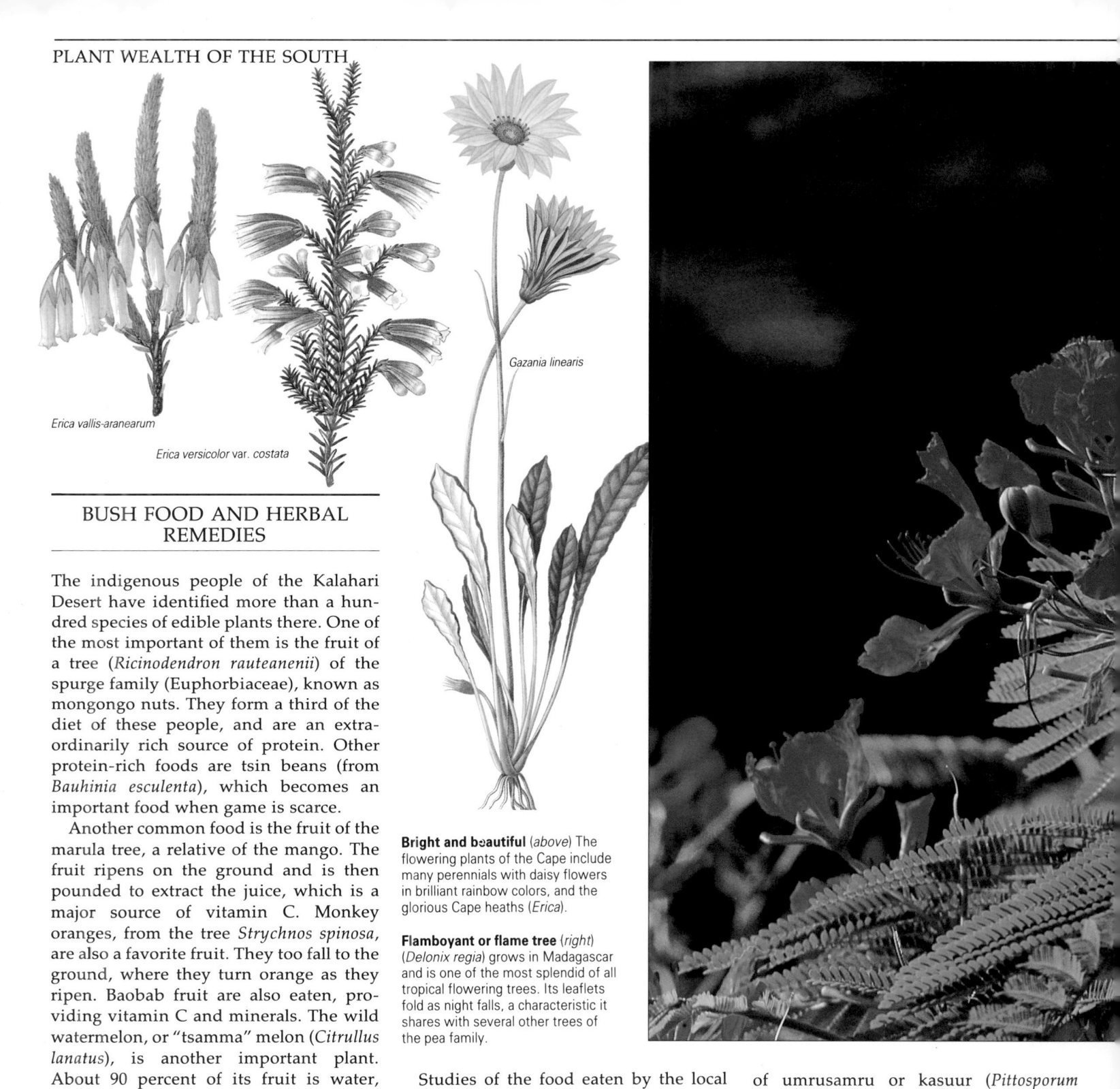

Erica vallis-aranearum

Erica versicolor var. *costata*

Gazania linearis

BUSH FOOD AND HERBAL REMEDIES

The indigenous people of the Kalahari Desert have identified more than a hundred species of edible plants there. One of the most important of them is the fruit of a tree (*Ricinodendron rauteanenii*) of the spurge family (Euphorbiaceae), known as mongongo nuts. They form a third of the diet of these people, and are an extraordinarily rich source of protein. Other protein-rich foods are tsin beans (from *Bauhinia esculenta*), which becomes an important food when game is scarce.

Another common food is the fruit of the marula tree, a relative of the mango. The fruit ripens on the ground and is then pounded to extract the juice, which is a major source of vitamin C. Monkey oranges, from the tree *Strychnos spinosa*, are also a favorite fruit. They too fall to the ground, where they turn orange as they ripen. Baobab fruit are also eaten, providing vitamin C and minerals. The wild watermelon, or "tsamma" melon (*Citrullus lanatus*), is another important plant. About 90 percent of its fruit is water, making it a valuable source of liquid for both people and animals in the dry months. The flesh is mashed to a watery pulp that can be eaten or drunk. This wild watermelon is the ancestor of the cultivated watermelon, though its fruits are smaller and tougher.

The corms of species of the iris family (Iridaceae), such as *Gladiolus* and *Moraea*, form a significant part of the diet, and the massive corky tubers of Hottentot bread (*Dioscorea elephantipes*), a type of yam, are also eaten. The herb plant *Sceletium strictum,* which contains mesembrine, a substance that is related to cocaine, is used as a narcotic; it is taken by chewing the dried leaves.

Bright and beautiful (*above*) The flowering plants of the Cape include many perennials with daisy flowers in brilliant rainbow colors, and the glorious Cape heaths (*Erica*).

Flamboyant or flame tree (*right*) (*Delonix regia*) grows in Madagascar and is one of the most splendid of all tropical flowering trees. Its leaflets fold as night falls, a characteristic it shares with several other trees of the pea family.

Studies of the food eaten by the local people are now being undertaken with a view to cultivating the plants as commercial crops as a new source of food, and as a means of greening desert areas.

Medicinal plants

Among the Bantu peoples of southern Africa herbal medicine is often by far the most important form of health care. In South Africa's Natal province some 89 percent of the urban black population consults traditional practitioners; there are 30 of them to every one doctor with medical training. Among the prescriptions, the usual herbal remedy for diarrhoea is the boiled root of impila, or ox-eye daisy (*Callilepis laureola*); the bark

of umrusamru or kasuur (*Pittosporum viridiflorum*) is used to treat a form of gastritis known locally as inyongo.

As herbal remedies are still usually gathered from wild plants, this has led to a serious conservation problem. Wild ginger (*Siphonochilus* (*Kaempheria*) *aethiopicus*) and *Warburgia salutaris* are both now critically endangered because of overcollecting. Attempts are being made to cultivate these and other important herbs, and pharmacological studies are being conducted to find the chemical reasons for their efficiency.

Plant exports and imports

The rich plant life of southern Africa has provided many species for gardens and

for agriculture. The orange barberton daisy (*Gerbera jamesonii*) was discovered by Robert Jameson in 1888. A hundred years later a range of colors from yellow to scarlet had been developed, and more than 8 million blooms were being sold each year in the Netherlands alone. Many species of *Pelargonium*, too, have been collected from this region. They have been hybridized and are widely grown as ornamentals throughout the world. The Rhodes grass (*Chloris gayana*) is now an important pasture and hay grass in warm areas outside southern Africa.

Many plants have also been imported, often to the detriment of the local plant life. The Western Cape lowland fynbos alone formerly covered nearly 1 million ha (2,471,000 acres) but is now reduced to some 15 percent of this by alien plants and development, and 270 species of this unique vegetation are threatened with extinction. One of the most invasive of the introduced plants is *Hakea sericea*, a member of the protea family that was introduced from Australia as a hedge plant. It has become all too successful, spreading rapidly and moving into some 480,000 ha (1,186,000 acres) of mountain fynbos in the Cape. The area affected has been reduced since the release in 1975 of the hakea fruit weevil (*Erytenna consputa*) from Australia, which has proved to be an effective biological control.

Another alien is the banana poka (*Passiflora mollissima*), from the Pacific Ocean island of Hawaii. In 1987 this vigorous climber was noticed smothering forest margins in the eastern Cape forest of Knysna, and an eradication program was subsequently initiated.

MADAGASCAR – A UNIQUE PLANT BANK

The island of Madagascar is divided into two distinct zones by the central highlands. The western half of the island is dry, and supports arid deciduous forest and scrub, with species of baobab (*Adansonia*) and densely thorny *Didieria* species. The eastern part was once largely covered with lush tropical rainforest, though much of this has been destroyed and transformed into secondary grassland, with grasses such as *Imperata cylindrica* and *Aristida similis*. What remains of the rainforest is the home of the unique traveler's palm (*Ravenala madagascariensis*), as well as many true palms.

The rainforest also contains a great many species that have some medicinal value. For example, the Madagascar periwinkle (*Catharanthus rosea*), now widespread in the dry tropics, produces the anticancer alkaloid leurocristine, used to treat childhood leukemia. The Tsimikety ("those who do not cut their hair") people of northern Madagascar, where one of the last extensive areas of rainforest is found, have considerable herbal knowledge. Among their remedies, *Grangeria* species are used for headaches. The plants of Madagascar are the focus of much research to identify those that are medicinally useful, as these may be the source of important cures in the future.

Wild watermelons are a precious source of water in the desert. As a bonus, the oily seeds can also be eaten. A related genus, *Cucumeropsis*, which is native to western Africa, is cultivated for its edible seeds, which also have a high food value.

Curiosities of the drylands

Southern Africa is well known for its dryland vegetation, where many plants are strikingly adapted to drought. A curious example is *Welwitschia mirabilis*, which grows in the coastal desert of Namibia, on flat stony ground between sand dunes. The plant was first discovered at Mossamedes in 1860 by the German botanist Dr Friedrich Welwitsch, after whom it was named. It has a swollen, rather tuberous stem that is usually only about 10 cm (4 in) high, though much larger specimens have been found, and it reaches 1.5 m (5 ft) in diameter. The stem may extend far below ground, and below that there is a massive taproot, which draws on the moisture that lies deep beneath the desert's surface.

The plant is curious because it produces only two true leaves after the seed leaves (cotyledons) have emerged. These two leaves continue to grow throughout the life of the plant, which may be up to 2,000 years, though radiocarbon dating has established that the average age of welwitschia plants is some 500–600 years. The leaves are produced on either side of the top of the stem.

The secret of the extraordinary longevity of welwitschia leaves is that they are continually renewed by cell division at the leaf base, while they shred and disintegrate at the leaf tip, despite being very tough. At a certain leaf length a

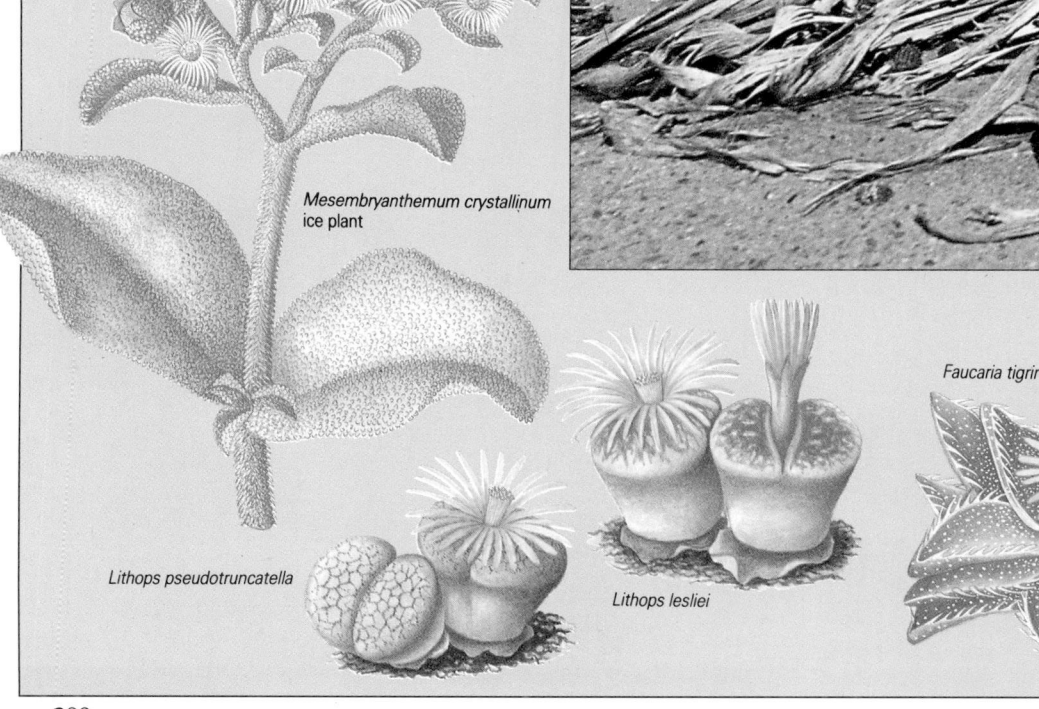

Mesembryanthemum crystallinum
ice plant

Lithops pseudotruncatella

Lithops lesliei

Faucaria tigrina

Enigma of the desert (*above*)
Although *Welwitschia* sets numerous seeds, many are infertile. Those fertile seeds that escape being eaten by desert animals can wait for a long time for rain to break their dormancy.

Living stones and succulents (*left*)
The annual ice plant, covered with glistening projections, trails on the sand. Its perennial relations are compact, and camouflaged except when in flower.

steady state of cell division is reached. The plants are either male or female. They are cone-bearing plants (gymnosperms), reproducing by means of cones that arise from pits beside the leaves rather than producing flowers, as angiosperms do.

Living stones

Another dryland adaptation is found in the succulents, of which the living stones (*Lithops*) are the most bizarre example. These plants, only a few centimeters high,

are perfectly camouflaged as stones, to prevent their being eaten by grazing mammals for their valuable water content. They achieve the mottled effect of natural stones by having orange-red chromoplasts and betalain pigments in their outer and inner (epidermal and subepidermal) layers of cells.

The body of the plant is made up of just two rounded, highly succulent leaves with a groove running between them. The leaves are attached to a short under-

ground stem. New leaves draw water from the old ones, whose withered remains lie at the base of the plant.

Some species have developed a further adaptation: they have transparent "windows" of clear tissue at the tip of the leaves, which allow light to penetrate to the photosynthetic tissues deep inside the leaf. The plants lose their camouflage when they produce their daisylike flowers, often brightly colored, from the groove between the leaves.

The extraordinary aloes

The quiver tree (*Aloe dichotoma*), with its thickset stem that can measure up to 1 m (3 ft) in diameter, is a giant succulent member of the lily family (Liliaceae). At the tips of its branches sprout rosettes of leaves from which grow short spikes of bright yellow flowers full of nectar. This is the main attraction to pollinating sunbirds. However, baboons often tear the flowers apart in their determination to get to this source of sweetness. The indigenous African San peoples used the soft branches as quivers for their arrows, from which the tree derives its name.

Aloe is a large genus of some 200 species, most of which grow in Africa. The quiver tree is not the only one of curious appearance. *A. plicatilis* forms a large shrub with dichotomously branched stems each topped by a fan of leaves. From the center of the fan arise deep pink flowers on a slender stem. The fleshy leaves of the rare spiral aloe (*A. polyphylla*) form a low, spiral-shaped rosette in the mountains of Lesotho. Another rarity, *A. haemanthifolia*, grows in the mountains of the southwestern Cape, where it survives winter snow.

Some species are very attractive. *A. candelabrum* has candles of bright red flowers, and both *A. arborescens* and *A. aristata* are grown in gardens for their striking flower spikes, which are scarlet and orange respectively.

Africans add the leaves of *A. aristata* to their porridge, and use the jellylike sap of many aloes to ease burns and stings. The old leaves of tree aloes are sometimes used for fuel. Perhaps the best known aloe is *A. vera*, a source of the bitter-tasting aloes long used as a purgative. Extracts from this plant were also used for cosmetic purposes by the ancient Egyptians. In recent years this use has been rediscovered, and *Aloe vera* has been marketed as a natural beauty treatment for hair and skin.

Quiver trees grow in the blistering heat of the rocky deserts of southern Africa.

FROM JUNGLE TREES TO DESERT SCRUB

FORESTS OF TEAK AND RHODODENDRON · CUSHION PLANTS AND FLOWERING TREES · PLANT USES AND ABUSES

Few regions can boast such a range of niches for plants as the Indian subcontinent. Woolly plants, hardy scrub and rhododendron forest grow on the world's highest mountains, the Himalayas, brilliant flowering trees light up dark deciduous forests, and the coasts of northwest India and Bangladesh are fringed with the mangrove swamps of the Ganges delta (the Sundarbans). There are dense rainforests on the Western Ghats in southwest India and in western Sri Lanka – rich in orchids and epiphytes – and thorn plants and succulents in the Thar Desert. Away from the mainland lie chains of palmclad islands: Lakshadweep, the Maldives, and farther south the Chagos Archipelago, including Diego Garcia. These ancient coral reefs have few plants, but some species are found nowhere else in the world.

COUNTRIES IN THE REGION

Bangladesh, Bhutan, India, Nepal, Pakistan, Sri Lanka

DIVERSITY

	Number of species	Endemism
Bangladesh and India	15,000	33%
Bhutan and Nepal	7,500	10%
Maldives	260	2%
Pakistan	5,500	5%
Sri Lanka	2,900	30%

PLANTS IN DANGER

	Threatened	Endangered	Extinct
Bangladesh and India	20	18	4
Sri Lanka	12	8	
Bhutan, Nepal, Pakistan	No current information		

Examples *Cycas beddomei* (Beddomes cycad); *Dioscorea deltoidea* (kin); *Diospyros oppositifolia* (opposite-leaved ebony); *Frerea indica* (frerea); *Lilium macklineae* (Shirhoy lily); *Paphiopedilum druryi* (Drury's slipper orchid); *Prunus himalaica* (Himalayan cherry); *Saussurea roylei* (Royle's saussurea); *Ulmus wallichiana* (Wallich's elm); *Vanda coerulea* (blue vanda)

USEFUL AND DANGEROUS NATIVE PLANTS

Crop plants *Cannabis sativa* (hemp); *Cinnamomum zeylandicum* (cinnamon); *Elettaria cardamomum* (cardamom); *Oryza sativa* (rice); *Piper nigrum* (pepper); *Saccharum spontaneau*; *Tectona grandis* (teak)
Garden plants *Buddleia crispa*; *Clematis montana*; *Corydalis cashmeriana*; *Delphinium* species; *Pleione humilis*; *Potentilla agryophylla*; *Rhodondendron* species
Poisonous plants *Antianis toxicaria* (chandla); *Atropa acuminata* (Indian belladonna); *Buscus wallichiana* (boxwood tree); *Datura stramonium* (purple thorn apple); *Gloriosa superba* (glory lily)

BOTANIC GARDENS

Calcutta (5,000 taxa); National Botanic Research Institute, Lucknow (6,000 taxa); Royal Botanic Gardens, Lalitpur (2,300 taxa); Royal Botanic Gardens, Peradeniya (4,000 taxa)

FORESTS OF TEAK AND RHODODENDRON

When the Indian subcontinent broke away from the ancient supercontinent of Gondwanaland, it probably brought with it tall conifer trees and podocarps that had evolved in largely temperate conditions. However, in response to the continent's new, largely tropical position, and to changes in the topography as the mountains eroded, a new plant life evolved. Plants came primarily from the east, and to a lesser extent from Africa. Today the subcontinent is dominated by families that are found in southeast Asia and probably made their way to the subcontinent through Assam in northeast India.

Mainland forest and scrubland

Most of the subcontinent receives an equable 500–2,000 mm (20–80 in) of rain each year. Where this falls relatively evenly throughout the year, as on the Western Ghats and in western Sri Lanka, rainforest flourishes.

These forests support a great diversity of species. The Western Ghats alone has about 1,500 species found nowhere else

Rhododendron forests in the Himalayas Tallgrowing rhododendrons like red *R. arborea* and *R. hodgsoni* grow in the forests alongside junipers and magnolias, while at high altitudes bordering the alpine meadows more scrubby species, such as yellow-flowered *R. anthopogon*, are common.

(endemics), including the spreading banyan tree (*Ficus benghalensis*). It also supports wild species of commercial crops, such as pepper (*Piper*), banana (*Musa*), cardamom (*Elettaria*), ginger (*Zingiber*), the beautiful ginger lilies (*Hedychium*) and several species of tuberous aroid, including the over-collected *Arisaema*. The wet tropical hills of Assam have about 2,000 endemic species; they represent a rich transition zone between the plants of Indochina and those of the subcontinent.

In most of the region there are one or two dry seasons separated by monsoon rains. Deciduous or monsoon forest thrives in these conditions. The deciduous forests contain many species of flowering trees, such as *Erythrina indica*, with its spectacular red flowers, the orchid tree (*Bauhinia purpurea*), and the fragrant-flowered *Gardenia*. Some of these forests are dominated by teak (*Tectona grandis*), particularly in Madhya Pradesh

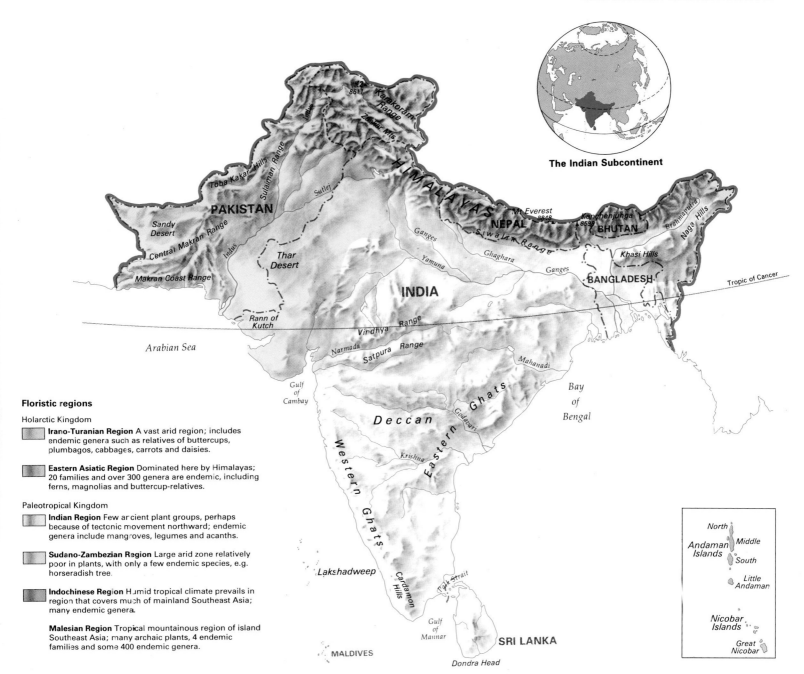

The Indian Subcontinent

Floristic regions

Holarctic Kingdom

Irano-Turanian Region A vast arid region; includes endemic genera such as relatives of buttercups, plumbagos, cabbages, carrots and daisies.

Eastern Asiatic Region Dominated here by Himalayas; 20 families and over 300 genera are endemic, including ferns, magnolias and buttercup-relatives.

Paleotropical Kingdom

Indian Region Few ancient plant groups, perhaps because of tectonic movement northward; endemic genera include mangroves, legumes and acanths.

Sudano-Zambezian Region Large arid zone relatively poor in plants, with only a few endemic species, e.g. horseradish tree.

Indochinese Region Humid tropical climate prevails in region that covers much of mainland Southeast Asia; many endemic genera.

Malesian Region Tropical mountainous region of island Southeast Asia; many archaic plants, 4 endemic families and some 400 endemic genera.

in central India, or sal (*Shorea robusta*) in the north. In areas of low seasonal rainfall, as in the arid Deccan peninsula, only savanna and scrub vegetation can grow, and *Acacia* thorn forests predominate.

In the far north of India, scrubby plants and succulents grow in the extremely dry, cold conditions. The rich plant life of the Himalayas totals about 9,000 plant species. Very small, squat, resilient plants, such as the woolly snowball (*Saussurea tridactyla*), and species of *Diapensia*, grow at the highest altitudes. Below these, in relatively arid areas, are temperate plants such as birch (*Betula*) and willow (*Salix*). The moister upper slopes support rhododendron forests and *Magnolia* species. Lower still coniferous forest, in which oaks (*Quercus*) also grow, clothes the mountain slopes.

The foothills of the Himalayas are covered with a wealth of coniferous trees such as juniper (*Juniperus*), pine (*Pinus*), larch (*Larix*), fir (*Abies*), cedar (*Cedrus*), spruce (*Picea*) and hemlock (*Tsuga*). In some areas these trees grow in single-species communities; elsewhere they are mixed with oaks and chestnuts (*Castanea*).

In northeast Pakistan, near the headwaters of the Indus and its tributaries, typical species include the deodar (*Cedrus deodar*), blue or Nepalese pine (*Pinus wallichiana*) and chir pines (*P. roxburghii*), Himalayan fir (*Abies pindrow*) and Himalayan spruce (*Picea smithiana*).

Plants of the islands

In the rainforest of southwestern Sri Lanka, the true cinnamon (*Cinnamomum zeylandicum*) and the kaluwara or Sri Lankan ebony (*Diospyros ebenum*) grow. This area is rich in endemics, including tree species of *Shorea* and *Ficus*. Epiphytes – plants that grow on other plants but are not parasitic – such as shrubby *Medinilla* species and orchids abound. The terrestrial foxglove orchid (*Phaius wallichii*) and the daffodil orchid (*Ipsea speciosa*) both occur nowhere else.

The natural plant life of Lakshadweep, the Maldives and the Chagos Archipelago comprises she-oak (*Casuarina*), mixed coconut or lettuce tree (*Pisonia*) woodland, *Scaevola taccada* scrub, marshland, and a few areas of broadleaf woodland, with *Ficus*, *Morinda citrifolia* and beach almond (*Terminalia catappa*). Much of the plant life on the larger islands has been replaced by coconut plantations, but the smaller islands retain some elements of their natural vegetation in a relatively undisturbed state.

Map of floristic regions The plant life of most of the Indian subcontinent reflects its largely tropical climate; the plants of the north have adapted to arid conditions.

CUSHION PLANTS AND
FLOWERING TREES

The plants of the Indian subcontinent have had to adapt to extreme conditions – notably very high altitudes and marked seasonal changes in climate. Two of the subcontinent's most remarkable plants are also among its smallest. They are a sandwort, *Arenaria bryophylla*, and a stitchwort, *Stellaria decumbens*, which contend for the title of the highest altitude plant in the world. They have been found at 6,100–6,200 m (20,000–20,340 ft), the sandwort growing on Mount Everest.

The environment where these plants grow is incredibly harsh, with freezing temperatures every night and most of the day; conditions are made worse by very strong winds and poor soils with few nutrients. However, the plants have adapted to overcome these difficulties in several ways. Both are cushion plants that grow in very dense, rounded mounds of short, many-branched stems. The wind passes over these cushions rather than through them, so less water is lost through the drying effect of the wind (desiccation). Because the plants are so compact, the temperature at the center of the cushions is slightly higher than that of the surrounding air; this makes it easier to withstand the intense cold. The plants also retain their dead leaves, which provide a source of nutrients in an otherwise nutrient-poor area.

Many other plants have remarkable ways of coping with life at high altitudes. Some have hairs on their leaves and stems – the snowball plants (*Saussurea*) have dense, long white hairs over their stems, leaves and flowers. These are thought to protect the plant against either high daytime temperatures, the intense ultra-violet radiation found at high altitudes or the effects of frost and freezing.

Flowers in the dry season

In many lowland areas there is a distinct hot, dry season. The trees adapted to this climate tend to be deciduous; they lose their leaves in the dry period, and produce their flowers after leaf-fall. This early flowering means that the fruits will begin

Epiphytic orchid *Coelogyne cristata* grows in the cloud forests of the Himalayas. It clings to bark or mossy rocks by means of specialized roots. It has a creeping rhizome from which arise pseudobulbs, succulent organs for storing water and nutrients.

Screwpines on the Maldives The lower trunk is surrounded by a cage of aerial prop roots, which give support and are also a response to periodic flooding by the sea. Screwpines are not pines or palms, but monocotyledonous trees.

PERADENIYA BOTANICAL GARDENS

The Peradeniya Botanical Gardens in Sri Lanka lie some 100 km (60 mi) east of the capital, Colombo, just outside the royal town of Kandy. The first gardens on the site were founded by King Kirthi Sri in the mid-18th century. The colonial Dutch and British governments later established the botanical gardens, which were subsequently developed into the striking formal gardens of today. Their setting, on rolling hills in the horseshoe curve of the Mahaweli, the longest river in Sri Lanka, is dramatic in itself.

In March and April the gardens become luminous with displays of flowering Sri Lankan trees, such as the murutha (*Lagerstroemia speciosa*), with its large candles of mauve flowers, the yellow poiciana (*Peltophorum inerme*), with its masses of yellow blooms, and the na or ironwood (*Mesua nagassarium*), which has scented white flowers and, at other times of the year, bright crimson new leaves.

The medicinal garden is of particular interest – it contains about 300 of the 500 or more species of plants used in the traditional medicine of Sri Lanka. Among them is the snakeroot (*Rauwolfia serpentina*), which has long been used to relieve the effects of snake bite. It is now used in Western medicine in the drug reserpine, which is derived mainly from the roots of the plant, and is used to lower blood pressure.

to ripen by the time the rains come; the seeds are then dispersed, and germinate in favorable growing conditions. As the seedlings grow they are shaded by the canopy of new leaves on surrounding trees, so the young plants have a chance to become established before the onset of the next dry season.

Firmiana colorata is a typical deciduous tree that has adapted to the dry season. Its clusters of bright coral red, tubular flowers, which produce copious nectar, are rendered even more conspicuous by appearing on leafless branches. Flowers of this kind are particularly attractive to birds, in this case to sunbirds and bulbuls, which act as pollinators.

The silk cotton tree (*Bombax ceiba*) is also pollinated by birds. It too produces its spectacular display of large red flowers on leafless branches; in common with many of these trees, the leaves are not invariably shed – they are retained if the trees grow in moist areas, dropping only when the new leaves form in March. The silk cotton tree has developed a number of other adaptations for its survival. Its fruits open before they fall, revealing seeds that are covered with silky hairs (giving the tree its common name), which are caught and dispersed by the wind. The bark of the young tree is covered with conical prickles that discourage predators – though they disappear as the tree grows older. The young branches of *Erythrina indica* also bear prickles that are dropped after two or three years.

The trees *Dillenia pentagyna*, of the northern sal forests, and karmal (*Sterculia guttata*) also flower before the new leaves unfurl. The flowers of karmal trees are a livid purple, and have a strong smell that attracts carrion flies and fruit flies, which perform the task of pollination.

Deciduous forests composed almost entirely of sal trees grow along the base of the Himalayas, and in the northeast and southeast of the Deccan. The hardwood trees are extremely resistant to fire – a constant threat in the hot, dry conditions. Like most trees that grow in areas with a distinct dry season they are deciduous, but they are never quite leafless; the young foliage appears in March along with the flowers. The seeds ripen three months later and sometimes germinate before they are released, so that once they land on the ground they can quickly become established.

In the drier parts of the western Deccan *Hardwickia binata* flourishes. Its seedlings cope with the difficulties of finding enough water by producing a taproot an amazing 3 m (10 ft) long.

PLANT USES AND ABUSES

The Indian subcontinent contains 18,000 or so species of flowering plants, a large number of which are used for a variety of purposes, or are of horticultural value all over the world. Many plants of the Himalayas, for example, are now so familiar to gardeners in temperate regions that their genus names have entered the vernacular. They include clematis, aster, delphinium, aquilegia, geranium, iris, primula, mahonia, and the Himalayan balsam and busy lizzie (*Impatiens*).

The forest trees are also used for a multitude of purposes. In India alone, minor products (everything except timber) such as pharmaceuticals, resins, gums, oils and bamboo (widely used in the construction industry and for furniture) probably account for roughly half of the country's net revenue from the forestry sector. However, the plant life of the region is very seriously threatened by the excessive requirements and rapid expansion of the human population.

Multipurpose plants

Among the many species with a number of uses are the palms, and in particular the talipot or umbrella palm (*Corypha umbraculifera*), so called because up to 15 people can shelter beneath a single leaf. The leaf bases of this tree provided the "paper" for important manuscripts and government documents; there are examples in museums that are over 1,000 years old. Twine from the dried young leaves can be made into fishing nets and mats. In addition, the roots are beneficial as a treatment for diarrhoea, sago flour can be extracted by rinsing the pith, and sugar can be tapped from the enormous flower stems (8 m/26 ft tall) just before the flowers open.

Another multipurpose plant widely used in rural areas is the soapnut (*Sapindus laurifolius*). As its common name suggests, the fruit contains the soapy alkaline substance saponin, which is a substitute for soap. But this is not all. The root can be taken as an expectorant to relieve catarrh, and the seed kernels are given as pessaries to stimulate contractions of the uterus in childbirth.

The sohnja or drumstick tree (*Moringa oleifera*) of the western Himalayas is widely cultivated for the gum that can be obtained from the fruits. (It is similar to gum tragacanth, obtained from species of *Astragalus*.) A clear oil (ben oil) is extracted from the seeds and used as a lubricant and in perfumery.

Timber trees under threat

Timber is the most valuable plant commodity in the region, with teak particularly in demand. It is one of the world's finest, heaviest timbers, and is

Mango trees grow wild in the forests of Assam. The conspicuous inflorescences of the musk-scented flowers are pollinated by bats. Mangifera indica is the only mango in cultivation, though other species bear luscious fruits that are eaten locally.

especially favored for boatbuilding because it resists certain burrowing marine mollusks. The timber of sal, which is highly resistant both to rotting and to attack from burrowing animals, is used to make railroad sleepers for the subcontinent's vast rail network.

Among the important Sri Lankan timber trees are burutha or satinwood (*Chloroxylon swietenia*), which has beautifully patterned yellow wood; the kaluwara or Sri Lankan ebony, with its black heartwood, which is greatly favored by wood carvers and cabinetmakers; and the true cinnamon, prized for its bark, which is exported mainly to Europe where it is used in cooking.

Despite the commercial importance of trees, in most countries of the region less than 15 percent of the land has any form of forest cover, and much of that is degraded. Roughly one percent of the remaining forests are being lost each year, generally to inappropriate forms of extensive agriculture. In areas with an abundance of species, such as the Western Ghats and southwestern Sri Lanka, steady loss of the forests, sometimes even within protected areas, threatens the longterm viability of some populations of endemic species, and extinction may be only a few decades away.

In some areas, such as Kashmir in the north, there are fine stands of deodar. However, the young trees will probably be the last generation to grow into stately trees. Uncontrolled cattle grazing and the collection of firewood have removed most of these saplings, so there is no regrowth.

A similar problem faces the forests of Baluchistan in Pakistan. The area was once covered in seemingly endless juniper forests, but now stands of stumps are all that remain of much of these forests. The trees were taken for fuelwood, which provides more than 50 percent of domestic energy needs for 75 percent of the people of Pakistan. When the population was small the trees were able to regenerate, but as it has increased, so has the demand for wood.

Uncontrolled cutting of the trees for fuel on the small islands off the south and southwest coasts of India has made fuelwood a scarce commodity there as well. In recognition of this problem, the Maldives government launched a largescale treeplanting program. Many more of these programs are needed if the forests are to survive into the future.

THE MEDICINAL PLANTS OF INDIA

As many as 2,500 species of plants in India are believed to have some medicinal properties. More are being discovered in the wake of a government program that is investigating the reasons for the cures these effect. The enormous task of documenting the uses of India's plants was first tackled at the end of the 19th century. The findings were published between 1889 and 1893 in the *Dictionary of the Economic Products of India*, a nine-part, 5,000-page work that still forms the basis of what is known today.

Much of this dictionary reads like an old herbal, with innumerable cures for headaches and gastric disorders. However, some concoctions are recommended for remarkably specific conditions; for example, maidenhair fern (*Adiantum*) prevents baldness; bamboo leaves mixed with black pepper stop diarrhoea in cattle; the milky sap of banyan trees softens cracked skin on the soles of the feet; delphinium roots kill maggots in goat wounds; and charred teak wood, soaked in poppy juice and pounded into a smooth paste, relieves swollen eyelids and improves the sight.

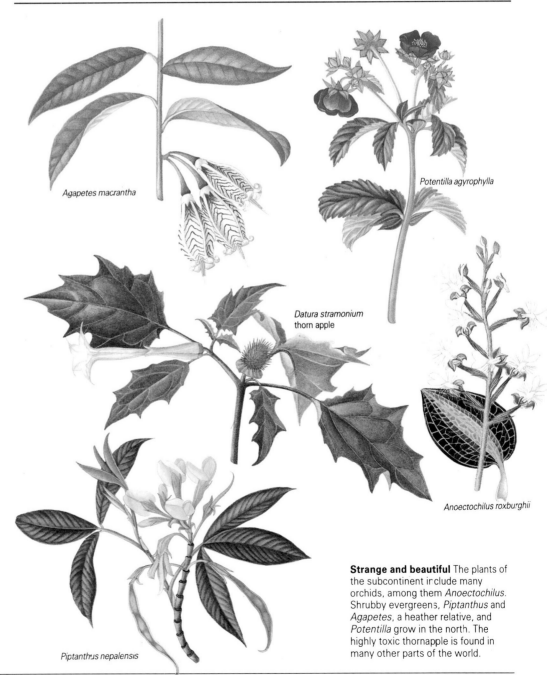

Agapetes macrantha

Potentilla agyrophylla

Datura stramonium
thorn apple

Anoectochilus roxburghii

Piptanthus nepalensis

Strange and beautiful The plants of the subcontinent include many orchids, among them *Anoectochilus*. Shrubby evergreens, *Piptanthus* and *Agapetes*, a heather relative, and *Potentilla* grow in the north. The highly toxic thornapple is found in many other parts of the world.

The Sundarbans

The mangrove forests of the Sundarbans are the most extensive in the world. They cover more than 4,000 sq km (1,540 sq mi) of the Ganges Delta in Bangladesh and northeast India, and have been protected for more than a century.

Swamp survival

The Sundarbans forest mainly consists of species typical of the less salty, landward fringes of most Asian mangrove swamps. It differs from most swamp forests, probably because of the enormous volume of fresh water that is brought down by the Ganges, Brahmaputra and lesser rivers, reducing the water's saltiness. It is similar in that it contains relatively few plants other than trees.

The main forest species is sundri (*Heritiera fomes*) – the origin of the name Sundarbans, previously "Sundribans". It is mixed in places with small trees such as

Where the forest meets the sea The Sundarbans covers much of the vast delta of the Ganges. The sacred river of India, the Ganges rises in the Himalayas, as does the Brahmaputra. These two, and a number of other lesser rivers and tributaries, bring an enormous volume of fresh water down from the mountains, giving a unique character to the mangrove forests of the Sundarbans. Here the trees have evolved vertical pneumatophores, which are exposed at low tide to "breathe"; when the tide comes in the breathing pores close to prevent the entry of water.

Inside the Sundarbans In the forest, marked by an absence of lianas and few epiphytes, sundri grows with *Nipa fruticans*. This stemless palm of tidal swamps has large fruits up to 30 cm (1 ft) in diameter, which float on the sea to be dispersed.

gewa (*Excoecaria agallocha*) and goran (*Ceriops decandra*). In the Sundarbans the sundri has vertical, woody extensions (pneumatophores) on its root system that project above the ground, though in other parts of its range these are absent. The pneumatophores have pores (lenticels) and specialized internal tissue that enable the plant to breathe (exchange gases with the surrounding air). The height of the pneumatophores varies from area to area, depending on the heights reached by the tides. The roots of gewa trees also project above the ground in small loops or "knees" of the horizontal roots below ground. (Gewa trees have copious white sap that is poisonous, and can cause blindness if it touches the eyes.)

Goran trees do not have specialized roots to cope with waterlogged soil and wave action, but they do have lenticels in their bark through which gases can be exchanged. They are members of the true mangrove family and, like other species, produce fruit with seeds that germinate while still on the tree. The young root (hypocotyl) that develops from the seed is robust, heavy and pointed; when the seed is released, the root therefore stands a good chance of penetrating the mud if the tide is out.

An endangered habitat

The devastating floods experienced in Bangladesh during the 1980s initiated a wide range of engineering projects to keep the water within the riverbanks and away from the densely populated flood-plains. As a result, the original pattern of flow of fresh water through the innumerable channels has changed, and it is possible that this has caused the sundri trees to die back in some areas. The trees, which are the source of a very hard timber, have been felled at a greater rate than their slow growth can sustain. Goran trees too are felled, and are used for firewood, boatbuilding, tanning and houseposts.

Given that these practices are continuing, it seem unavoidable that there will be changes in the forest ecosystem. These are likely to be compounded by the effects of global warming and the predicted rise in sea levels. In recognition of these threats, there are moves to have the whole of the area covered by mangrove forest declared an International Peace Park, which would be administered jointly by Bangladesh and India.

A PARADISE FOR PLANTS

A HAVEN FROM THE ICE · ADAPTING TO MOUNTAIN TERRAIN · DEPENDING ON NATIVE PLANTS

No less than 30,000 species of higher plants grow somewhere within the borders of China, some 10 percent of the world's total. Great topographic and climatic diversity provide a wide range of different habitats that support many types of plants. Alpine meadow and high altitude desert plants grow on the vast snow-capped mountain ranges that surround the Plateau of Tibet. To the north lies the scrub of the Takla Makan desert. In the extreme northeast the steppe grasslands resemble those of neighoring Siberia, while Hainan island in the extreme southwest of the region luxuriates in tropical heat and humidity, and supports lush tropical plants and rainforest. China's plant life has been enhanced by the limited effects of the ice ages there, allowing many ancient plants to survive that became extinct elsewhere.

COUNTRIES IN THE REGION

The People's Republic of China, Taiwan

DIVERSITY

	Number of species	Endemism
China	30,000	7%
Hong Kong	2,500	1%
Taiwan	3,500	25%

PLANTS IN DANGER

Information limited; at least 350 species threatened in China and 300 in Taiwan

Examples *Abies beshanzhuensis* (Baishanzhu mountain fir); *Burretiodendron hsienmu* (xianmu); *Camellia granthamiana*; *Coptis teeta*; *Cycas taiwaniana*; *Davidia involucrata* (dove tree); *Ginkgo biloba* (maidenhair tree); *Kirengeshoma palmata*; *Metasequoia glyptostroboides* (dawn redwood); *Panax ginseng* (ginseng)

USEFUL AND DANGEROUS NATIVE PLANTS

Crop plants *Actinidia chinensis* (kiwi fruit); *Citrus medica* (citron); *Citrus sinensis* (sweet orange); *Eleocharis tuberosa* (Chinese water chestnut); *Glycine mas* (soybean); *Litchi chinensis* (litchi/lychee)
Garden plants *Berberis calliantha*; *Camellia reticulata*; *Cotoneaster* species; *Ginkgo biloba*; *Mahonia lomariifolia*; *Pleione praecox*; *Primula sinensis*; *Rhododendron* species; *Wisteria sinensis*
Poisonous plants *Nerium odora*; *Physalis alkekengi*; *Veratrum wilsonii*; *Wisteria sinensis*

BOTANIC GARDENS

Beijing (2,500 taxa); Canton (3,200 taxa); Hong Kong (1,200 taxa); Ping-Tung (1,200 taxa); Shanghai (4,000 taxa)

A HAVEN FROM THE ICE

China has a remarkably high number of plant genera, and even families, that include only a single species. These are often endemic to China (found only there), and quite frequently restricted to small areas. Two well-known examples are the dove, or handkerchief, tree (*Davidia involucrata*) and the ginkgo or maidenhair tree (*Ginkgo biloba*). Several major Eurasian or more widely distributed genera are also exceptionally well represented. These include the rhododendrons (*Rhododendron*), with 850 or so species worldwide, of which some 500 occur in China.

A wealth of species

The richness and diversity of China's plant life is due partly to the survival of many ancient plants that became extinct in other parts of the world. A major reason for their survival is thought to be

On top of the world *Rhododendron yunnense*, seen here growing at over 3,000 m (9,850 ft), surveys the thickly wooded landscape typical of much of western China. A thicket-forming shrub, its range extends into Tibet and Burma.

the fact that large areas of China escaped glaciation during the ice ages. Plants were able to retreat before the slowly advancing icesheets to areas free of ice, where the climate favored their survival. After the ice thawed, diversity was encouraged by the wide range of habitats available. The uplifting of the Plateau of Tibet and its associated mountain systems in western China created a multitude of different microclimates.

From tropical rainforest to desert

Tropical plants are found in only the most southerly parts of the mainland and on Taiwan, Hainan and the smaller islands of the South China Sea. Tropical rainforest grows in parts of these areas; it is composed mainly of evergreen broadleaf trees such as dipterocarps (*Dipterocarpus*),

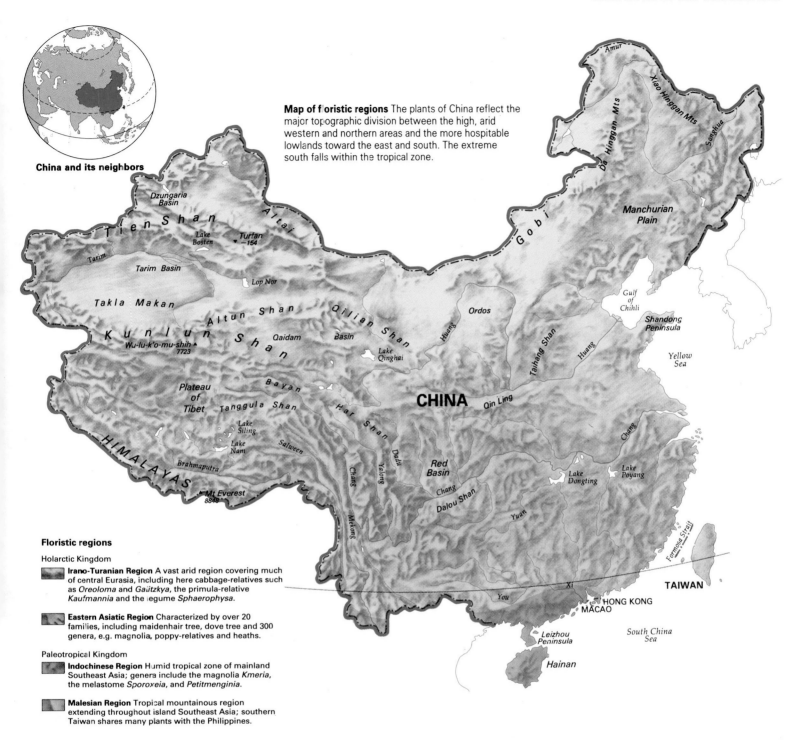

China and its neighbors

Map of floristic regions The plants of China reflect the major topographic division between the high, arid western and northern areas and the more hospitable lowlands toward the east and south. The extreme south falls within the tropical zone.

Floristic regions

Holarctic Kingdom

Irano-Turanian Region A vast arid region covering much of central Eurasia, including here cabbage-relatives such as *Oreoloma* and *Gaützkya*, the primula-relative *Kaufmannia* and the legume *Sphaerophysa*.

Eastern Asiatic Region Characterized by over 20 families, including maidenhair tree, dove tree and 300 genera, e.g. magnolia, poppy-relatives and heaths.

Paleotropical Kingdom

Indochinese Region Humid tropical zone of mainland Southeast Asia; genera include the magnolia *Kmeria*, the melastome *Sporoxeia*, and *Petitmenginia*.

Malesian Region Tropical mountainous region extending throughout island Southeast Asia; southern Taiwan shares many plants with the Philippines.

Parashorea and banyans (*Ficus*). However, much of the original forest has been cleared and replaced by scrub or cultivated land.

Most of China, from Manchuria in the north, southward to northern Guangdong (including Hong Kong and Macao), and from the east coast inland to the Tibetan borderlands, supports temperate to subtropical forests. There are more than 20,000 species, the most typical being deciduous or evergreen oaks (*Quercus*, *Lithocarpus*, *Castanopsis*). Conifers are also well represented, with several Chinese specialties such as golden larch (*Pseudolarix*) and China fir (*Cunninghamia*). Little trace remains of the original plant life in lowland areas, however; almost all

the available fertile land has been under intensive cultivation for many hundreds of years, and the area is also extremely densely populated.

The Himalayan area of southeastern Tibet and southwestern China is extremely rich in plant species – there are some 12,000 in northern Yunnan province alone. Damp mountain slopes are clothed with forest, giving way at high altitudes to alpine scrub and meadows. Many different rhododendrons, gentians and lilies grow here.

The Plateau of Tibet was formed only comparatively recently and has a cold, dry climate in which plants find it difficult to survive. Its distinctive vegetation consists of only some 500 species, varying from

low scrub and meadow through grassland to high-altitude desert with specialized cushion plants.

In the extreme north of China plant life is similar to that of neighoring areas of Soviet Central Asia, Mongolia and Siberia. Desert plants grow in the arid parts of Xinjiang province in the northwest and in western Inner Mongolia. Eastern Inner Mongolia and adjoining parts of Manchuria are covered with drought-resistant grasses. The forests are dominated by conifers such as spruce (*Picea*), pine (*Pinus*) and fir (*Abies*), which grow in the mountains of northern Xinjiang (the Tien Shan and Altai) and the Da Hinggan range on the Mongolian–Manchurian border.

ADAPTING TO MOUNTAIN TERRAIN

Southwest China is a land of high mountains and deep valleys, most of which run roughly north to south. The rugged topography of the region and its generally damp climate offer an extraordinarily wide range of ecological niches that have favored the development of a very rich and varied plant life. Northwest Yunnan and western Sichuan support an assortment of more than 15,000 plant species, many of them local endemics. Some grow only on one mountain or group of mountains – *Gentiana omeiensis, Magnolia dawsoniana* and *Picea aurantiaca* for example.

China's rhododendrons

The largest genus of higher plants in China is *Rhododendron*. Its species grow virtually throughout China, except in the far northwest. Only a few species grow in the north; they are much more numerous south of the Chang river, and are found predominantly in the mountainous regions of west and southwest China. A single mountain in Sichuan may support 15 or more species, and no less than 277 occur in Yunnan province.

The rhododendrons of southwest China vary widely in both size and growth habit. Some are trees that reach heights of 9–12 m (30–40 ft). Others are shrubs of various sizes, including tiny, lowgrowing, matlike forms. Their characteristics reflect their adaptation to the widely varying habitats that they occupy, from forest at low and medium altitudes to alpine scrub and mountain heathland at the upper limits of woody vegetation. Some species are epiphytic, growing on the branches of trees. It is even possible to find small rhododendron species growing on larger ones. All rhododendrons need plenty of moisture, so none grow in arid areas.

Most Chinese rhododendrons are evergreen, and many of the bigger species have large, leathery leaves. *Rhododendron giganteum* has been found to grow to a height of 24 m (80 ft), with leaves up to 40 cm (16 in) long. China's monsoon climate means that there is copious rainfall in summer, but long dry periods in winter, when moisture in the soil is often frozen. Moisture loss from large evergreen leaves would normally be high during dry spells, but the rhododendrons have

developed ways to reduce the amount of water they lose. When they cannot take up much water because the ground is dry or frozen, the edges of the leaves roll under to cover their lower surfaces, and the leaves also fold down toward the stem. This protects the lower surfaces of the leaves, from which water loss is usually highest, and significantly reduces the total leaf area exposed to drying winds. The lower surfaces of the leaves of many rhododendron species have a mass of brown or reddish, feltlike scales or hairs (called an indumentum) that also help to reduce water loss.

Rhododendrons have fibrous root systems that spread widely but shallowly, which enables them to absorb rainwater quickly before it runs off the slopes on which they frequently grow. In cultivation rhododendrons are lime-hating plants, but in the wild many grow on limestone mountains. It is possible that their shallow roots do not penetrate the acid upper layer of soil down to the limy rock below. It is also likely that the magnesian limestone of many Chinese mountains causes them less problems than carboniferous limestone.

ENRICHING THE WORLD'S GARDENS

The Chinese have been making gardens for more than 3,000 years. Some of their ornamental plants have been selected and hybridized for 2,000 years. The exact origins of the chrysanthemum *Dendranthema morifolium*, for example, are still obscure because the cultivated plants are complex hybrids.

Europeans, who reached China by sea during the 16th century, were soon attracted to the ornamental plants they found there. In 1656 the first Dutch embassy to Beijing returned full of praise for Chinese tree peonies (*Paeonia suffruticosa* cultivars). During the following century and a half a slow trickle of plants acquired from Chinese gardens began to arrive in Europe. From the mid-19th century plant collectors were able to travel inland for the first time across remote parts of China. They sent back vast numbers of plants from the wild to Europe and North America, where gardens were greatly enriched by the previously unknown rhododendrons, camellias, primulas, lilies, roses, cotoneasters, magnolias and other fine ornamentals.

Golden straplike petals (*above*) Witchhazel (*Hammamelis mollis*) is a small tree that flowers in winter before its leaves unfurl. This allows the modest but fragrant flowers to be shown to maximum advantage to pollinating insects.

The blue poppy of Yunnan (*left*) *Meconopsis betonicifolia* is a tall herb of cool, moist, sheltered places in Tibet and western China. Here in well-drained soil it maintains a perennial way of life, though it also sets seed freely.

Folding leaves in winter (*below*) Growing in the woodlands of western China, the large tree rhododendron *R. calophytum* has to adapt to a wide range of conditions, including icy winters.

Other mountain plants

Much of the high Plateau of Tibet is so cold and arid that it supports no shrubs at all. Winters there are very long and cold, and plants must be able to flower and seed quickly in the short growing season. Even in July and August the temperature often drops below freezing at night, and the little precipitation there is may fall as hail, sleet or snow.

The plants are either entirely herbaceous, dying down to spend the winter underground, or grow into tight cushions with small, often hairy leaves that lose water only very slowly. *Arenaria roborowskii*, a typical cushion plant of the area, has a long, woody tap root that penetrates deep into rocky soil to find moisture, and short, densely clustered stems that bear tiny, needlelike leaves.

The genus *Nomocharis*, which has only some six to eight species, is very closely related to the lily genus, *Lilium*. It grows in the mountains of southwest China and southeast Tibet (and in northeast Burma). All the species (with the possible exception of one) include parts of Yunnan province within their range of distribution, which suggests that they evolved fairly recently from the true lilies in the mountains of northwest Yunnan.

Nomocharis is a good example of how species diversify in response to local topography: different species have evolved on neighboring mountain ranges, separated by deep valleys where conditions are unsuitable for them. A series of closely related species of *Nomocharis* grows on the mountains that form the border between China's Yunnan province and Burma. These are replaced on the mountain ranges to the east by a different group of species.

DEPENDING ON NATIVE PLANTS

China is a very ancient agricultural area, where people probably began the transition from hunting and gathering to cultivating food between 10,000 and 8,000 years ago. It was once thought that Chinese agriculture was originally based on plants introduced from elsewhere, but it is now becoming clear that the earliest domesticated species were in fact local. Thousands of years ago there were plenty of wild plants available in China for domestication as crops, and these original species are still found growing wild in China today.

The earliest date for which there is evidence of Chinese agriculture is about 5,000 BC. At that time there were three quite distinct centers of agriculture in China, each dependent on different crops. One was situated on the loess plateau of central north China, where there is little rain and the winters are cold. The principal grain crops were foxtail millet (*Setaria italica*), and later, broom-corn millet (*Panicum miliaceum*) and gaoliang (*Andropogon sorghum*).

Another area lay in the warm, humid lands of the lower Chang river valley and adjacent parts of the east China coast. Here the major grain crop was rice (*Oryza*

sativa). Rice cultivation had reached an advanced stage here by 5,000 BC, earlier than at any other known site in the world, so it is highly likely that this was the first area in which rice was domesticated. Other important food crops here were aquatic lotus (*Nelumbo nucifera*), which was grown both for its seeds and its roots, and water-caltrop (*Trapa*).

The third agricultural zone, in the far south of China, is based on the cultivation of tropical root crops as staples rather than grains.

Crop plants that originated in China are now widely grown around the world. Some of the most important include soya bean (*Glycine max*), hemp (*Cannabis sativa*), peach (*Prunus persica*), apricot (*Prunus armeniaca*) and persimmon (*Diospyros kaki*) from northern China; and the tangerine (*Citrus reticulata* var. *deliciosa*), loquat (*Eriobotrya japonica*), tea (*Camellia sinensis*), lychee (*Litchi chinensis*) and longan (*Dimocarpus longan*) from the south.

Many beautiful herbs There are an enormous number of flowering herbs of horticultural value to be found in China. Although a species may be endemic (found nowhere else), it may be a representative of a widespread genus such as *Allium*, *Primula* or *Leontopodium*, which includes the famous edelweiss of the European Alps.

Stongly textured foliage, creamy flowers and the shiny berries of evergreen *Vibernum rhytidophyllum* have made it a popular garden shrub. Native to central and western China, it is one of several viburnums now widely used in horticulture.

Leontopodium haplophylloides

Allium cyaneum

Primula veitchii

Cotoneaster salicifolia

Staphylea holocarpa

Decorative shrubs China is very rich in attractive flowering shrubs and trees. It is, therefore, no wonder that plant collectors from Europe and North America have risked life and limb to take home specimens from the wild.

Chaenomeles speciosa flowering quince

Chinese medicine

The Chinese have long had an extensive knowledge of the medicinal properties of plants. One of their legendary god-ancestors, Shen Nong (the Divine Husbandman), is credited not only with the origination of agriculture, but also with testing numerous wild herbs on himself to determine their medicinal qualities. Books listing medicinal plants and their uses were already in existence more than 2,000 years ago. They were continuously revised and expanded; one of the most famous and comprehensive, dating from 1578, lists 1,892 medicinal materials, most of them derived from plants.

In modern China traditional medicines are still prescribed, and a great deal of research has been carried out to provide a scientific basis for their use. They are very often successful, even in cases where orthodox medicine can offer no effective treatment. The root of the yellow-flowered peony (*Paeonia lutea*), known as *Shaoyao* in Chinese, is used to treat a variety of disorders including headache with vertigo, painful stomach spasms and diarrhoea, irregular or painful menstruation and certain kinds of fever. The bulbs of several lily species, called *Bai He* in Chinese, are prescribed for chronic coughing, certain disorders of the blood, neurosis and insomnia.

Deforestation and pollution

China has the largest population of any country in the world, with more than 1,100 million people. Despite its large size, the pressures of population are severe and seriously affect the vegetation, mostly because of deforestation and pollution. Wood is still extensively used as fuel, especially in rural areas where it is often the only available material for heating and cooking. It is also in demand for industrial use.

The rapid development of industry in China since the establishment of the People's Republic in 1949, combined with few effective controls on pollution, has led to severe problems from acid rain and the contamination of rivers. Such problems were often ignored during the earlier years of the People's Republic, but official attempts are now being made to reduce them. Major tree-planting campaigns have produced good results in some areas, though there is still pressure on natural forest resources. Legislation on pollution has been passed, but it is not always enforced.

Great strides have been made, however, in the creation of nature reserves for the protection and conservation of at least some areas of natural vegetation with rare plants and animals. The first such reserve was established in 1956 at Dinghu Shan in Guangdong province in southeastern China. Since then some 120 such reserves have been set up around the country, several of them covering hundreds of thousands of hectares.

THE VERSATILE BAMBOO

Bamboos are some of the most versatile of Chinese plants. They belong to several genera within the family Gramineae. These woody stemmed grasses come in many sizes, from dwarf species only centimeters high to giants that grow to tens of meters. Their versatility lies in their canes, which are both flexible and immensely strong. For thousands of years they have been used in the construction of buildings and for scaffolding and ladders, for making simple rafts and boats, as fishing rods, water-pipes and musical instruments.

Cut to length and split, the canes can be made into chopsticks, the standard eating implement of the Chinese. Split more finely, they can be woven into baskets, matting, hats and many other useful objects. Many have fibers suitable for making paper. Apart from the multitude of structures, objects and products that their canes provide, bamboos have other uses. Some species produce edible shoots and others have medicinal value. They are also very popular as ornamentals in Chinese gardens. With their wide range of uses, bamboos have made a significant contribution to Chinese culture.

Elegant striped stems *Bambusa vulgaris* "Vittata" has long been grown in Chinese gardens; it is probably the mostly widely grown of all bamboos throughout the tropics. It is also used in the manufacture of paper pulp.

China's living fossils

In the Tianmu Mountains in Zhejiang province on the southeast coast of China grow what may be the last surviving, truly wild, ginkgo trees, remnants of a whole class of plants that was once widespread throughout the northern hemisphere. During the last ice age adverse climatic conditions wiped out every species of this class except the one that now remains in China, *Ginkgo biloba*.

Although only a few wild ginkgos still exist in one small location, the tree, also known as the maidenhair tree, has been widely cultivated in China for a very long time. It is slow growing and long lived; ancient trees can still be seen in many places around the country, especially near old temples. The tree was introduced to Japan and Korea long ago, and reached Europe during the early 18th century.

The ginkgo is not the only ancient tree surviving in China. There are also the plants of the genus *Metasequoia*, conifers that are closely related to the North American swamp cypresses (*Taxodium*). The genus was known from fossil remains about 100 million years old, but until 1941 was believed to be extinct. In that year a Chinese forestry scientist discovered three trees that appeared to belong to an unknown species growing near the border between the provinces of Sichuan and Hubei in central China. In 1944 more trees of the same species were found, and study confirmed that they did indeed belong to a previously unknown species of the genus *Metasequoia*. This species was described and named in 1948 as *Metasequoia glyptostroboides* – commonly known as the dawn redwood.

Other survivals from the past

The exceptional conditions in China during the glacial periods that enabled the ginkgo and metasequoia to survive also favored other plants. The golden larch (*Pseudolarix amabilis*) is the only surviving species of its genus, which is possibly 200 million years old. It is found only in a few mountainous areas south of the Chang river. In a natural state it grows among mixed coniferous and broadleaf woodland, becoming conspicuous in fall when its deciduous needles turn golden. A 400-year-old specimen in the Tianmu Mountains is 53 m (174 ft) tall and has a trunk 3.6 m (12 ft) in diameter. A reserve has been established in this mountain area to protect naturally occurring stands of golden larch.

The evergreen conifer *Cathaya argyrophylla* is one of China's rarest trees, and was discovered only as recently as 1955. At first known from one locality in the mountains of Guangxi province in central southern China, it has now been found at several scattered sites. It is resistant to drought and likes plentiful sunlight, but grows slowly and is difficult to propagate. Before the ice ages *Cathaya* was widespread throughout Eurasia, but came close to extinction three million years ago because of glacial action.

Another fascinating relict tree is the broadleaf *Davidia involucrata*, the only surviving member of its genus. In spring this beautiful, medium-sized tree produces heads of small flowers each with a pair of conspicuous large white bracts, which hang from the branches and have given rise to its common names of dove and handkerchief tree.

Fan-shaped leaves (*left*), the distinctive foliage of the gingko or maidenhair tree. A tree carries either male or female flowers. After fertilization the female trees bear fruit, the flesh of which has an evil smell though the kernel is edible.

Ethereal dove tree (*right*) White bracts do not usually appear in abundance until the tree is mature, when they form a canopy above an inconspicuous flower. The solitary fruits turn brown on ripening. The young foliage is pleasantly scented.

TROPICAL PLANTS IN PROFUSION

LUXURIANT FORESTS · ADAPTATION IN THE FOREST · SPICES, EBONY AND ORCHIDS

Southeast Asia is particularly rich in plant species, with a total of about 20,000 flowering plants. Almost all the region was once covered by dense forest, but logging for valuable timber trees has substantially reduced their extent. In both the north and the south there is a moderately dry season for part of each year, and plant growth depends on the heavy monsoon rains. Here semi-deciduous, seasonal tropical forests grow. Nearer the Equator the climate remains hot and wet all year round, and tropical rainforests flourish. The highest rainforests in the world are on the Malay Peninsula, Sumatra and Borneo. The conditions are perfect for luxuriant plant growth. There is also great variety – more than 2,000 species of orchid grow on Borneo alone. Many important food plants also originated in this region.

Widespread palms Members of the palm family can be found in many parts of Southeast Asia. Most produce a huge inflorescence below the crown of leaves; this sealing-wax palm (*Cyrtostachys renda*), a palm of peaty soil, is no exception.

COUNTRIES IN THE REGION

Brunei, Burma, Cambodia, Indonesia, Laos, Malaysia, Philippines, Singapore, Thailand, Vietnam

EXAMPLES OF DIVERSITY

	Number of species	Endemism
Burma, Cambodia, Laos Thailand and Vietnam	17,000	20%
Malaysia and Singapore	8,000	high
Borneo	10,000	high

PLANTS IN DANGER

Little information; many species threatened or lost because of extensive deforestation

Examples *Allobunkillia* species; *Amorphopallus titanum* (titan arum); *Johannesteijsmannia lanceolata* (umbrella leaf palm) *Maingaya* species; *Maxburretia rupicola*; *Nepenthes* – Mount Kinabulu species (pitcher plant); *Paphiopedilum rothschildianum* (Rothschild's slipper orchid); *Phyllagathis magnifica*; *Rafflesia* species; *Strongylodon macrobotrys* (jade vine)

USEFUL AND DANGEROUS NATIVE PLANTS

Crop plants *Artocarpus altilis* (breadfruit); *Camellia sinensis* (tea); *Citrus limon* (lemon); *Cucumis sativa* (cucumber); *Durio zibethinus* (durian); *Eugenia caryophyllus* (cloves); *Mangifera indica* (mango); *Musa* species (banana); *Myristica fragans* (nutmeg)
Garden plants *Cyrtostachys renda* (sealing wax palm); *Ficus benjaminia* (weeping fig); *Pandanus sanderi*; *Paphiopedilum barbatum*
Poisonous plants *Antiaris toxicaria*; *Croton tiglium*; *Gloriosa superba* (glory lily); *Strychnos nux-vomica*

BOTANIC GARDENS

Bogor; Manila (10,000 taxa); Penang (1,000 taxa); Rangoon; Singapore (3,000 taxa)

LUXURIANT FORESTS

The type of forest found in Southeast Asia is not only determined by the climate; soil also influences the distribution and growth of plants. In areas with a strongly seasonal climate, such as central Burma and the Indochina peninsula, scrubby savanna forests or thorn forests grow. On the flat coastal plains there are extensive swamplands dominated by mangroves. Mangroves also grow in the brackish tidal swamps along the muddy coasts, supported by sturdy prop roots that rise up out of the murky water. Inland, where the soil is fertile, many of the former fresh-water swamps have been converted into rice paddies, producing the most important staple food crop of Southeast Asia. A distinctive kind of swamp is also found over areas of acidic inland peat, which is extensive in Sumatra, the Malay Peninsula and Borneo.

In the high mountain ranges throughout Southeast Asia the climate is cooler than on the lowland plains, as it is in mountainous zones throughout the world; plant growth here may be more limited. Close to the Equator, however, the climate even of mountain altitudes

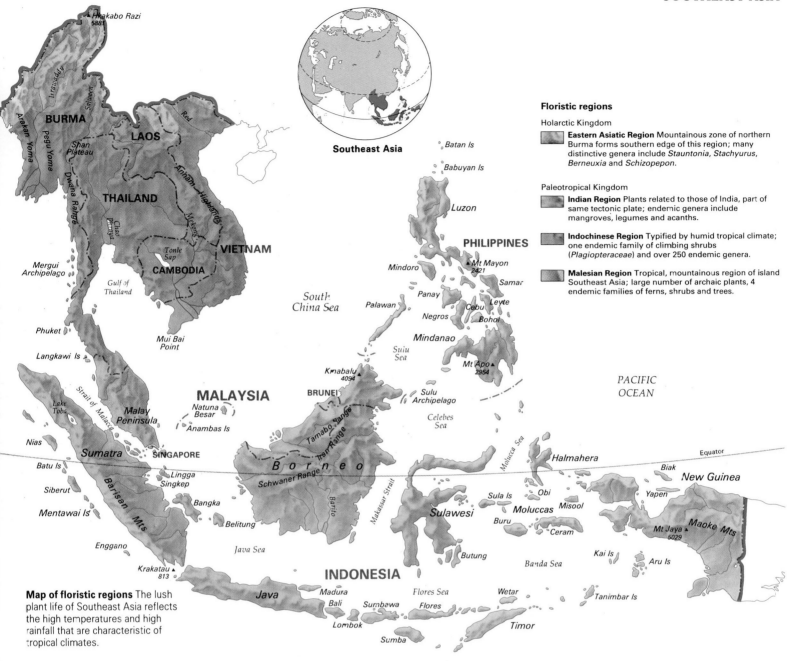

Southeast Asia

SOUTHEAST ASIA

Floristic regions

Holarctic Kingdom

Eastern Asiatic Region Mountainous zone of northern Burma forms southern edge of this region; many distinctive genera include *Stauntonia*, *Stachyurus*, *Berneuxia* and *Schizopepon*.

Paleotropical Kingdom

Indian Region Plants related to those of India, part of same tectonic plate; endemic genera include mangroves, legumes and acanths.

Indochinese Region Typified by humid tropical climate; one endemic family of climbing shrubs (*Plagiopteraceae*) and over 250 endemic genera.

Malesian Region Tropical, mountainous region of island Southeast Asia; large number of archaic plants, 4 endemic families of ferns, shrubs and trees.

Map of floristic regions The lush plant life of Southeast Asia reflects the high temperatures and high rainfall that are characteristic of tropical climates.

A glorious lily *Gloriosa superba* is a climbing lily, seeking support from trees to which it clings by its tendril-like leaftips. This region has a profusion of flowering herbs, brilliant in color and often remarkable in appearance.

tends to remain fairly constant and relatively humid, conditions that support the growth of rainforest.

An abundance of species

The lowland Malaysian rainforests, which are warm and humid for most of the year, support a vast number of species. The most distinctive family is the Dipterocarpaceae. These are largely lofty, timber-producing giants that are the dominant trees in most kinds of lowland forest. The seasonal tropical (monsoon) forests to the north and south support fewer species. The most famous timber tree here is teak (*Tectona grandis*). Bamboo also grows abundantly, and particularly in areas where tree felling or land disturbance has allowed these highly opportunistic plants to invade, colonizing any open spaces.

To the north of the region, where the climate is less humid, temperate genera become increasingly common, such as

ash (*Fraxinus*) and elm (*Ulmus*). Some of the more temperate genera, such as *Rhododendron* and *Vaccinium*, extend toward the Equator in the cooler, more mountainous rainforests. Above the treeline alpine plants such as gentians (*Gentiana*), primroses (*Primula*) and buttercups (*Ranunculus*) grow. On the high Himalayan ranges temperate plants are widespread. Among the trees are included fir (*Abies*), spruce (*Picea*) and pine (*Pinus*).

In Malaysia a few Australian families are found, such as she-oaks (Casuarinaceae) and Australian heaths (Epacridaceae). These are especially abundant in the heath forests that grow – most extensively on Borneo – on silica-rich sands in which the topsoil has been leached of minerals such as iron. Some temperate Australasian plants also grow among the high mountain herbs, such as the mat-forming species of *Gunnera* and *Nertera*, for example.

221

ADAPTATION IN THE FOREST

The Dipterocarpaceae are a family of trees strongly centered in Southeast Asia. They are most numerous in the lowland rain-forest, where they are the dominant family of large trees. The trees are unique in several respects. In an area of a few square kilometers up to five or six genera, comprising 30 or 40 different species, can be found growing together; the genera *Dipterocarpus*, *Hopea* and *Shorea* have particularly large numbers of species. They are the commonest large canopy-top trees of the rainforest, growing in dense stands and developing straight, unbranched trunks that frequently reach 40 m (130 ft), and occasionally 50 or even 60 m (165 or 200 ft), in height, with a diameter of up to 1.5 m (5 ft), as they compete for light. When a giant grows successfully and emerges through the canopy, it produces a huge spreading crown in the bright sunlight.

Trees of the Dipterocarpaceae flower at about the same time over large areas. When this happens, once or twice a decade, the whole of the upper canopy bursts spectacularly into white, pink or red blossoms. Different species flower in sequence over a period of several weeks; the air becomes fragrant with their scent and is later filled with falling petals. The fruits of the various species mature at slightly different rates, so they all ripen simultaneously and fall to the ground together to provide a feast for the forest animals. Not all the harvest is eaten, however; the fruits that escape germinate immediately, rapidly forming crowded carpets of several million seedlings per hectare. Almost all these will die over the next few years; only when a tree falls, leaving a sunlit gap, can a young sapling grow to form a replacement canopy.

This mass production of seedlings can be exploited, because large numbers of the young plants can easily be persuaded to grow up to form a new forest if care is taken not to damage them while logging, and provided that the gaps in the overhead canopy are not allowed to become too large, which would destroy the warm, moist atmosphere they require for growth.

Prickly palms
Throughout the region, and especially in the lowland rainforests of Malaya and

Slipper orchid (*above*) *Paphiopedilum sukhakulii* is a ground dweller, endemic to northeastern Thailand. Behind the central column, above the pouched lip, are concealed two pollinia (pollen balls). These stick on pollinating insects as they investigate the flower, attracted by its lurid markings.

Mangrove swamps (*below*) Mangroves fringe many low muddy coastlines in Southeast Asia. To overcome the problem of living in waterlogged soil all mangroves possess specialized roots. These develop kneelike growths (pneumatophores) capable of absorbing air when they are exposed at low tide.

Borneo, climbing prickly palms called rattans are abundant. These palms belong to the subfamily Calamoideae, all of which bear scaly fruit. The rattan stem varies from pencil-thin to the thickness of a human arm. It is covered in very sharp, spiny leaf sheaths. These palms scramble through the forest canopy in their search for light, hanging on by means of long grappling hooks that are either modified flowering parts or the prolonged tips of their feather-shaped leaves.

Sun-loving pioneers
Macaranga is a huge genus of small to medium-sized trees. The genus comprises 280 species, with a range that extends from Africa to the Pacific, and is strongly concentrated in Southeast Asia: Borneo has 44 species and New Guinea 73, though between them the two islands

Sonneratia pokes up pneumatophores through the mud at the outer edge of the mangrove swamp, where the trees are regularly inundated by the sea

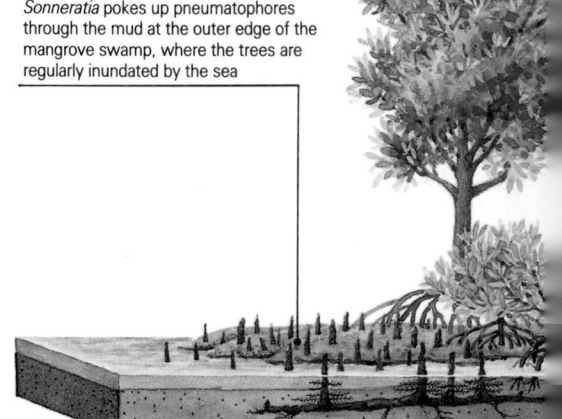

have only a handful of species in common. *Macaranga* are gap-opportunists – sun-loving trees that grow to form closely packed stands in large forest clearings or where other plants have been swept away by landslides, and along river banks. They bear all the hallmarks of pioneer trees – rapid growth, pale, low-density timber and the production, for most of the year, of copious small seeds. These are dispersed by birds, which are further attracted by the fleshy extra seed covering of some species.

Ants and scale insects inhabit the hollows within the twig stems of some *Macaranga* species. The trees produce starch grains that provide the ants with food, while the scale insects feed on the trees' sap. In the same way that herders guard a herd of goats, the ants keep the scale insects inside the twig hollows, protecting them against predators and periodically "milking" them to collect the sweet secretions they produce.

This relationship benefits the trees as well as the ants. A *Macaranga* that is free of ants will have its leaves damaged by herbivorous insects; ants have been seen removing would-be leaf eaters and insect eggs, and also the tips and tendrils of climbers that might otherwise eventually smother the tree.

Ants scurry busily around the protective hollows on the twigs of *Macaranga*. These trees are not alone in forming close associations with animals. In tropical zones competition between plants is fierce, and an animal partner can help in the fight for survival.

TORRENT-LOVING PLANTS

Throughout the tropics the swiftly flowing rocky streams are liable to flood when they are swollen by sudden heavy rain. Their banks are often fringed by wiry-stemmed shrubs and small trees, such as the evergreen *Garcinia cataractalis*, herbs and even ferns like *Dipteris lobbiana* of Borneo. These plants are known as rheophytes. They have narrow, willow-like leaves and form a very distinctive community in places where swift, powerful currents of water would damage broader, less tough leaves. Botanists have not yet established why rheophytes are far more abundant in some parts of the world than in others; they are extremely common in Southeast Asia, and especially in Borneo.

Rheophytic species are found within many different families of plants. Their narrow, leathery leaves are often so similar, even in plants that are totally unrelated, that the species can only be positively identified by studying their flowers or fruits.

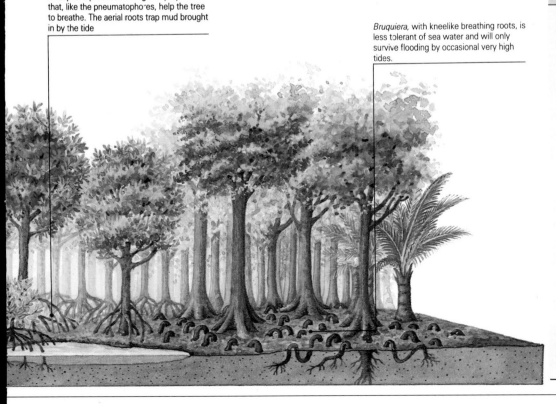

Rhizophora produces a tangle of prop roots that, like the pneumatophores, help the tree to breathe. The aerial roots trap mud brought in by the tide

Bruquiera, with kneelike breathing roots, is less tolerant of sea water and will only survive flooding by occasional very high tides.

SPICES, EBONY AND ORCHIDS

Some of the most important food plants have come from Southeast Asia. Archeological findings have suggested that agriculture, based on a type of rice (a grass called *Oryza sativa*), may have developed here earlier than anywhere else in the world. Citrus fruits such as oranges and lemons all originated in this region. Other export crops from the rainforest include bananas (*Musa*), of which many species grow wild in wet open places, and mangoes (*Mangifera*), which grow on medium-sized forest trees. Numerous other fruits are widely grown, such as the delicately flavored mangosteen (*Garginia mangostana*) and the rambutan (*Nephelium lappaceum*). Most famous of all, perhaps, is the durian (*Durio zibethinus*), notorious for its dreadful smell and delicious taste.

In the past several European powers laid claim to territories in the archipelago, drawn by the riches of the spice trade, principally nutmeg (*Myristica fragrans*) and cloves (*Syzygium aromaticum*) from the Moluccas in eastern Indonesia (often known as the Spice Islands). Cardamom (*Elettaria cardamomum*), cinnamon (*Cinnamomum rerum*) and ginger (*Zingiber officinalis*) were also highly prized.

Resin and timber
Resins taken from the giant coniferous kauri tree (*Agathis*) and from Dipterocarpaceae and Burseraceae species were once widely used in the manufacture of paints and varnishes. Together with a yellow gum known as gambodge, obtained from various *Garcinia* species, and named after Cambodia, which is its principal source, these resins have been important items of trade for some 2,000 years.

An extraordinary flowerhead The banana is a huge herb with a stem made up of the bases of the enormous leaves. The fruit develops in large bunches called hands; after fruiting the plant dies, to be replaced by suckers arising from its base.

In the 19th century a major market developed for the solidified latex gutta-percha (*Palaquium*), which was once used to insulate underwater cables and for making golf balls, and is still used today to make temporary dental fillings. In the 1920s an international trade developed in jelutong (*Dyera costulata*), a tasteless latex that formed the basic ingredient of chewing gum. The market for chewing gum in the United States increased greatly during this period as a result of the prohibition of alcoholic drink.

Fine timbers exported from the region include teak, especially from Burma and formerly from Java and Thailand, where the forests have been so heavily depleted that logging activities were suspended at the end of 1988. Ironwood grows in Sumatra and Borneo. This heavy, durable timber was traditionally used to make houseposts; it can also easily be split into the shingles that are still widely used as a roofing material.

The rainforests support numerous species of *Diospyros*. Most are small trees of the gloomy undergrowth, but a few reach medium size and develop the black heartwood known as ebony, of which Sulawesi is a major exporter. Despite the large number of species that grow in this region (155 in Malaya; 267 in Borneo), the timber they produce is so similar that it can be grouped for sale into only a handful of categories.

Dipterocarp trees that produce heavy, dark, durable timber have long been valued, especially the chengal (*Neobalano capus*) of the Malay Peninsula. Recently dipterocarp species that produce softer, paler, non-durable timbers have become more important. These include light red meranti, which comes from certain species of *Shorea*, keruing (*Dipterocarpus*) and kapur (*Dryobalanops*). Most of these woods are of low and medium density, excellent for plywood or furniture.

Other important timber species are pines (*Pinus kesiya* and *P. merkusii*), which also give resin for turpentine production. Yemane, a relative of teak, grows wild in slightly seasonal forests and is one of the few species that has proved successful when grown on timber plantations.

Two sizeable trees that are common throughout the region and are now well-known houseplants everywhere are the banyan or weeping fig (*Ficus benjamina*) and the rubber tree (*Ficus elastica*).

Orchids and palms

The most important Southeast Asian plants in horticulture are the orchids and the palms. With the Singapore Botanic Garden as an important innovator, a major industry has developed, based on cultivating native orchids and their hybrids and exporting cut flowers by air to many parts of the world.

In contrast, only a few of the wide range of native palms have been cultivated; many have great potential, such as the sealing-wax palm (*Cyrtostachys renda*) of peatswamp forests, which develops slow-growing clumps and has attractive red leaf sheaths. The fishtail palms, such as *Caryota mitis*, are sometimes seen in the

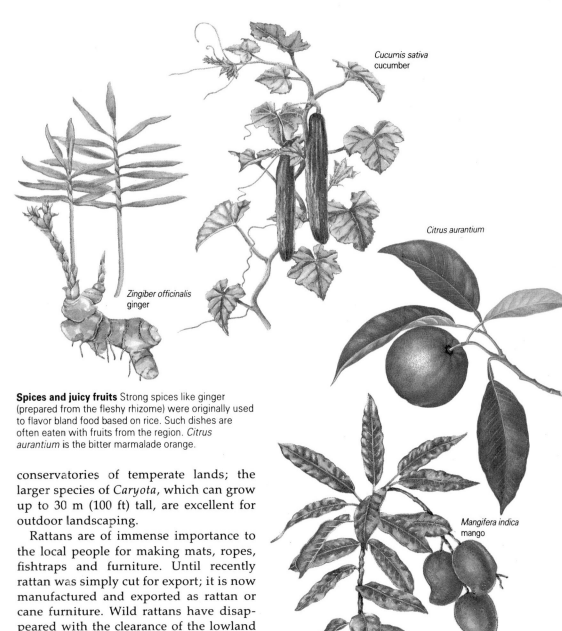

Cucumis sativa
cucumber

Citrus aurantium

Zingiber officinalis
ginger

Mangifera indica
mango

Spices and juicy fruits Strong spices like ginger (prepared from the fleshy rhizome) were originally used to flavor bland food based on rice. Such dishes are often eaten with fruits from the region. *Citrus aurantium* is the bitter marmalade orange.

conservatories of temperate lands; the larger species of *Caryota*, which can grow up to 30 m (100 ft) tall, are excellent for outdoor landscaping.

Rattans are of immense importance to the local people for making mats, ropes, fishtraps and furniture. Until recently rattan was simply cut for export; it is now manufactured and exported as rattan or cane furniture. Wild rattans have disappeared with the clearance of the lowland rainforests in the last fifty years, but attempts to cultivate them have recently been successful.

THREATS TO THE FOREST

The rapid rise in population during the 20th century has placed enormous pressure on the forests of Southeast Asia. Large areas of once permanent forest have been cleared to make way for agriculture or plantation crops (mainly oil palms from Africa and rubber trees from Brazil). Such valuable export crops are essential to generate the hard currency needed to import industrial goods. Another threat has been timber extraction – of teak from the seasonal tropical forests and of dipterocarps from the rainforests. As the forests diminish, some species will become extinct and others will be endangered.

These threats and pressures are now being countered by a growing understanding of conservation. All Southeast Asian nations have established networks of national parks and other protected areas. This is a good beginning, but much more is needed to give longlasting protection. This can only be achieved by policing, a widespread education program, and effective population control, which would reduce the pressure to utilize more and more land for farming in order to feed the growing population. There is also a need to develop higher-yielding crops and to improve farming systems.

Parasites and carnivores

Rafflesias, the most distinctive of parasites, obtain nutrition entirely from the living host. The genus contains about 14 species. The largest of these, *Rafflesia arnoldii*, has the biggest flower in the world: it can measure up to 80 cm (32 in) across and weigh as much as 7 kg (15 lb). The tiny seed germinates on *Tetrastigma*, a stout liana that is a member of the vine family. Inside its host the *Rafflesia* develops fungal-like threads, but grows neither a stem nor leaves; instead it produces only a huge flower, with blotched red and white petals, that forms on the stem of the vine. The blossoms smell of rotting meat, attracting flies that act as pollinators by carrying pollen from male to female flowers. Local people have long believed that the flower has special medicinal properties, and this has contributed to its rarity in the wild.

Saprophytes
Another group of plants that are dependent on a host for their nutrition are saprophytes, which take sustenance from the rotting plant material on which they live. They do not themselves cause this material to decompose, but live in close association with the fungi that do. Most saprophytes are small and inconspicuous, growing among the dark forest undergrowth. Like other parasites, they lack the green pigment chlorophyll that is used by most plants to make their food by the process known as photosynthesis.

The tropical forests of Southeast Asia also harbor partial or hemiparasites, including plants of the family Viscaceae and a related family, the Santalaceae. Many members of these families grow as epiphytic shrubs, tapping their host tree for water and mineral nutrients; however, they do possess chlorophyll and are able to manufacture their own food. The Santalaceae contain ground-growing bush and tree hemiparasites that take moisture and nutrients from the roots of other tree species. This group includes sandalwood (*Santalum album*), much prized for its perfumed wood.

Carnivorous pitcher plants
In Southeast Asia the most spectacular carnivorous plants are the Old World

The biggest flower in the world (*right*) The center of the evil-smelling *Rafflesia arnoldii* flower contains a spiky disk, which conceals the stamens in the male flowers and the ovary in the female flowers. Large, squashy fruits contain the tiny seeds.

Pitcher plant (*below*) Once an insect has crawled over the slippery overhanging lip of *Nepenthes maclarlanei*, just one of the many striking species of pitcher plant, there is no escape. The lid prevents rain from diluting the digestive fluid at the base.

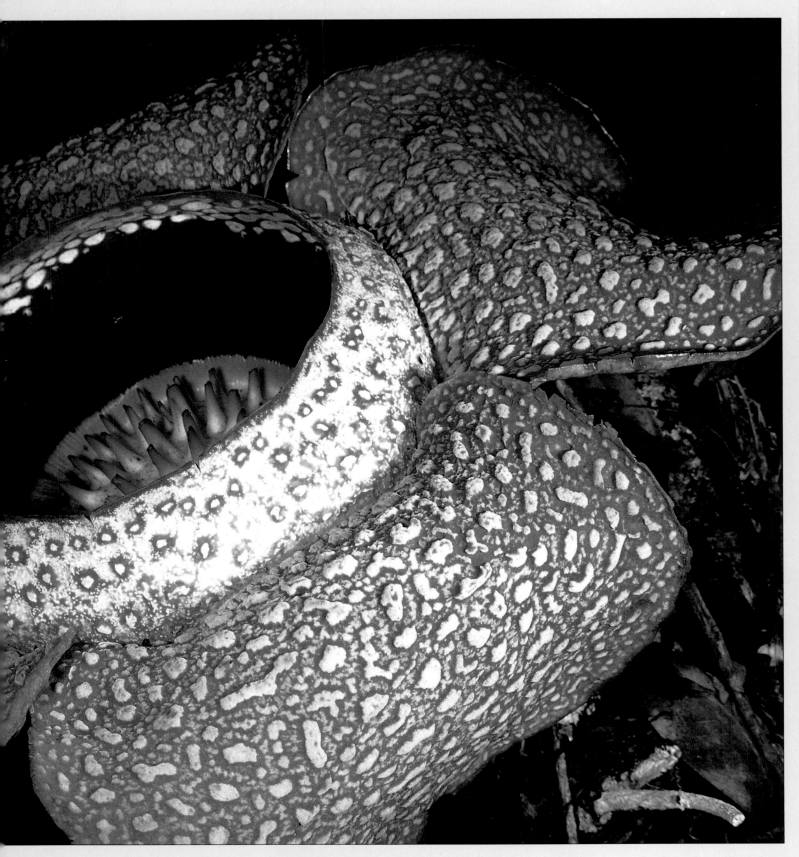

pitcher plants, *Nepenthes*, a genus of 70 species. The stems and leaves of *Nepenthes* bear nectaries, and the leaves also produce terminal tendrils that aid the plant in climbing. Behind these tendrils most leaves develop a pitcher-shaped pouch with an overhanging lid. These pitchers are often bright red, green, yellow or brown; digestive fluids collect in the base of each one. Unsuspecting insects are attracted by the nectaries and the brightly colored pitchers; they walk down on to the slippery pitcher walls and fall to the base, where they are eventually dissolved in the juices.

Nepenthes species are especially common on poor rocky or sandy soils, and it is thought that the insects they devour help to supply them with essential nutrients that they are unable to find in the soil. *Nepenthes rajah* has the largest pitchers, 30 cm (12 in) long, which hold up to 2 liters (3.5 pints) of liquid. The stems of the scrambling and climbing species provide tough cords that are traditionally used in Borneo for binding longhouse floors and to make matting.

The sacred lotus

The sacred lotus (*Nelumbo nucifera*) is an ancient plant. Fossil evidence of it growing 160 million years ago shows that is has changed little over eons of time. Yet, like the waterlily to which it is related, its blossoms are sophisticated for a plant that evolved so long ago.

It lives in still water. Some of the circular leaves float on the water, but as the plant matures the leaves become more like an inverted cone and are held above the surface. When young, these are rolled tightly to emerge between the floating leaves. In the center of each exquisite flower is a strange receptacle whose circular holes are the stigmas.

In the east almost all parts of the sacred lotus are eaten, and this may have a bearing on its religious significance. The leaf stalks provide a salad, the rootstock is roasted and the pearlike seeds taste of almonds.

It is a sacred plant in many lands – from Egypt to Japan, where it is planted for devotional purposes. In India it is Padma, symbol of the Ganges; from there it traveled eastward with the spread of Buddhism, across southern Asia to China, where it was depicted as the seat of the Buddha and became a symbol of feminine beauty.

The sacred lotus, a plant of great beauty revered for many hundreds of years.

FLORAL CROSSROADS OF THE ORIENT

WIDESPREAD AND UNIQUE PLANTS · SUCCESSFUL STRATEGIES · THE SEVEN HERBS OF SPRING

Japan and Korea are sited at a crossroads for plants. In the past, Japan was linked to what is now the Soviet Union through the island of Sakhalin, and to the southeast Asian mainland through Korea, providing routes for plants pressured by climatic changes. Today the mountainous islands of the region isolate and protect plant populations. Many plants known only as fossils in other parts of Eurasia still survive in Korea and Japan. The plant life ranges from subtropical forest in the south, through deciduous hardwoods to coniferous softwoods in the north, with alpine scrub and snowline plants. The number of species increases farther inland. In the spring the mountains are covered in the soft pink and white petals of flowering cherries, while the fall produces a riot of color from golden birches and flame maples.

WIDESPREAD AND UNIQUE PLANTS

The plants in the north of Japan and Korea are related to many northern hemisphere plants elsewhere in the world, Japan having been linked in the past to the Eurasian landmass. The spruce *Picea hondoensis*, for example, is closely related to the sitka spruce (*Picea sitchensis*) of Pacific northwest America, having similar papery cones. Familiar northern hemisphere trees such as birch (*Betula*), willow (*Salix*), alder (*Alnus*) and poplar (*Populus*) grow at increasingly higher altitudes as they extend southward from the northern island of Hokkaido.

Japan and Korea both support many species of temperate woodland trees found elsewhere in the hemisphere, such as temperate oak (*Quercus*), beech (*Fagus*), chestnut (*Castanea*), hornbeam (*Carpinus*), maple (*Acer*), cherry (*Prunus*) and apple (*Malus*). The other temperate woodland species found here are less common; they were able to survive the effects of glaciation in mountain refuges such as the Caucasus in the south of the Soviet Union, the Appalachians in the eastern United States, and in parts of China. They include species of magnolia, *Zelkova*, hackberries (*Celtis*), buckeyes and horse chestnuts (*Aesculus*), walnuts (*Juglans*), snowbells (*Styrax*) and fringe trees (*Chionanthus*).

Mixed forests of the rugged north Broadleaf evergreen trees and conifers clothe the slopes of Hokkaido and of North Korea. The gold and red foliage indicates the presence of deciduous trees.

COUNTRIES IN THE REGION
Japan, North Korea, South Korea

DIVERSITY

	Number of species	Endemism
Japan	4,022	27%
North Korea, South Korea	2,900	14%

PLANTS IN DANGER

	Threatened	Endangered	Extinct
Japan	8	4	–
North Korea, South Korea	10	8	1

Examples *Abies koreana* (Korean fir); *Arisaema heterocephalum*; *Chrysanthemum zawadskii*; *Cyclobalanopsis hondae*; *Cymbidium koran*; *Euphrasia omiensis*; *Fritillaria shikokiana*; *Gentiana yakusimensis*; *Magnolia pseudokobus*; *Rhododendron mucronulatum*

USEFUL AND DANGEROUS NATIVE PLANTS

Crop plants *Cryptomeria japonica* (Japanese cedar); *Larix kaemferi*
Garden plants *Berberis seiboldii*; *Camellia japonica*; *Hamamelis japonica*; *Kerria japonica*; *Lilium auratum*; *Magnolia stellata*; *Paeonia japonica*; *Rodgersia podophylla*; *Rosa multiflora*; *Viburnum tomentosum*
Poisonous plants *Nerium odora*; *Taxus cuspidata*; *Wisteria floribunda*

BOTANIC GARDENS

Chollipo Arboretum, Sosan (7,000 taxa); Hiroshima (8,000 taxa); Kyoto (5,500 taxa); Pyongyang (2,000 taxa); Tokyo Metropolitan (10,000 taxa)

Japan and Korea

Floristic provinces

Holarctic Kingdom/Eastern Asiatic Region

North Chinese Province Area in northwest Korea that approximately coincides with the temperate deciduous forest zone.

Manchurian Province Extends north of Korea into Manchuria; here includes bellflower *Hanabusaya*.

Japanese-Korean Province Largest province, distinguished by endemic parasol pines (*Sciadopitys*) and the peony-relative *Glaucidium*.

Sakhalin-Hokkaido Province A plant-rich province that includes endemic species of fir, globe-flower and aconite.

Map of floristic provinces Japan and Korea are both included in the Eastern Asiatic Region; much of it is mountainous, with the large number of endemic plants typical of isolated mountain environments.

Many plants that are widespread in the northern hemisphere grow in the shrub layer of the woods; *Euonymus*, *Viburnum*, *Berberis* and *Rubus* are examples. There are others that grow only in more restricted areas; these include *Clethra*, *Enkianthus* and *Pieris*.

At ground level too there are a number of familiar herbs such as pansy (*Viola*), *Primula* and *Anemone*. The ground layer of much of the woodland is a smothering, lowgrowing bamboo, but there are also many ferns. Temperate mixed woodland thrives in the valleys of Hokkaido; farther south it grows on hillsides above ground that has been cleared for cultivation.

Some Japanese plants belong to a group found only in eastern Asia and the Himalayas. Many represent a single genus, and several are the only species in their family. Among the best known,

because they are now popular as garden plants, are sacred bamboo (*Nandina domestica*) and the red-berried shrub *Aucuba japonica*. The Japanese cedar (*Cryptomeria japonica*) is the most familiar tree in Japan; together with the larch, it is the basis of the forestry industry. The climbers *Stauntonia* and *Akebia*, and the *Aspidistra*, *Liriope*, lily turf (*Ophiopogon*) and the giant lily (*Cardiocrinum*) – all members of the lily family (Liliaceae) – also belong in this group.

Links with the south

A number of the plants of southern Japan and the coastal region of South Korea belong to families that are distributed throughout the southern hemisphere. These include two species of the evergreen *Pittosporum* and two of the coniferous *Podocarpus*. As in Taiwan, China and

Malaysia, the broadleaf evergreen forest is dominated by the pink-flowered *Machilus*, a relative of laurel, *Castanopsis*, with its chestnutlike prickly fruits and orange and silver foliage, and *Neolitsea*. Numerous evergreen oaks and many other tree relatives of the laurel, including the camphor laurel (*Cinnamomum*), form a highly diverse and colorful mixture. *Camellia* also grows as a forest tree here. Along the coast fig trees are common, together with the mainly tropical *Michelia*, a relative of magnolia. Fall brings with it a distinctive riot of color produced by golden birches, flame maples and the soft yellow of *Cercidiphyllum*.

The most exceptional plant of all grows marooned on the southern cliffs of Japan. This is the primitive cycad, *Cycas revoluta*, which has related species in Australia and Polynesia.

SUCCESSFUL STRATEGIES

In a mountainous region such as this, plants are frequently exposed to winds, the cold temperatures of high altitudes, snow and frost. Where forest predominates, the major problem for plants is to obtain sufficient light. Some plants take advantage of the lack of competition, evolving like alpines and desert plants to tolerate extremes of deprivation.

Among the successful snowline plants, Japan and Korea have an exceptional number of ground-hugging heath shrubs, which tolerate acid soils and have strong, deep roots that secure them on steep scree slopes. There are 20 genera from the heath family (Ericaceae), including the woody plants *Lyonia*, *Phyllodoce*, *Cassiope*, *Vaccinium*, *Tripetaleia* and *Andromeda*. All these plants are lowgrowing, which both reduces the extent to which they are buffeted by desiccating winds in exposed areas and also protects them from the intense cold.

The low pine scrub in Japan and Korea is dominated by a dwarf, five-needled pine, *Pinus pumila*. In some places it grows with dwarf *Sorbus* species and shrubby alders; other mountainous scrub communities consist of a dwarf holly (*Ilex sugeroki*), with dwarf juniper, willow, *Acer tschonoskii* and *Vaccinium* species. The scrubby oak *Quercus dentata* survives the harsh conditions of the mountain slopes in northern Honshu that face the winds and snow that sweep in across the Sea of Japan.

The fight for light

In the forests, plants compete for light by scrambling up and flopping among other plants, or by climbing and twining up them. Others, like those of the woodland floor, tolerate deep shade or complete their year's activity in the spring before the tree canopy closes.

In mountainous areas lax shrubs (shrubs with arching branches) that use trees and other surfaces to raise their branches to the light can be seen everywhere. They grow in the cracks and ledges of ravines, and up the sides of cliffs that border new roads built to hairpin up mountains. One of them, now a favorite garden plant, is yamabuki (*Kerria japonica*), with its gold flowers that burst into bloom at the beginning of spring. Other lax shrubs include shrubby honeysuckle (*Lonicera*), the chalk-white *Deutzia*, crimson *Rhododendron kaempferi*, fragrant mock orange or syringa (*Philadelphus*) and *Weigela*; all of them insinuate themselves among other plants.

Among the climbers are some of the most widely known of Japan's plant exports. The Japanese honeysuckle (*Lonicera japonica*) forms sheets on the ground and twines its way up the stems of other plants. The long mauve racemes of *Wisteria* cascade down ravines or in shrine gardens. Kuzu (*Pueraria lobata*), a member of the pea family (Leguminosae)

Winter in the mountains The pyramidal shape of coniferous trees is well suited to shed snow. The branches of a more spreading crown would tend to snap under its increasing weight during heavy and prolonged snowfalls.

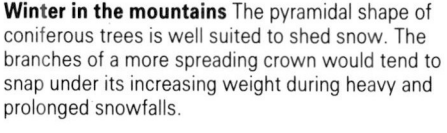

with hairy leaves, commonly grows in thickets. At the edge of woodland the tough twining vine *Akebia* is widespread, trailing across grassy banks. Its stalked leaflets are a soft gray green, and it produces purplish flowers followed by edible fruit.

In the evergreen woodlands of southwestern Japan climbers such as *Stauntonia hexaphylla* send out meter-long growths that seek a support to twine round. Other climbers emerge from the dense shade into the canopy and drape themselves across the treetops in liana loops. In the shrine forest at Nara on Honshu scrambling stems of roses hook themselves by their thorns to the great dark trees.

Arum of the northern swamps (*left*) *Lysichiton camtschatcensis* is an aquatic plant. The huge white spathes enclosing the flower cluster act as flags to attract pollinating insects to the columnar spadix, which is studded with tiny green flowers.

On the forest floor (*above*) In the still air the large, thin leaves of *Rodgersia podophylla* are in no danger of being torn by the wind. Like many other shade-loving plants, this species spreads by means of a creeping underground stem.

On the woodland floor

Woodland floor plants must either be able to survive in shade or grow rapidly in spring, when sunlight penetrates to the ground. In the dense woodland of valley floors, where large trees such as *Pterocarya rhoifolia*, the silver-leaved *Magnolia obovata* and the ash *Fraxinus spaethiana* flourish, much of the ground cover consists of shade-tolerant ferns such as *Polystichum* and *Dryopteris* species. Where it is damp and there is little light, herbaceous plants with large thin leaves, such as *Rodgersia podophylla* and *Diphylleia grayi*, make the best of the subdued light that filters through the foliage.

In mixed beech or maple woods *Anemone*, *Trillium* and *Erythronium* all push their way through the leaf litter in early spring, to open their foliage and flowers before the tree canopy unfurls. Among the ferns on shady banks grow clumps of peonies (*Paeonia japonica*) and the pleated leaves of *Veratrum*.

Most exciting of all the plants of dappled shade are species of *Arisaema*. About fifty Japanese species were described originally, though this number has recently been revised to 25. The single stout stalk is often mottled and striped with wine-colored stains. The arum-like flowers have the inflorescence (spadix) and caplike spathe of all aroids; in some species these have long tails that are held in the center fold of the leaf and later released to lie on the ground as a lure for pollinating beetles. *Arisaema* species are being overcollected, and are sadly endangered by their own beauty.

A THOUSAND YEARS TO MAKE A FOREST

Japan lies on a belt of intense tectonic activity, and suffers from both earthquakes and volcanic eruptions. All over the region plants are colonizing land that has been devasted by mud and lava flows. Mount Fuji, Japan's most famous volcano, erupted in the year 804 and again in 1707. More than 250 years later, the lava flow supports only a twiggy low scrub and the Fuji cherry, which is now quite common.

Oshima, an island to the southwest of Tokyo, has been built up by repeated lava flows, volcanic eruptions having taken place in 684, 1777, 1876 and 1950. The order in which the competing plants become established (plant succession), and the length of time it takes for the plant community to become stabilized (climax vegetation), can be worked out by examining the plant life on the various lava flows. On the earliest flow of 684, broadleaf evergreen forest grows; woody species take hold in the crevices and shelter in cracks, but they remain dwarfed. On the flow of 1777 alder, cherry and *Cornus* (dogwood) trees have become established, while the flow of 1876 supports a woody, lowgrowing scrub with alder and *Weigela* species. Only a few species have successfully colonized the lava flow in the 40 years since the 1950 eruption. These include alder and *Reynoutria japonica* (also known as *Polygonatum cuspidatum*), a herb related to dock. It may be another thousand years before a true forest community successfully establishes itself in this vulnerable environment.

THE SEVEN HERBS OF SPRING

Plants play an important part in Japanese life. In daily life they are cultivated in gardens, eaten, bred to create new varieties and skillfully crafted into useful and beautiful objects. On a deeper level, they are also recognized as universal symbols: rice, for example, is seen as the source of life itself, and the rice-growing cycle provides a measure of the seasons and years.

Plants in daily life

The arrival of a new season in Japan is a cause for celebration; in both art and literature it is most often depicted in a language that reflects the plant calendar. In the *Manyoshu*, the earliest anthology of Japanese poetry (AD 745), there are references to the Seven Herbs of Spring and the Seven Plants of Autumn. The herbs show green in winter and are eaten on 7 January, after the New Year. Other spring plants include pine, bamboo and plum blossom. The fall plants include bush clover, wild pinks, and the plumes of the *Miscanthus* grass, along with cherry, wisteria, iris, tree peony, chrysanthemum and autumn maple. Both groups herald the arrival of the seasons.

In daily life plants are primarily important as a source of food. For those unfamiliar with Japanese cuisine, the range of plants used can be baffling and every meal requires a botanical interpreter; for example, the fiery fibers in the rice under raw fish are wasabi root; the vegetable components of the dish *sukiyaki* are oak mushrooms, chrysanthemum greens, sliced lotus root and eggplant; in the bar food called *oden*, the gelatinous lumps are derived from the edible tubers of *kon-nyaku*, a member of the arum family; *ume boshi* is the name for the sharp pickled plums colored with the herb *perilla* that are added to breakfast rice; and sweet bean cakes are wrapped in cherry or oak leaves, depending on the season.

Control and selection

The idea of striving toward perfection is an integral part of Japanese culture. It is also applied to methods of forestry, which has been practiced since the 16th century when Japanese cedar or sugi was first cultivated for pitprops. Northwest of Kyoto, in the south of Honshu island, the trees are pruned to produce knot-free timber, and form what is regarded throughout Japan as a historic landscape. Pines for gardens are also trained to become the ideal tree – simplified, layered and densely needled; they are highly prized, and also highly priced.

Japanese horticulturists are skilled in perpetuating exceptional plants through selection, and they also specialize in breeding new varieties. Among them are conifers such as the juvenile-leaved *Cryptomeria japonica* "Elegans", *Chamaecyparis pisifera* "Plumosa", and abnormalities such as *Chamaecyparis pisifera* "Filifera", with its threadlike twigs.

Plant breeding has produced hundreds of variations on many of the native plants, such as azaleas (*Rhododendron*), irises, tree peonies, chrysanthemums, camellias and maples. Variegated plants, which have green foliage mottled or streaked with

Japanese gardens are regarded as a symbolic representation of the wild landscape. The plants reflect the changing of the seasons, and some bushes, often of azalea, may be clipped to represent stones or the waves of the sea.

Arundinaria japonica
Japanese bamboo

Lilium auratum
golden-rayed lily

Kerria japonica

Garden flowers in Japan Bamboos are seen in many gardens, but flowers are rarely planted; yamabuki (*Kerria*) is one of the few. Flowers are important in *ikebana*, skillful arrangements for quiet contemplation.

cream because of a lack of chlorophyll, have been widely collected. Among them dwarf palms and selaginellas, golden striped bamboos, tiger's-eye pines, mottled maples and dappled asarum are highly sought-after.

Traditional crafts using materials such as paper, wood and bamboo are still widely practiced in Japan. Handmade paper, a translucent web made from the fibrous inner bark of the paper plants known as kozo, mitsumata and gampi, is still regarded as the perfect covering for sliding screens. The timber of the ornamental tree *Paulownia* is used specifically to make chests in which bridal clothing is stored, and also to make boxes for hanging scrolls; keyaki (*Zelkova serrata*) is used to make traditional furniture such as chests; cherry wood is carved into blocks for print making, and boxwood (*Buxus*) is used to make combs. The long canes of many species of bamboo make excellent builders' scaffolding, and the versatile wood is also used to make whisks for the tea ceremony, flutes, canisters, wooden spoons and writing brushes.

THE ART OF BONSAI

Bonsai is the ancient Japanese art of growing miniaturized trees in containers, shaping them to become tiny, exact replicas of the normal adult plant. Good bonsai trees never look dwarfed or deformed; the first priority is the tree's health and growth, responding to its need for light, food and water.

The bonsai gardener trains the trees over a considerable period of time, using a range of highly skilled techniques. These include root and shoot pruning, needle trimming, bark abrasion, chemical antiquing and wiring to bend branches, resulting in a tree that is perfectly proportioned and looks as if it is growing in a natural landscape. The plants can be grown from seeds or from wild seedlings.

Common trees used for bonsai are juniper (*Juniperus rigida*), Hinoki cypress (*Chamaecyparis obtusa*) and small-leaved deciduous trees such as elm, zelkova and Japanese maple. Among the flowering trees, hawthorn, cherry, apple, Chinese quince and azalea are favorites. Much the most popular plant is pine – the New Year symbol of longevity and strength, trained to convey endurance. As with their rugged counterparts in nature, many bonsai trees are centuries old; these specimens are priceless.

The pine style *Pinus mugo* trained informally with two harmonious curves of the trunk. Bonsai can be trained in a number of styles, but the informal upright or *moyogi* is the most common.

Flowering cherries

Cherry blossom marks the arrival of spring in Japan, the start of the annual pilgrimage of flower viewing called O-hana-mi, which has been practiced for more than a thousand years. In ancient times only wild single-flowered cherries were known. These were brought under cultivation in the gardens of the nobility at the old capital, Nara, and by 1800 a large collection had been established at the Heian court in Kyoto, in the north-west of the Kii Peninsula.

Cultivating double flowers

The origin of the cultivated Japanese cherries, known throughout the world as Sato Zakura, is obscure. It is thought that they were derived from the Oshima cherry (*Prunus speciosa*) and *P. serrulata* var. *spontanea*. Double-flowered cherry trees must first have arisen by chance, but they have been cultivated for centuries, propagated by grafting – the technique of uniting a shoot from one plant with the rootstock of another. In the earliest written work on cherries, *Kadan Komoku* (1681), 40 cultivars are mentioned, and about 20 of these are still grown. In both Japan and Korea the beauty of the flower has been the main influence in the breeding of these trees, rather than the quality of the fruit.

Toward the end of the 19th and in the early part of the 20th century, interest in cultivating plants declined, and many cherry varieties were lost, including the great white cherry called "Tai Haku". Fortunately many cherry trees are long lived, and some survived this period of neglect long enough for botanists and plant collectors from Europe and America to buy dormant shoots, which they could propagate. Many cultivars, including "Tai Haku", that are extinct in Japan now flourish in the gardens of Europe and North America.

A national pursuit

Flower viewing is enjoyed by the entire population of Japan, and tourists travel great distances to see the best sites. The first cherry trees come into flower from mid-March on Kyushu, the southernmost island. Moving north, they flower progressively later, those of the plantations on southwestern Hokkaido coming into bloom about 6 May. Takao, northwest of Tokyo, became a major cherry-viewing location in the last 25 years. The trees in this garden range from old species such as

P. kurilensis, which comes from the Kurile archipelago south of Kyushu, to the latest, highly bred cultivars. Grafts from national treasures grow next to drifts of wild species such as the Fuji cherry (*P. incisa*), which grows in thickets on the slopes of Mount Fuji. The best-loved cherry in Japan, "Somei Yoshino", also grows here; planted together in great numbers, the trees have single flowers, tinged with the palest pink that fades to white.

The cherry is the national flower of Japan, and flower viewing is a national pursuit. The cherry is also regarded as a martial flower, because its transitoriness symbolizes the lightness with which the warrior (*samurai*) holds his life in the

O-hana-mi, flower viewing (*above*) Spontaneously, as the warm weather brings the cherries into bloom, enormous numbers of people come out to enjoy them. Famous sites attract large crowds and a party atmosphere prevails.

Cherry blossom clothes the Yoshino Hills (*right*) The blossoms appear with the first leaves of spring, but their beauty is transitory. They will soon fall, to be blown away like spirits on the wind.

service of his lord. In the Way of the Warrior (*bushido*), the code of honor by which the *samurai* lived, death was identified with cleanness and innocence, characteristics symbolized by the cherry. However, cherry watching as a popular festival is primarily a celebration of the coming of spring and sunshine.

PLANTS IN ISOLATION

A RICH AND VARIED FLORA · ADAPTING TO THE ENVIRONMENT
TIMBER TREES, FRUIT CROPS AND ORNAMENTALS

Over one-tenth of the world's flowering plants are found in Australia, New Zealand, New Guinea and the Pacific islands. From tropical rainforests near the Equator to the alpine herbfields of Mt Kosciusko in the southeast, Australia has a colorful and fascinating range of plants. Eucalypts and wattles of numerous varieties, gaudy flowers, and peculiar growth forms adapted to dry climates, fires and poor soils are all characteristic. By contrast, New Guinea, with its lush tropical forests and great diversity of plants, and New Zealand are much wetter. The plant life of New Zealand ranges from subtropical to temperate, with forested lowlands and many alpine species. The numerous islands of the region vary from coconut-fringed tropical gardens to bleak, windswept, grass-covered subantarctic rocks.

COUNTRIES IN THE REGION

Australia, Fiji, Kiribati, Nauru, New Zealand, Papua New Guinea, Solomon Islands, Tonga, Tuvalu, Vanuatu, Western Samoa

EXAMPLES OF DIVERSITY

	Number of species	Endemism
Antarctica	2	0%
Australia and New Zealand	26,000	90%
Papua New Guinea	11,000	55%
Pacific Islands	4,000	50%

PLANTS IN DANGER

	Threatened	Endangered	Extinct
Australia and New Zealand	1,755	256	121
Papua New Guinea and Pacific Islands	No information available		

Examples *Argyroxiphium kauense* (Kau silverwood); *Banksia goodii* (Good's banksia); *Clianthus puniceus* (lobster claw); *Dacrydium franklinii* (Huon pine); *Eucalyptus caesia* (caesia); *Hibiscus insularis* (Philip Island hibiscus); *Hibiscus kokio* (kokio); *Myosotium hortensia* (Chatham Island forget-me-not); *Neoveitchia storckii* (vuleito); *Platycerium grande* (stag's horn fern); *Sophora toromiro* (toromiro); *Tecomanthe speciosa*

USEFUL AND DANGEROUS NATIVE PLANTS

Crop plants *Artocarpus altilis* (breadfruit); *Cocus nucifera* (coconut); *Colocasia esculenta* (taro); *Eucalyptus species* (gum trees); *Macadamia integrifolia*
Garden plants *Acacia dealbata* (mimosa); *Callistemon citrinus* (crimson bottlebrush plant); *Eucalyptus ficifolia*; *Hebe* species; *Parahebe catarractae*; *Platycerium bifurcatum* (stag's horn fern)
Poisonous plants *Coriaria* species (tutu); *Cycas media*; *Grevillea viscidula*; *Solanum aviculare*; *Urtica ferox* (tree nettle)

BOTANIC GARDENS

Adelaide (4,500 taxa); Royal Botanic Gardens, Melbourne (15,000 taxa); Royal Botanic Gardens, Sydney (6,000 taxa)

A RICH AND VARIED FLORA

Tropical forest remnants in the north of Australia, where the rainfall is high, contain a great variety of trees, ferns and creepers, while the forests of the east and southwest are dominated by gum trees, or eucalypts (*Eucalyptus*). In the temperate forests of the southeast and Tasmania, southern beeches (*Nothofagus*) predominate, and southern heaths of the family Epacridaceae are common in the south and southwest. Mangrove forests or mangrove scrub fringe most of the northern and eastern coasts.

In the semiarid Australian interior there are savannas of scattered eucalypts and grasses, while in the central desert areas the eucalypts are replaced by mulga scrub of lowgrowing acacia trees (*Acacia*). Many of the species have developed water-storing roots and flattened green stems instead of true leaves.

New Guinea lies to the north of Australia and just south of the Equator. Its coasts are fringed with mangroves. The rich tropical lowlands support rainforests and swamps. Upland areas have elements of the plant life of the ancient supercontinent of Gondwanaland, in particular southern beeches and tree ferns. Plants of the high alpine zones, such as the shrubby coprosmas (*Coprosma*) and the lowgrowing woody veronicas (*Parahebe*), have relatives in Australia and New Zealand. So too do the perching lilies (*Astelia*), with a distribution that extends to southern South America and Hawaii.

In New Zealand forests of conifers and broadleaf trees grow on both North and South Island. The forest trees of the subtropical north often have affinities with those of Southeast Asia and Polynesia. The noble kauri (*Agathis australis*), for example, which grows up to 60 m (200 ft) in height, has close relatives northward to the Philippines. The natural vegetation of South Island is mostly southern beech forest, together with subalpine to alpine grassland and herbfields.

Icebound Antarctica supports only two species of flowering plants: the grass *Deschampsia antarctica* and the pearlwort *Colobanthus quitensis* grow where soil is exposed on the climatically less severe Antarctic Peninsula which reaches toward the southern tip of South America. Mosses and lichens are able to survive on snow-free rocks and gravels.

The columnar flowering spikes of blackboys (*Xanthorrhoea*) appear only after a fire has passed, stimulated into growth by the heat. With their distinctive topknot of grassy leaves sprouting from a trunk composed of fibrous leafbases, these treelike plants dominate many of the drier areas of the Australian interior.

Plants of the Pacific islands

Stretching eastward from Australasia, the vast Pacific Ocean is dotted with thousands of islands. In general, plant diversity on the islands decreases eastward with increasing distance from the rich plant sources of Australia and Southeast Asia. The plants of the continental fragments of Fiji and New Caledonia are tropical, with some relicts from Gondwanaland such as southern beeches, proteas, members of the monkeypuzzle family (Araucariaceae) and coniferous podocarps (Podocarpaceae). The other Pacific islands fall into two groups. Elevated islands such as

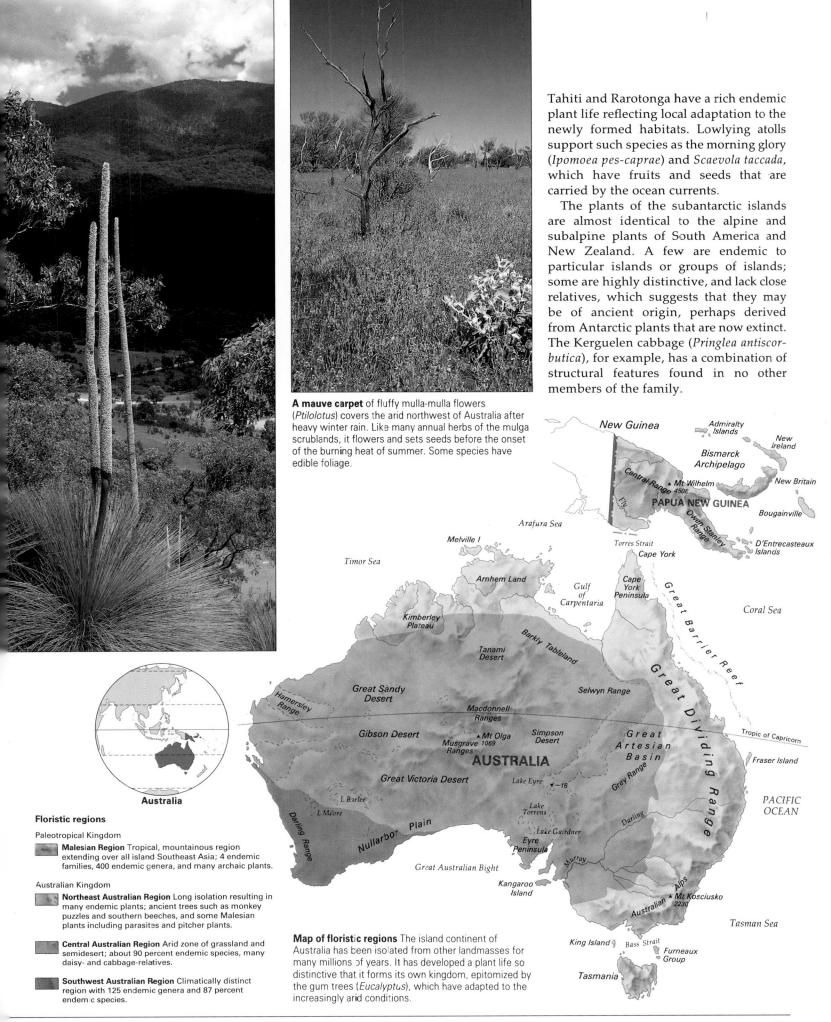

Tahiti and Rarotonga have a rich endemic plant life reflecting local adaptation to the newly formed habitats. Lowlying atolls support such species as the morning glory (*Ipomoea pes-caprae*) and *Scaevola taccada*, which have fruits and seeds that are carried by the ocean currents.

The plants of the subantarctic islands are almost identical to the alpine and subalpine plants of South America and New Zealand. A few are endemic to particular islands or groups of islands; some are highly distinctive, and lack close relatives, which suggests that they may be of ancient origin, perhaps derived from Antarctic plants that are now extinct. The Kerguelen cabbage (*Pringlea antiscorbutica*), for example, has a combination of structural features found in no other members of the family.

A mauve carpet of fluffy mulla-mulla flowers (*Ptilolotus*) covers the arid northwest of Australia after heavy winter rain. Like many annual herbs of the mulga scrublands, it flowers and sets seeds before the onset of the burning heat of summer. Some species have edible foliage.

Australia

Floristic regions

Paleotropical Kingdom

Malesian Region Tropical, mountainous region extending over all island Southeast Asia; 4 endemic families, 400 endemic genera, and many archaic plants.

Australian Kingdom

Northeast Australian Region Long isolation resulting in many endemic plants; ancient trees such as monkey puzzles and southern beeches, and some Malesian plants including parasites and pitcher plants.

Central Australian Region Arid zone of grassland and semidesert; about 90 percent endemic species, many daisy- and cabbage-relatives.

Southwest Australian Region Climatically distinct region with 125 endemic genera and 87 percent endemic species.

Map of floristic regions The island continent of Australia has been isolated from other landmasses for many millions of years. It has developed a plant life so distinctive that it forms its own kingdom, epitomized by the gum trees (*Eucalyptus*), which have adapted to the increasingly arid conditions.

ADAPTING TO THE ENVIRONMENT

Many of the plants of Australasia and the Pacific islands have characteristic and distinctive growth forms and flower types. They have evolved in isolation in response to unique pressures of natural selection, and make this region a natural laboratory for students of plant evolution.

Pollination by animals

The intricate web of the relationship between plants and animals is nowhere more evident than in methods of pollination. The brightly colored flowers of Australia demonstrate that many plant groups have adapted to take advantage of the huge diversity of available insect, bird and mammal pollinators.

Typical bird-pollinated flowers include the Australian "fuchsia" *Correa pulchella* with its tubular red flowers, *Eremophila maculata* with its red or orange gullet flowers, and the red brush blossoms of many eucalypts. Such flowers usually have no scent, but offer copious sugary nectar. The flowers of banksias are often small, and grouped together in large numbers near ground level on a stout supporting stalk. The thin nectar flows to the ground, where it attracts small mammals to the flowers, including the tiny honey possum.

New Zealand presents a marked contrast, having numerous plants with small greenish, white or yellow flowers that are pollinated by flies, moths and short-tongued bees. The flowers are often flat or dish-shaped, with short tubes that make it easy for insects to reach the nectar. Many of New Zealand's plants produce male and female flowers on separate individuals so that self-pollination is impossible; prominent examples of this are the Spaniards (*Aciphylla*), *Coprosma* and the coniferous podocarps.

Living with fire

When bushfires take hold in Australia their smoke can dim the sun 1,600 km (1,000 mi) away in New Zealand. In this dry climate many plants have evolved ways to cope with fire. Resins and oils, common in the leaves and wood of eucalypts and other plants, encourage fires to burn hot and quickly. Such fires do less damage to the woody and underground parts of plants. Mallees and many

The brushlike flowers of *Banksia* are pollinated by the honey possum. This tiny mammal lives exclusively on nectar, and depends for its survival on a range of plants that produce similar flowers in every month of the year.

other trees possess lobed, woody swellings (lignotubers) found at the base of the stem on the top of the root crown. These store carbohydrates and other nutrients, and contain numerous dormant buds borne at or near their surface. Damage by fire triggers these buds to grow and produce new stems.

Tufted trees such as the blackboys (*Xanthorrhoea*) of Australia and the cabbage tree (*Cordyline*) of New Zealand are very well adapted to regrowth following a

fire; although the heat may scorch the outside of the trunk, the protected inner tissue remains undamaged.

Adapting to poor soil

Many Australian plants are scleromorphic, which means they have small, hard leaves with thick-walled cells and a very thick cuticle or skin. It has been assumed that this helps plants resist drought, but scleromorphy is not confined to dry regions. It is now widely accepted that scleromorphic leaves enable plants to cope with the low levels of nutrients in most Australian soils. The plants grow slowly, and the leaves are long-lived, conserving precious nutrients.

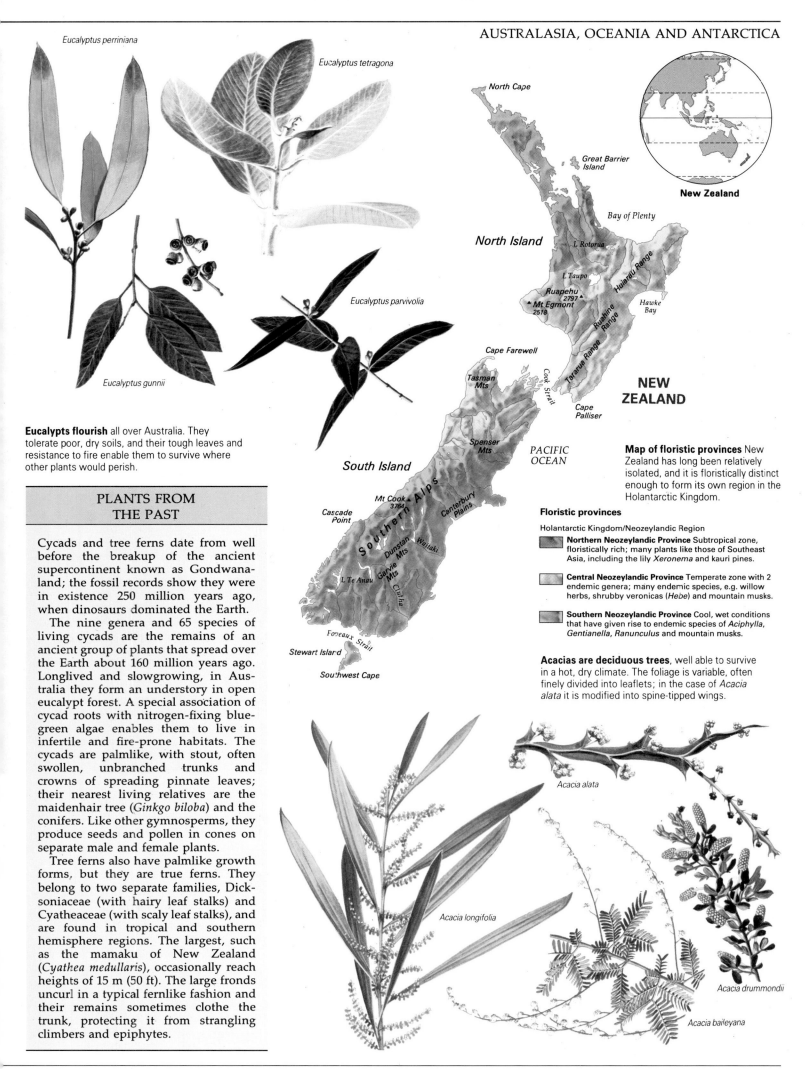

Eucalyptus perriniana

Eucalyptus tetragona

Eucalyptus parvivolia

Eucalyptus gunnii

Eucalypts flourish all over Australia. They tolerate poor, dry soils, and their tough leaves and resistance to fire enable them to survive where other plants would perish.

New Zealand

North Cape

Great Barrier Island

Bay of Plenty

North Island

L. Rotorua

L. Taupo

Ruapehu 2797

Mt Egmont 2518

Huiarau Range

Ruahine Range

Tararua Range

Hawke Bay

Cape Farewell

Tasman Mts

Cook Strait

Cape Palliser

NEW ZEALAND

Spenser Mts

South Island

PACIFIC OCEAN

Mt Cook 3764

Cascade Point

Southern Alps

Canterbury Plains

Dunstan Mts

Waitaki

L Te Anau

Gervie Mts

Clutha

Foveaux Strait

Stewart Island

Southwest Cape

Map of floristic provinces New Zealand has long been relatively isolated, and it is floristically distinct enough to form its own region in the Holantarctic Kingdom.

Floristic provinces

Holantarctic Kingdom/Neozeylandic Region

Northern Neozeylandic Province Subtropical zone, floristically rich; many plants like those of Southeast Asia, including the lily *Xeronema* and kauri pines.

Central Neozeylandic Province Temperate zone with 2 endemic genera; many endemic species, e.g. willow herbs, shrubby veronicas (*Hebe*) and mountain musks.

Southern Neozeylandic Province Cool, wet conditions that have given rise to endemic species of *Aciphylla*, *Gentianella*, *Ranunculus* and mountain musks.

Acacias are deciduous trees, well able to survive in a hot, dry climate. The foliage is variable, often finely divided into leaflets; in the case of *Acacia alata* it is modified into spine-tipped wings.

Acacia alata

Acacia longifolia

Acacia drummondii

Acacia baileyana

PLANTS FROM THE PAST

Cycads and tree ferns date from well before the breakup of the ancient supercontinent known as Gondwanaland; the fossil records show they were in existence 250 million years ago, when dinosaurs dominated the Earth.

The nine genera and 65 species of living cycads are the remains of an ancient group of plants that spread over the Earth about 160 million years ago. Longlived and slowgrowing, in Australia they form an understory in open eucalypt forest. A special association of cycad roots with nitrogen-fixing blue-green algae enables them to live in infertile and fire-prone habitats. The cycads are palmlike, with stout, often swollen, unbranched trunks and crowns of spreading pinnate leaves; their nearest living relatives are the maidenhair tree (*Ginkgo biloba*) and the conifers. Like other gymnosperms, they produce seeds and pollen in cones on separate male and female plants.

Tree ferns also have palmlike growth forms, but they are true ferns. They belong to two separate families, Dicksoniaceae (with hairy leaf stalks) and Cyatheaceae (with scaly leaf stalks), and are found in tropical and southern hemisphere regions. The largest, such as the mamaku of New Zealand (*Cyathea medullaris*), occasionally reach heights of 15 m (50 ft). The large fronds uncurl in a typical fernlike fashion and their remains sometimes clothe the trunk, protecting it from strangling climbers and epiphytes.

The Mount Cook lily (*Ranunculus lyallii*) was once prominent in the snowy herbfields of the southern Alps of New Zealand. Now it is a rarity that can be found only in remote, moist crevices, where it is inaccessible to grazing animals.

Symbiosis offers another solution to the problem. Some plants, such as members of the pea family (Leguminosae), the she-oaks (*Casuarina*), some podocarps and cycads, have developed associations with bacteria or blue-green algae in their swollen root-nodules. These fix atmospheric nitrogen, and pass the nutrients to the plants. Associations between fungi and plant roots (mycorrhiza) occur in many plants. This helps the roots take up the trace elements and phosphorous essential for growth.

In many shrubs and young trees from New Zealand the stems branch at right angles and bear small leaves in an interlacing growth. Plants that branch, or divaricate, in this way, such as the mountain wineberry (*Aristotelia fruticosa*) and the kowhai (*Sophora microphylla*), are not confined to poor soils; they are particularly common on sites exposed to wind and frost. Such a growth form may originally have been a defence against browsing by moas, giant flightless birds. It may also be a response to the equable but unpredictable climate, which can bring frost in summer and warm growing days in winter. Evergreen species that have buds protected inside the bush are well adapted to cope with these extremes.

TIMBER TREES, FRUIT CROPS AND ORNAMENTALS

European immigrants to Australia and New Zealand brought with them their own way of life and many of their useful plants. Except for timber, they had little need to exploit the native plants of the region. With the increasing dominance of European culture, many traditional uses of these plants have been forgotten by the indigenous people.

The major threats to the plant life of Australasia come from the milling of the forests for timber and wood chips, and the conversion of species-rich native grassland to pasture dominated by introduced (exotic), often European species. Mountain and desert areas are largely free from disturbance, except locally where tourism and mining operate.

Exploiting the forests

The Maoris of New Zealand and the aboriginal people of Australia used only small amounts of timber for fuel, building, weapons, tools and canoes. Many trees such as ironwood (from *Choricarpia* and *Casuarina* species) were highly prized. However, there is evidence that substantial areas of forest were destroyed by fire. The earliest white explorers saw huge potential in the remaining tracts of virgin forest that they found. Conifers, particularly kauri and the podocarps, and the taller gum trees such as jarrah (*Eucalyptus marginata*) and karri (*E. diversicolor*), have tall, straight trunks that make them valuable timber trees.

Australasian timber remains in demand in the growing economies of Southeast Asia, both as pulp for paper and as sawn timber; this exploitation is still one of the greatest ecological threats to the region. Exotic pine forests, particularly of Monterey pine (*Pinus radiata*), have been planted, and these are satisfying much of the demand for timber and pulp. However, such monocultural crops are always at risk from disease, and they have replaced native vegetation.

Commercial potential

Although wild plants were widely used by the indigenous people of the various islands, their role in modern commerce is very small and the economic potential and medicinal value of the huge plant life has barely begun to be realized. One

The pineapple-like fruits of the screwpine were once a staple food for the people of the tropical islands of the Pacific. It is a characteristic tree of sandy shores, stabilized by the often numerous aerial prop roots that arise from the trunk.

exception in the tropics is the coconut, which has long provided food, drink, wood, thatching materials, fibers and utensils, and is a valuable commodity in international trade.

A few Australian species have recently acquired some importance as fruit and crop plants. The Queensland nut (*Macadamia integrifolia*) is now cultivated for its edible fat-rich nut in Australia, and outside the region also in the United States in California and Hawaii. The evergreen tree billygoat plum (*Terminalia ferdinandiana*) produces small fruit that were once a traditional aboriginal food and have recently been shown to have very high concentrations of vitamin C; they are now being sold as a health food in a number of countries.

The diverse Australian plant life is very rich in attractive ornamental plants. The pea, citrus, myrtle and other families contain hundreds of popular species, though most are unknown in cultivation outside their native region. The New Zealand genus *Hebe*, on the other hand, with about a hundred shrubby species from which several hundred cultivars and hybrids have been developed, is often to be seen in the gardens of North America and Europe. Sadly, the picking of native wildflowers for both the local and the export flower trades has become a threat to several spectacular and showy species such as the smoke bushes (*Conospermum*) of Western Australia and the red mistletoe (*Peraxilla*) of New Zealand, neither of which has yet been raised in cultivation.

Vulnerable communities

Islands have particularly fragile and vulnerable communities of often unique plants. One of the greatest threats comes from the introduction of exotic plants and animals. Grazing by goats has caused huge damage on subtropical islands, almost exterminating the shrub *Hebe breviracemosa*, which is endemic to the Kermadec Islands lying to the northeast of New Zealand, while woody weeds such as guava (*Psidium guajava*) and mysore thorn (*Caesalpina decapetala*) often invade native vegetation.

On some subantarctic islands goats, sheep and rabbits have severely modified the vegetation, and have even driven some species to the brink of extinction. On Macquarie Island, for example, rabbits have so heavily grazed the dominant tussock grass (*Poa foliosa*) that it no longer stabilizes the steep coastal slopes. As a result, bare soil lies exposed or the land has been stripped to the underlying rock. On Chatham Island the blue-flowered endemic giant forget-me-not (*Myosotidium hortensia*), which once formed a broad belt on the seashore just above the highwater mark, has been much reduced by pigs and sheep, though it survives as a garden plant in Europe and North America.

In the wealthier countries of the region, prospects for conservation are steadily improving. Both Australia and New Zealand now have government departments responsible for conservation, and large areas are secure in national parks and reserves. Small nations, too, are coming to see the cultural, aesthetic and economic value of reserves, though they are also under pressure to develop their economies by exploiting their resources.

Relicts of a bygone age

Some of the first evidence for the once controversial theory of continental drift came from the distributions of southern hemisphere plants. How, for example, did the southern beeches and their associated fungi and flightless insects come to be found in South America, Australia, New Zealand, New Guinea and New Caledonia, and as fossils in Antarctica? Had these distant lands once been joined in some way? Nowadays many plant groups with the same distribution as the southern beeches are assumed to have been present across the ancient supercontinent of Gondwanaland before it fragmented as the constituent tectonic plates moved apart.

When the Gondwanan and the Laurasian supercontinents collided 15 million years ago New Guinea and the other islands to the north of Australia became a bridge for the plant and animal life of these two great landmasses, though few crossed it. The plants and animals of these ancient lands now meet along Wallace's line, which runs between Borneo and Sulawesi in Southeast Asia and is named after its 19th century discoverer, Alfred Russel Wallace. Asian plants such as pines, magnolia and dipterocarps are absent or rare in New Guinea, but plants of Gondwanan origin abound – southern beeches, podocarps, monkey puzzles and members of the family Winteraceae. The islands are rich in palms; those of Asian origin are found mostly west of Wallace's line, while most of those from the south grow to the east of it.

Most of the conifers of Australasia belong to two Gondwanan families: the primitive-looking monkey puzzles, with their tall straight trunks and whorled branches, and the podocarps, whose seeds have a swollen, fleshy base. Most are forest trees, but some are small alpine shrubs and one, the New Caledonian *Parasitaxus ustus*, which grows on species of *Falcatifolium* (also a podocarp), is the only parasitic conifer known.

Some families of flowering plants, for example the Proteaceae and the rushlike

The fall colors of deciduous species of southern beech (*right*) distinguish them from their evergreen cousins. Dotted among them are cabbage or ti trees (*Cordyline*), which underline the affinities of New Zealand plants with those of Southeast Asia.

Flowerbuds and young fruit of the tree *Degeneria vitiensis* (*below*). This is an ancient species, and the solitary flowers display a combination of primitive characteristics including numerous fleshy stamens.

An ancient evergreen tree lives again. *Idiospermum australiense* was long thought to be extinct, but was rediscovered some 20 years ago. Like *Degeneria*, its floral structure is similar to that of the primitive flowers of the northern magnolias.

Restionaceae, share the distribution of the podocarps and monkey-puzzles, which suggests that they also originated in Gondwanaland. In Australia the Proteaceae have evolved into a spectacular array of species with a great variety of flower and fruit forms. The leaves of many Proteaceae, however, are similar to those found as fossil deposits (*Banksieaephyllum*) in the Oligocene epoch (38–25 million years ago).

Remnants of the most primitive flowering plants of Gondwanaland are still found in the rainforests of Australia, New Zealand, New Caledonia and Fiji. The family Winteraceae is one of the best known. These small trees or shrubs have primitive wood; like the wood of conifers it has no vessels. The flowers have spirally arranged parts, flat leaflike stamens, and carpels that resemble folded leaves with the stigma running along the joined edges. Other plants of the region have similar ancient features. *Degeneria*, with two species in Fiji, has carpels that are open when immature; *Idiospermum australiense*, an evergreen rainforest tree of Queensland, which was only rediscovered as recently as 1972, has seeds the size of tennis balls. Many other flowering plants rapidly evolved, while these probably persisted virtually unchanged after Gondwanaland broke up some 12 million years ago.

A jewel in the desert

In the true deserts of Australia only ephemeral herbs, tough grasses and shrubs such as saltbushes (*Atriplex*) will grow on the driest sites. It is here that Sturt's desert pea (*Clianthus formosus*), with its spectacular scarlet flowers, is found. This trailing herb, in common with other ephemerals, grows rapidly after rain. The finely divided foliage is covered with hairs; this reduces evaporation and enables the plant to survive long enough to complete its life cycle in the scorching heat. The stems scramble over a distance of more than a meter, but grow erect to support the large flowers that seem to survey the landscape with their purple-black eyes. Where the rainfall is more reliable the plant may become biennial, and flower in its second year.

This brilliant desert flower has one close relative, the glory pea (*C. puniceus*), found only in New Zealand. The flowers are similar, but the glory pea lacks the distinctive eye and, like so many plants in New Zealand, has developed a shrubby growth form. It is a handsome plant, first cultivated by the Maoris and now grown widely in gardens. However, its position in the wild is very different. Even in the 19th century it was a rarity, and now it is almost extinct, confined only to the Bay of Islands and Lake Waikoremoana region of North Island. Here, unlike the desert pea, it lives on the coast. A white-flowered form is also known.

Sturt's desert pea flowering in the sandy soils of Australia's arid interior.

GLOSSARY

Abscission The detachment of one structure from another, such as leaves or fruit falling from a tree.
Acid soil Soil that has a pH of less than 7; it is often peaty.
Acid rain Popular term used to describe PRECIPITATION that has become acid through reactions between moisture and chemicals in the atmosphere. Excessive acidity is usually caused by polluting emissions of sulfur dioxide and nitrogen oxides from burning fossil fuels.
Adaptation The process by which a plant changes genetically so that it becomes better fitted to cope with its environment.
Advanced No longer resembling the ancestral type; the opposite of PRIMITIVE.
Adventitious A term used to describe organs that arise from unusual parts of the plant, e.g. roots growing from aerial shoots rather than the base of the stem, or buds occurring on leaves.
Aerobic Requiring oxygen for life.
Afforestation The planting of trees to produce a forest.
Algae A group of very simple plants that range in size from single cells to immense seaweeds many meters long.
Alien A plant that has been introduced to an area in which it would not naturally occur.
Alkaline soil Soil that has a pH of more than 7; chalk or limestone soils.
Alpine A plant that is adapted to grow in the TUNDRA-like environment of mountain areas, between the treeline and the permanent snowline.
Angiosperm A plant producing seeds enclosed in an ovary; a flowering plant. (See also GYMNOSPERM.)
Animal An organism with sensation and voluntary movement, not capable of photosynthesis.
Annual A plant that grows from seed, flowers, sets seed and dies within one year.
Anther In flowers the terminal part of the male organs (stamen), usually borne on a stalk (filament) and developing to contain POLLEN.
Aquatic Plants and animals that both grow and are able to live in water.
Arctic The northern POLAR region. In biological terms it also refers to the northern region of the globe where the mean temperature of the warmest month does not exceed 10°C (50°F). Its boundary roughly follows the northern TREELINE.
Arid An area that is hot and dry. Arid areas are rarely rainless, but rainfall is intermittent and quickly evaporates or sinks into the ground. Little moisture remains in the soil, so plant life is sparse.
Aroid A plant belonging to the family Araceae, characterized by having a large number of small flowers arranged on a spike (spadix) that has a large leaf (spathe) at the base.
Awn A stiff, bristlelike extension to a plant part, usually near the tip.

Basal Borne at or near the base.
Biennial A plant that lives for two years only, usually producing foliage in the first year and flowers and fruit in the second year.
Biomass The total mass of all the living organisms in a defined area or ecological system.
Bog An area of permanently wet ground with a plant community of mosses, heaths and rushes.
Boreal Northern cold winter climates lying between the ARCTIC and latitude 50°N, dominated by coniferous forest (TAIGA).
Bract A leaflike structure borne directly below a flower or INFLORESCENCE.
Broadleaf A type of tree with broad, flat leaves that are often shed for the dry season, e.g. beeches (*Fagus*) and most oaks (*Quercus*). In temperate regions winter is the dry season, and many trees shed their leaves in the fall.
Bulb An underground organ comprising a very short stem and a number of tightly packed leaves. The inner leaves are used to store food, while the outer leaves give protection. As bulbs grow they often divide and thus act as organs of vegetative reproduction. Onions are bulbs.
Bulbil A small bulb or bulblike organ.
Buttress roots Roots that emerge from the stem above soil level, and slope to the ground at an angle to support the plant.

Calyx Collective term for all the SEPALS of a flower.
Cambium A layer of dividing cells that provides new cells for the specialized tissues either side of it. (See PHLOEM, XYLEM.)
Canopy The uppermost layer of a forest, where the branches spread out and the foliage restricts the amount of light reaching the forest floor.
Carnivorous Flesheating. Sundews (*Drosera*) are carnivorous plants, trapping and digesting insects.
Carpel The female part of a flower. There is usually more than one carpel present in each flower.
Chalk soil A shallow, well-drained alkaline soil consisting of only one layer, rich in organic matter and invertebrates.
Chaparral The scrub vegetation growing in the areas of California, in the United States, where there is a MEDITERRANEAN climate.
Chlorophyll The green pigment of plant cells, contained in the CHOROPLAST; essential for PHOTOSYNTHESIS.
Chloroplast The microscopic structure within plant cells in which PHOTOSYNTHESIS takes place.
Climax vegetation The final stage in the development of an area of vegetation.
Conifer A cone-bearing tree, e.g. pine, larch, fir, etc. (See GYMNOSPERM.)
Continental climate The type of climate found in the interior of continents, with a wide temperature range both daily and seasonally, and with relatively low rainfall.
Continental drift The theory that today's continents, formed by the breakup of prehistoric supercontinents, have slowly drifted to their present positions. (See also GONDWANALAND, LAURASIA, PANGEA.)
Corm A fleshy, usually underground, stem. A corm often resembles a BULB, but it is solid, not formed in layers.
Corolla Collective name referring to all the PETALS of a flower.
Cotyledon The first leaves of a plant embryo.
Crasslacea acid metabolism (CAM) A process first discovered in the plant family Crassulaceae, where the stomata are closed during the day to prevent water loss and opened at night to take in the carbon dioxide (CO_2) needed for PHOTOSYNTHESIS. The CO_2 is stored as malic acid, and released when needed the following day. Most plants open their stomata during the day and do not store CO_2.
Crown The point on the trunk of a tree from which most of the major branches arise.
Cultiver A named cultivated form of a species.
Cushion plant A lowgrowing plant, densely branched to form a hemispherical mound similar to a cushion.
Cuticle The waxy or fatty layer on the outer surface of epidermal cells (see EPIDERMIS).
Cycad Any member of the group Cycadopsida, a group of palmlike trees.

Deciduous Plants that shed their leaves seasonally, usually for the dry season.
Deforestation The permanent removal of forests, usually to provide cultivated land or fuelwood.
Dehiscent A plant structure that splits or opens to release seeds or pollen.
Desert A very ARID area with less than 25 cm (10 in) rainfall a year. In hot deserts the rate of evaporation is greater than the rate of PRECIPITATION, and there is little vegetation.
Dicotyledon A plant belonging to one of the two subclasses of flowering plants; its seeds contain two COTYLEDONS. (See also MONOCOTYLEDONS.)

Dioecious A species of plant in which the male and female flowers are borne on separate plants.
Distribution The way that plants are spread over the Earth.
Divaricating Splitting apart or turning in different directions.
Dominant The prevailing species in a plant community.
Dormancy The stage in the life cycle of a plant when there is no active growth, usually to withstand difficult environmental conditions such as cold or drought.
Dwarf A plant of small size compared to that of its close relatives.

Ecosystem A community of plants and animals and the environment (habitat) in which they live.
Endemic A plant group (FAMILY, GENUS or SPECIES) that is NATIVE to one specific area, and found nowhere else in the world.
Enzyme A complex protein that promotes a chemical reaction without being used up or changed; a biological catalyst.
Ephemeral Shortlived; the term is often applied to plants that flourish only during a brief wet period in arid zones.
Epidermis The outer protective layer, usually single-celled, of many plant organs, particularly leaves and HERBACEOUS stems.
Epiphyte A plant that grows on the surface of another, without depending on it for nutrition.
Equatorial Close to the Equator, the longest line of latitude, farthest from the poles.
Erosion The process by which exposed land surfaces are broken down into smaller particles or worn away by water, wind or ice.
Escape A plant that has spread from cultivation into the surrounding uncultivated land.
Evergreen A plant that retains its leaves all year round, unlike DECIDUOUS plants.
Exotic A plant that is not NATIVE to an area but has become established after being introduced from elsewhere.
Extinct No longer in existence.

Family A rank in the classification of plants; a group of genera (see GENUS). All the flowering plants are arranged into between 200 and 300 families.
Fasciation A bundle or band composed of a number of stems growing together as one.
Fastigiate Borne parallel and close together, as in the branches of *Populus fastigiata*.
Fen Area of moist peaty land, rich in organic matter.
Fern Any plant belonging to the class of plants Filices, a group of plants with vascular tissue but not producing flowers or seed.
Fertilization The fusion of male and female reproductive cells (gametes). Cross-fertilization occurs between gametes from separate plants; self-fertilization occurs between gametes from the same plant.
Field layer The lowest level of plants in a forest, below the CANOPY and UNDERSTORY. It contains HERBACEOUS plants, LICHENS, MOSSES, and small woody plants.
Filament The narrow basal portion of the stamen that bears the anther.
Flora The plants of a specific area; also a book that lists and describes the plants of an area.
Floristic region A region supporting a characteristic flora. Floristic regions are grouped within kingdoms, and further subdivided into provinces.
Flower The structure concerned with the sexual reproduction of a plant. It usually consists of carpels, petals, sepals and stamens, though any of these may be missing or highly modified.
Forma A division of plants ranking below VARIETY, used to distinguish between plants with trivial differences.

Fossil Remains or traces of a plant that have been turned to stone, leaving a permanent record of its structure in the rock.

Fruit The ripened ovary of a seed plant and its contents; the name is often loosely applied to the part of a plant that contains the ripe seeds.

Fungus A plantlike organism that lacks chlorophyll and does not manufacture its own food, but lives off other organisms or their remains.

Gametophyte The stage in the life cycle of a plant that has only half the total complement of chromosomes. In flowering plants the gametophyte is not visible to the naked eye. (See also SPOROPHYTE.)

Genus A rank in the classification of plants that groups together more or less closely related SPECIES. The first of the two words in a Latin plant name is the name of the genus to which that plant belongs. (See also SPECIES.)

Germination The first stage in the growth of a seed, spore or pollen grain.

Gondwanaland The southern supercontinent composed of present-day Africa, Australia, Antarctica, India and South America. It began to break up 200 million years ago. (See also CONTINENTAL DRIFT.)

Grass Any member of the plant family Gramineae.

Growing season The period during the year when a plant is actively growing. (See also DORMANCY.)

Gymnosperm A seed plant in which the seeds are not enclosed in an ovary; conifers are the most familiar example. (See also ANGIOSPERM.)

Habit The characteristic mode of growth or occurrence of a plant; the form and shape of a plant.

Habitat The locality within which a particular plant is found, usually including some description of its character, e.g. grassland habitat.

Halophyte A plant that has adapted to grow in soil affected by salt water.

Hardy Not harmed by extremes of climate; usually applied to plants in temperate areas that can withstand frost.

Heath An area of open land covered with lowgrowing shrubs such as heather and ling.

Herbaceous A plant that does not develop persistent woody tissue above ground. Herbs either die at the end of the growing season or overwinter by means of underground organs such as bulbs, corms and rhizomes.

Humus Decomposing organic matter in the soil.

Hybrid The offspring of two plants that are not of the same species.

Inflorescence An arrangement of more than one flower in a single structure, e.g. corymb, panicle, spike and umbel.

Insectivorous Trapping and digesting insects to absorb nutrients from them.

Internode The part of a stem between two nodes.

Introduction Allowing a plant to grow in an area in which it is not NATIVE.

Irradiance The amount of light energy striking the Earth's surface at any particular point.

Latosol A type of soil, lacking in minerals and with a clay subsoil, usually found in tropical and warm temperate regions.

Laurasia The ancient northern supercontinent, which broke up and formed present-day North America, Europe and Asia. (See also CONTINENTAL DRIFT.)

Leaching The process by which water moves soluble minerals from one layer of soil to another or out of the soil into streams, etc.

Leaf Aerial and lateral outgrowth from a stem – the foliage of a plant. Its prime function is the manufacture of food by PHOTOSYNTHESIS.

Leaflets Small leaf parts each of which resembles a complete leaf.

Legume Any member of the plant family Leguminosae, noted for their ability to fix nitrogen because of their symbiotic relationship (see SYMBIOSIS) with a MYCORRHIZA.

Lenticel A pore through bark that allows gaseous exchange to take place.

Liana A woody climbing vine.

Lichen A close association (see SYMBIOSIS) between a FUNGUS and an alga (see ALGAE). The algae produce food for the fungus while the fungus provides protection for the algae.

Lignotuber A distended woody stem base or rootstock used for water storage.

Limestone soil Similar to CHALK SOIL but deeper, often dividing into two layers and having a greater moisture-holding capacity.

Littoral An area along the sea shore.

Liverwort Any plant belonging to the group Hepaticae, a group of simple plants closely related to MOSSES.

Mangrove Shrubs and trees that grow on tidal coastal mudflats and estuaries throughout the TROPICS. Many have aerial roots.

Maquis The typical vegetation of the Mediterranean coast, consisting of aromatic shrubs, laurel, myrtle, rock rose, broom and small trees such as olive, fig and holm oak.

Maritime climate An area where the (generally moist) climate is determined mainly by its proximity to the sea. The sea heats up and cools down more slowly than the land, reducing variations in temperature so the climate is equable.

Meadow An area of grassland artificially maintained by grazing or mowing.

Mediterranean climate Climate similar to that of the Mediterranean region: wet winters and hot, dry summers.

Meristem A group of cells capable of dividing indefinitely, providing new cells for the surrounding tissues.

Microclimate The climate of a very small area, e.g. below the canopy of a forest or between the buttress roots of a tree.

Mixed woodland Woodland containing both BROADLEAF trees and CONIFERS.

Mesophyte A plant with moderate moisture requirements.

Mesophyll The internal tissues of a leaf that are concerned with PHOTOSYNTHESIS.

Monocotyledon A plant belonging to one of two subclasses of flowering plants whose seeds contain a single COTYLEDON. The grasses are monocotyledons. (See also DICOTYLEDON.)

Monoculture An agricultural system in which only a single species or variety of crop plant is grown.

Montane The zone at middle altitudes on the slopes of mountains, below the ALPINE zone.

Moss Any plant belonging to the group Musci, a group of simple small plants lacking vascular tissue; closely related to the LIVERWORTS.

Mutation An inheritable change in a plant's genetic makeup.

Mycorrhiza An underground fungus living in association with the roots of higher plants. It enables the plant to extract nutrients from the humus in the soil and fix nitrogen from the air. (See also LEGUME, SYMBIOSIS.)

Native A plant that occurs naturally in a particular area.

Node The point on a stem where a leaf or branch emerges.

Ovary The hollow basal area of a CARPEL, containing one or more ovules and surmounted by the STYLE and STIGMA. It is made up of one or more carpels, which may fuse together.

Ovule The structure in an OVARY containing the egg cell. The ovule develops into the SEED after fertilization.

Palm Any member of the plant family Palmae, a group of evergreen trees.

Palmate A leaf with more than three segments or leaflets arising from a single point, as in the fingers of a hand.

Pangea The supercontinent that was composed of all the present-day continents and therefore included both GONDWANALAND and LAURASIA. It existed between 250 and 200 million years ago. (See also CONTINENTAL DRIFT.)

Parasite A plant that obtains its food from another living plant to which it is attached.

Peat Soil formed by an accumulation of plant material incompletely decomposed due to lack of oxygen, usually as a result of waterlogged conditions.

Pedicel A stalk bearing a single flower; the final branch of an INFLORESCENCE.

Peduncle A stalk that bears an INFLORESCENCE.

Perennate To live over from season to season.

Perennial A plant that lives for more than two years, and normally flowers annually.

Permafrost Ground that remains permanently frozen, typically in the polar regions. A layer of soil at the surface may melt in summer, but the water that is released is unable to drain away through the frozen subsoil.

Petal One of the leafy structures that make up a flower; usually present as one of a number in a whorl called a corolla, often highly adapted to attract pollinators (see POLLINATION).

Phloem Plant tissues that conduct food material. In woody stems it is the innermost layer of the bark. (See also XYLEM.)

Photosynthesis The making of organic compounds, primarily sugars, from carbon dioxide and water, using sunlight for energy and chlorophyll or related pigment for trapping it.

Pinnate A leaf made up of a number of LEAFLETS in pairs on either side of a central midrib. Ash trees (*Fraxinus*) have pinnate leaves.

Pioneer A plant that is among the first to colonize a new area.

pH measurement on the scale 0–14 of the acidity or alkalinity of soil. Neutral soils are pH7; soils with a pH of less than 7 are acid, those with a pH of more than 7 are alkaline.

Plant An organism capable of carrying out PHOTOSYNTHESIS, not capable of sensation or voluntary movement.

Pneumatophore A vessel in plant tissues used to transport air.

Podzol A type of soil deficient in HUMUS.

Polar The regions that lie within the lines of latitude known as the Arctic and Antarctic circles, 66° 32' north and south of the Equator; the lowest latitude at which the sun does not set in midsummer or rise in midwinter.

Polder An area of lowlying ground that has been reclaimed from the sea or other body of water.

Pollen Collective name for pollen grains, the minute spores produced in the ANTHERS of a flower.

Pollination The fertilization of plants by the transfer of pollen grains from the male anthers to the female pistil. Pollen is usually transferred by wind or insects, but animals such as hummingbirds also act as pollinators.

Pollinium A mass of pollen grains produced by one anther-lobe, cohering together and transported as a single unit during pollination, as in the orchids.

Prairie North American STEPPE grassland between 30°N and 55°N.

Polyploid A plant that has more than the normal two sets of chromosomes in its nonreproductive cells.

Precipitation Moisture that reaches the Earth from the atmosphere, including rain, snow, sleet, mist and hail.

Primary vegetation Vegetation that has not recently

been disturbed but has long remained in its natural CLIMAX state.

Primitive A plant characteristic that has been inherited largely unchanged from the plant's distant ancestors.

Propagule A unit of vegetative reproduction, e.g. a spore or bulbil.

Prostrate Lying flat, growing along the ground.

Prothallus The structure that results from the germination of the spore of a fern and that bears the sexual organs; the GAMETOPHYTE of a fern.

Rainforest Forest where there is abundant rainfall all year round. The term usually refers to tropical rainforests, which are particularly rich in plant and animal species and where growth is continuous, lush and rapid.

Rare Plants with a small total population that are consequently vulnerable to changes in the environment.

Reduction A plant structure that does not develop to the size or complexity it had in the plant's ancestors.

Refugia Habitats that have escaped drastic changes in climate, enabling species and populations to survive, often in isolation.

Relict A species that is one of a few survivors of a FLORA that has been dramatically altered by climatic or other change.

Rheophyte A plant that grows in running water.

Rhizome A horizontally creeping underground stem that lives over from season to season (perennates) and bears roots and leafy shoots.

Riparian Growing by rivers or streams.

Riverine Living in or near a river.

Root The lower, usually underground, part of a plant. It anchors the plant in the soil and absorbs water and mineral nutrients from it through the root hairs

Saprophyte A plant that cannot live on its own, but needs decaying organic material as a source of nutrition.

Savanna A habitat of open grassland with scattered trees in tropical areas. There is a marked dry season each year, and too little rain to support large areas of forest.

Sclerophyll Plants with leathery leaves that help to reduce water loss.

Scree An unsorted mass of rocks and boulders that accumulates at the bottom of a slope, having been broken off the rocks higher up by WEATHERING processes.

Scrub An area of shrubs and trees of stunted growth, the stunting typically caused by lack of water.

Secondary vegetation Vegetation that has taken the place of primary vegetation, for example after deforestation has cleared the land.

Sedge A grasslike plant typically from the genus *Carex*, though the name is also used for plants of other genera in the family Cyperaceae.

Seed A unit of sexual reproduction developed from a fertilized ovule, either exposed as in GYMNOSPERMS (including conifers) or enclosed in the fruit as in ANGIOSPERMS (flowering plants).

Semidesert A very arid area where there is only slightly more moisture available to support vegetation than in a desert.

Sepal A floral, usually green leaf forming the outer ring of structures on a flower, usually present as

one of several collectively called the calyx; sometimes absent.

Serpentine soil A soil that contains a large amount of magnesium and a number of toxic heavy metals such as chrome, copper and nickel.

Shrub A perennial woody plant with well-developed side branches that appear near the base, so there is no trunk. They grow to less than 10 m (30 ft).

Spadix An INFLORESCENCE with a fleshy central axis onto which the flowers are directly attached, as in the AROIDS.

Spathe A large leaflike structure at the base of a SPADIX or other INFLORESCENCE.

Speciation The process by which a new species evolves, usually occurring where a barrier prevents further cross-pollination between two groups of plants of the same species.

Species The basic unit of classification of plants. Species are grouped into genera; variations may be categorized into subspecies, variety and forma (form) in descending order of hierarchy. The second of the two words in a scientific plant name is the name of the species to which that plant belongs. (See also GENUS.)

Spike An INFLORESCENCE in which the flowers are attached directly to an unbranched central axis.

Spore A single-celled unit of vegetative propagation.

Sporophyte The stage in the life cycle of a plant that has the full complement of two sets of chromosomes. In flowering plants and ferns this is the dominant stage; in the mosses and liverworts it is represented by the spore-producing structures.

Stamen The male structure of sexual reproduction, consisting of a filament that supports the ANTHER, which produces the pollen.

Steppe Open grassy plains with few trees or shrubs. Steppe is characterized by low and sporadic rainfall, and experiences wide variations in temperature during the year.

Stigma The receptive part of the female reproductive organs of flowering plants, on which the pollen grains germinate; the apex of a CARPEL.

Stipule A leafy appendage, often paired, and usually at the base of the leaf stalk.

Stomata Pores that occur in large numbers in the EPIDERMIS of plants, particularly leaves, and through which gaseous exchange takes place.

Style The narrow part of the CARPEL between OVARY and the STIGMA.

Subsoil The lower levels of soil below those containing significant amounts of humus.

Subspecies A rank in the classification of plants between SPECIES and VARIETY. It is often used to denote a geographical variation of a species.

Subtropical The climatic zone between the TROPICS and TEMPERATE zones. There are marked seasonal changes of temperature but it is never very cold.

Succession The sequence of changes in the vegetation of an area, from the initial colonization to the development of CLIMAX communities.

Succulent A plant that stores water in soft tissues of the leaves or stem to withstand drought.

Symbiosis Two living things of different SPECIES that derive mutual benefit from their close association with each other.

Taiga Russian name given to the coniferous forest belt of the Soviet Union, bordering TUNDRA in the north and mixed forests and grasslands in the

south. The name is generally applied to the northern coniferous forest belt as a whole.

Taproot The main descending root of a plant.

Taxon (taxa) Any taxonomic group, such as a species, genus, family, etc. (see TAXONOMY).

Taxonomy The study of the ordering and classification of plants into taxa (see TAXON).

Temperate The climatic zones in mid latitudes, with a mild climate. They cover areas between the warm TROPICS and cold POLAR regions.

Terrestrial Plants whose entire life cycle is carried out on land.

Transpiration The loss of water from leaves that draws a column of water up the stem from the roots, where it is absorbed from the soil. It is controlled by the opening and closing of small pores called STOMATA.

Tree A large perennial plant with a single branched and woody trunk and with few or no branches arising from the base. (See also SHRUB.)

Tree fern A treelike fern supported by a distinct trunk with strengthened cell walls.

Treeline The limit of tree growth beyond which the growing season each year is not long enough for trees to grow.

Tropics The area lying between the Tropic of Cancer and the Tropic of Capricorn, which mark the latitude farthest from the Equator where the sun is directly overhead at midday in midsummer.

Tuber An underground stem or root that lives over from season to season (perennates) and is swollen with food reserves.

Tundra The level, treeless land lying in the very cold northern regions of Europe, Asia and North America, where winters are long and the ground beneath the surface is permanently frozen. (See also PERMAFROST.)

Tussock A tuft of grass or grasslike plants. The soil level below older tussocks is often higher than that of the surrounding ground, forming a mound.

Understory The layer of shrubs and small trees under the forest CANOPY.

Variety A named subdivision of a SPECIES, denoting minor differences between plants.

Vascular system The system of vessels that conduct water and nutrients around a plant. ANGIOSPERMS and GYMNOSPERMS are vascular plants.

Vegetation A mass of growing plants.

Vegetative reproduction Production of offspring without sexual reproduction or use of sexual apparatus.

Vessel A duct or tube.

Vicariance The occupation of geographically separate areas by closely related species.

Viviparous Seeds that germinate before becoming detached from the parent plant.

Weed A plant growing in cultivated ground where it is not wanted.

Wetland A HABITAT that is waterlogged all or enough of the time to support vegetation adapted to those conditions.

Whorl The arrangement of structures in a circle around a central axis.

Xerophyte A plant able to withstand long periods of drought.

Xylem Plant tissues that conduct water; see also PHLOEM; VASCULAR SYSTEM.

Further reading

Bellamy, David *Botanic Man* (Hamlyn, London h/b 1978, p/b 1984)Morrison **80b**
Bell, P. and Woodcock, C. *The Diversity of Green Plants* 3rd edn (Edward Arnold, London 1983)
Collison, A.S. *Introduction to World Vegetation* (George Allen & Unwin, London, 1977)
Davis, S.D. et al. *Plants in Danger – What Do We Know?* (IUCN, Switzerland 1986)
Everard, N. and Morley, B.D. *Wild Flowers of the World* (Ebury Press and Michael Joseph, London 1970)
Grime, J.P. *Plant Strategies and Vegetation Processes* (John Wiley and Sons, London 1981)
Heywood, V.H. (ed.) *Flowering Plants of the World* (Croom Helm, London 1985)
Hora, Bayard (ed.) *The Oxford Encyclopedia of Trees of the World* (Oxford University Press, Oxford 1981)
Hortus Third: A Concise Dictionary of Plants Cultivated in the United States and Canada (Macmillan Publishing Co., New York 1976)
Huxley, A. (ed.) *The Encyclopedia of the Plant Kingdom* (Hamlyn, London 1977)
Huxley, A. *Green Inheritance* (Collins/Harvill, London 1984)
Lotschert, W. and Beese, G. *Collins Guide to Tropical Plants* (Collins, London 1983)
Masefield, G.B., Wallace, M. et al. *The Oxford Book of Food Plants* (Oxford University Press, Oxford 1969)
Moggi, Guido *The Macdonald Encyclopedia of Mountain Flowers* (Macdonald, London 1985)
Moore, D.M. (ed.) *Green Planet – The Story of Plant Life on Earth* (Cambridge University Press, Cambridge 1982)
Perry, Frances *Flowers of the World* (Hamlyn, London 1972)
Raven, P.H., Evert, R.F. and Curtis, H. *Biology of Plants* 3rd edn (Worth Pub. Inc, New York 1981)
Royal Horticultural Society Dictionary of Gardening 2nd edn (Oxford University Press, Oxford, 1956; new, completely revised edition to be published by Macmillan, London in 1992)
Takhtajan, Armen *Floristic Regions of the World* (University of California Press, California 1986)
The Times Atlas of the World (Times Books, London 1988)
Wilkins, M. *Plant Watching – How plants live, feel and work* (Macmillan, London 1988)
There are also many Floras and Field Guides covering the various regions of the world.

Acknowledgments

Picture credits
Key to abbreviations: ANT Australasian Nature Transparencies, Heidelberg, Victoria, Australia; **BCL** Bruce Coleman Ltd, Uxbridge, UK; **J** Jacana, Paris, France; **NF** Naturfotografernas Bildbyra, Osterbybruk, Sweden; **NHPA** Natural History Photographic Agency, Ardingly, UK; **NI** Natural Image, Ringwood, UK; **OSF** Oxford Scientific Films, Long Hanborough, UK; **RHPL** Robert Harding Picture Library, London, UK.

b=bottom, c=center, l=left, t=top.

6–7 NHPA/G.I. Bernard 8–9 NHPA/John Shaw 10–11 OSF/Star Osolinski 11 OSF/Alastair Shay 12–13 Courtesy of the Illinois State Museum 16–17 Bob Gibbons 17t NHPA/Roger Tidman 17b J/Michel Viard 18 RHPL/Bill O'Connor 19 OSF/Jack Dermid 20 OSF/Jim Clare/Partridge Productions Ltd 20–21 J/M. Denis-Huot 22 BCL/Kim Taylor 23 NI/Peter Wilson 24 ANT/Jonathan Chester 24–25 OSF/Richard Packwood 26 OSF/Edward Parker 26–27 Zefa Picture Library 27 Hutchison Picture Library 28–29 BCL/Michel Viard 29 Still Pictures/Mark Edwards 30 NI/Liz Gibbons 30–31 Andrew Lawson 31 RHPL/British Library 32 Harry Smith Collection 32/33 NHPA/Martin Wendler 33 The Environmental Picture Library/Herbert Girardet 34 BCL/Peter Davey 34–35 NHPA/David Woodfall 35 NHPA/Michael Leach 36 OSF/Michael Leach 36–37 NHPA/Stephen Krasemann 37 OSF/G.A. Maclean 38 Still Pictures/Mark Edwards 38–39t OSF/Adrienne T. Gibson 38–39b BCL/Adrian Davies 40 ICCE Photo Library/Mark Boulton 40–41t BCL/Frans Lanting 40–41b BCL/Melinda Berge 44 NI/Tom Leach 46 NHPA/Stephen Krasemann 47 NI/Tom Leach 48 Fred Bruemmer 48–49 BCL/Bob and Clara Calhoun 50 J/René Volot 51t, 51b Fred Bruemmer 52 BCL/Bruce Coleman 53 BCL/Leonard Lee Rue 54–55 BCL/Leonard Lee Rue 56 OSF/John Gerlach 58t, 58b, 59 NHPA/John Shaw 60 J/François Gohier 61 BCL/Norman Tomalin 62 NHPA/John Shaw 62–63 OSF/Zig Leszczyski 64–65 BCL/Bob and Clara Calhoun 66t, 66b OSF/Deni Brown 66–67 OSF/Stan Osolinski 68–69 BCL/Jeff Foott 70 OSF/Kjell B. Sandved 71 NHPA/John Shaw 72t M.P.L. Fogden 72b NHPA/Stephen Krasemann 73 M.P.L. Fogden 74 J/R. König 76 OSF/M.P.L. Fogden 76–77 OSF/Scott Camazine 77 OSF/Dr J.A.L. Cooke 78–79 South American Pictures/Dr Peter Francis 80t South American Pictures/Tony

OSF/Doug Allan 83 OSF/Deni Brown 84t OSF/D.H. Thompson 84b, 84–85, 86–87 OSF/M.P.L. Fogden 88, 90 NF/Klas Rune 90–91 NF/Ake W. Engman 92 NF/Bruno Helgesson 93 NF/Jan Elmelid 94l BCL/N. Schwirtz 94r, 95 Karin Eriksson 96–97 BCL/Bob and Clara Calhoun 98–99 Linda Proud 100–101, 102, 103t Bob Gibbons 103b NHPA/Laurie Campbell 104, 105 Bob Gibbons 106–107 OSF/Tom Leach 108 NI/Paul Davies 110t NHPA/Ferrero/Agence Nature 110b Bob Gibbons 111 J/R. König 112 J/Prigent 112–113 Bob Gibbons 113 J/L. Lacoste 114 J/Pierre Pilloud 115 NHPA/J.P. Ferrero/Agence Nature 117t NHPA/Picture Box 117b J/Claude Nardin 118–119, 119 J/N. Le Roy 121 Bob Gibbons 122 J/R. König 123 BCL/Udo Hirsch 124 Bob Gibbons 126 AGE Fotostock 127 Bob Gibbons 128–129 NHPA/N. Callow 129 Bob Gibbons 130 Dr Alan Beaumont 130–131 NHPA/Brian Hawkes 131 OSF/Tom Leach 132 OSF/Raymond Blythe 133 NHPA/David Woodfall 134t NI/Paul Davies 134b Dr Alan Beaumont 135 Roy Lancaster 136 J/Michel Viard 138–139 NI/Paul Davies 140–141 Harry Smith Collection/I. Polunin 142 NHPA/Brian Hawkes 144 BCL/Norbert Rosing 145 NHPA/Nigel Dennis 146t NHPA/David Woodfall 146b Bob Gibbons 148 BCL/Hans Reinhard 149 BCL/Werner Layer 150–151 Vajda Janos 152 NHPA/Laurie Campbell 153 J/R. König 154t NI/Liz Gibbons 154b NHPA/Laurie Campbell 156t OSF/Deni Brown 156b, 157 Vajda Janos 158–159 OSF/Frithjof Skibbe 160 NHPA/David Tomlinson 162 BCL/Konrad Wothe 163 R.B. Burbidge 164 NHPA/G.J. Cambridge 165 NHPA/Laurie Campbell 166, 167 A. Bureokypol 168 RHPL 170 OSF/Fredrik Ehrenstrom 171 Harry Smith Collection/I. Polunin 172 Harry Smith Collection 173 The Royal Botanic Garden, Edinburgh 174 BCL/Michel Viard 175t OSF/Deni Brown 175b BCL/Leonard Lee Rue 176–177 Brian Matthew 178–179 BCL/Adrian Deere-Jones 180t J/Ruffier-Lanche 180b BCL/Peter Davey 181, 182 OSF/P. and W. Ward 183 J/Michel Viard 184l, 184r, 185 Susan Carter 186 BCL/Gunter Ziesler 188–189 NHPA/Ivan Polunin 189 NHPA/Anthony Bannister 190–191, 190 J/R. König 191 J/S. Cordier 192l Dr J.M. Lock 192r NHPA/David Woodfall 193 BCL/Peter Davey 194 NHPA/Anthony Bannister 196 J. Stewart 196–197t NHPA/Anthony Bannister 196–197b BCL/John Visser 198–199 Harry Smith Collection/I. Polunin 199 NHPA/Nigel Dennis 200–201 NHPA/Peter Johnson 202–203 BCL/M.P.L. Fogden 204 NHPA/N.A. Callow 206–207 BCL/M. Timothy O'Keefe 206 NHPA/N.A. Callow 208 Bob Gibbons 210, 210–211 Terese McGregor 212 Biofotos/Heather Angel 214–215 A.D. Schilling 214 J/N. Le Roy 215 Stephen Haw 216 J/Hervé Berthoule 217 Stephen Haw 218, 219 Bob Gibbons 220, 221 Harry Smith Collection/I. Polunin 222–223 NHPA/James Carmichael 223 NHPA/Ivan Polunin 224 Bob Gibbons 226, 226–227 Robert and Linda Mitchell 228–229 Stephen Haw 230 BCL/Orion Press 232 RHPL/Carol Jopp 232–233, 233 Peter Barnes 234–235 RHPL 235 BCL/Michel Viard 236 NHPA/Orion Press 237 RHPL/Carol Jopp 238–239 ANT/J. and E.S. Baker 239 NHPA/ANT/Jan Taylor 240 NHPA/ANT/Ken Griffiths 242 BCL/John Fennell 243t J/R. König 243b ANT/P. and M. Walton 244 ANT/Paddy Ryan 244–245 ANT/Rob Blakers 245 ANT/Andrew Dennis 246–247 ANT/Otto Rogge

Editorial, research and administrative assistance
Nick Allen, Roger Hyam, Shirley Jamieson, Susan Kennedy, Amanda Kirkby, Sandra Shepherd, Lesley Young

Artists
Julian Baker and Stephen Capsey (Maltings Partnership), Wendy Brammall (Artist Partners), A. Davies, J. Davies, Judith Dunkley, B. Harvey, R. Jennings, Christabel King, M. MacGregor, Liz Peperell (Garden Studio), C. Roberts, John Woodcock

Cartography
Sarah Rhodes
Maps drafted by Euromap, Pangbourne

Index
Barbara James

Production
Clive Sparling

Typesetting
Brian Blackmore, Catherine Boyd, Peter MacDonald Associates

Color origination
Scantrans pte Limited, Singapore

INDEX

Page numbers in *italics* refer to captions, illustrations, maps or tables